The Principles
of
Interferometric Spectroscopy

John Chamberlain 1937–1974

The Principles
of
Interferometric Spectroscopy

John Chamberlain

Completed, collated and edited by
G. W. Chantry and N. W. B. Stone
National Physical Laboratory,
Teddington, Middlesex.

A Wiley–Interscience Publication

JOHN WILEY & SONS
Chichester · New York · · Brisbane · Toronto

Library of Congress Cataloging in Publication Data:

Chamberlain, John Ernest.
  The principles of interferometric spectroscopy.
  'A Wiley–Interscience publication'.
  1. Fourier transform spectroscopy. I. Chantry, G. W.
II. Stone, Norman Walford Bavin. III. Title.
QC454.F7C47   538.3   78–13206
ISBN 0 471 99719 6

Typeset in Northern Ireland at The Universities Press, Belfast
and printed in Great Britain at The Pitman Press, Bath

Contents

Preface

When John Chamberlain died suddenly in October 1974, his coworkers lost not only a much-loved friend but also a most able and diligent colleague. During his brief career John had made major contributions to spectroscopy and seemed set to achieve an eminent position and to play a senior role in international science. His main field was asymmetric or dispersive Fourier transform spectrometry and he was, in fact, together with E. E. Bell in the USA, mainly responsible for the development of this branch of the art. It is indeed sad that these two, who had known each other for such a brief time, should die within a year or two of each other. John had known that his life was at risk from the high blood pressure to which he was subject, but he never allowed this to influence him in any way and was always a cheerful and considerate companion whose sensitivity and regard for the feelings of others was common knowledge.

Outside his scientific work, John's great passion was music and he built up a large collection of records, together with a superb reproduction system to indulge it. However, much of his spare time at home was taken up with a major project to which he was devoted—the writing of a comprehensive textbook on Fourier transform spectrometry. This he had been painstakingly assembling and at the time of his death there were twenty chapters in various stages of completion. His colleagues felt that the best memorial they could raise to him, as a scientist, would be to ensure that this book was published. However the text, as it stood, required considerable modification and pruning. Moreover there were very few diagrams and those mostly consisted of thumbnail sketches.

We decided therefore to edit the book down to manageable proportions, to supply the necessary diagrams, and to revise the text in accord with modern advances and with the newer ideas that had emerged since 1974. To do all this was a major task and we soon realized that we could not do a simple excision and grafting job. In the end we decided to redraft several sections of the book completely, but guided by John's original text and by the concepts which he had obviously chosen. We have done our best to remain faithful to his original conception of the book within these limitations and we hope that, were he alive now, he would approve of what we have done. We must however point out that Chapter 10 is entirely ours. It was a chapter that we felt was necessary, but the advances that have occurred in the field of computation made John's original sketches completely out of date. In the light of this, a completely new chapter, *ab initio*, seemed in order, but we take sole responsibility for it. Throughout the period when we were redrafting and revising the text we received help from several of our colleagues, but we would especially like to acknowledge the assistance of J. R. Birch and J. W. Fleming whose careful reading and criticism of each section of the book as it appeared helped to eliminate many errors and obscurities.

We dedicate this book therefore to our late much-missed colleague in the hope that it will prove helpful to present and future generations of interferometric spectroscopists and in some small way replace the personal help which John would undoubtedly have given had he lived.

Summer, 1978

G. W. Chantry
N. W. B. Stone

CHAPTER 1

Introduction

1.1. SPECTROSCOPY

Spectroscopy is the term describing the study of spectra; *spectrometry* is the quantitative measurement of the form of a spectrum. Fellgett[1] has proposed the term *spectrology* as a possibly more precise description of the subject. Strictly speaking, spectroscopy implies the visual observation of a spectrum, but the term is so widely understood in the more general sense to describe the subject as a whole, that it will be used with that meaning here. Spectrometry will have the meaning given above; spectrology will not be used.

1.1.1. Absorption spectrometry

The quantitative investigation of spectra involves the determination of precise information about the energies (that is, frequencies or wavelengths) at which systems absorb or emit electromagnetic radiation and the extent to which the absorption or emission occurs (that is the intensity). Infrared and long-wave spectroscopists deal, for the most part, with passive or absorbing systems on account of the feeble spontaneous emission at low frequencies. The specimen under study is probed, therefore, with radiation, the exact state of which before and after interaction with the system yields the spectral information required. The optical properties of a specimen are completely specified when the complex refractive index $\hat{n}(\sigma)$ at wavenumber σ is known.† This quantity may be expressed in either of the alternative forms

$$\hat{n}(\sigma) = n(\sigma)[1 - i\kappa(\sigma)] = n(\sigma) - ik(\sigma) \qquad (1.1a)$$

or

$$\hat{n}(\sigma) = n(\sigma) - i(\alpha(\sigma)/4\pi\sigma), \qquad (1.1b)$$

† Note: the spectral energy variable will be generally measured by the wavenumber σ having the dimensions of reciprocal length and the other variables, i.e. frequency ν (Hz), wavelength (*in vacuo*) λ (m), circular frequency ω (rad s^{-1}), are related to this by

$$\sigma = \frac{1}{\lambda} = \frac{\nu}{c} = \frac{\omega}{2\pi c},$$

where c is the speed of light (299.8×10^6 m s^{-1}). The SI unit for σ is m^{-1}, but the cm^{-1} persists as the practical unit. In the literature the symbols $\bar{\nu}$ and sometimes ν will be encountered with the same meaning as σ used here.

2

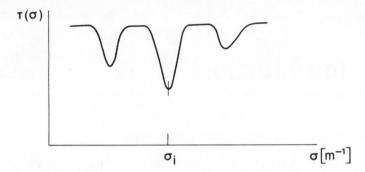

Figure 1.1. Example of a transmission spectrum $\tau(\sigma)$ showing the wavenumbers such as σ_i that are of interest to the spectroscopist

where $n(\sigma)$ is the real refractive index, $\kappa(\sigma)$ is the extinction coefficient, $k(\sigma) = n(\sigma)\kappa(\sigma)$ is the absorption index and $\alpha(\sigma)$ is the power absorption coefficient (neper m^{-1}).†

The real refractive index is related to the speed $v(\sigma)$ of the waves in the medium according to the relation

$$n(\sigma) = c/v(\sigma) \tag{1.2}$$

given by Maxwell; the absorption coefficient measures the intensity (or power) attenuation of the electromagnetic waves after travelling unit length within the (isotropic) medium.

The transmissivity of a bounded specimen of thickness d is given in terms of the incident and transmitted powers $I_0(\sigma)$ and $I(\sigma)$, respectively, by

$$\tau(\sigma) = I(\sigma)/I_0(\sigma) \tag{1.3}$$

and has values lying in the range $0 \leqslant \tau(\sigma) < 1$. The spectroscopist frequently measures the variation of $\tau(\sigma)$ with σ so that he can locate the positions of minima or points of inflection in $\tau(\sigma)$—Figure 1.1. These two types of turning point are taken to represent the locations of absorption bands or lines. The information obtainable from measurements of the positions of absorption features is adequately discussed in textbooks dealing explicitly with spectra and their interpretation.[3] However, to derive the maximum amount of information from a spectroscopic experiment it is desirable to determine an absolute measure of the absorption. In practice, the apparent loss of energy which the beam experiences in traversing the specimen arises from single and multiple reflection effects at the interfaces in addition to the true absorption mechanisms. One may write, therefore,[4]

$$\tau(\sigma) = \tau_A(\sigma)\tau_R(\sigma, d), \tag{1.4}$$

† The use of the neper (Np) to measure *power* absorption is discussed by Chamberlain and Chantry.[2]

where

$$\tau_A = \exp\left[-\alpha(\sigma)d\right] \tag{1.5}$$

is the transmissivity corresponding to the purely absorptive loss and $\tau_R(\sigma, d)$ represents the transmissivity ascribable to all causes other than pure absorption, principally the reflection effects (see ref. 4 for detailed treatment). Equation (1.5) is the mathematical formulation of the Bouger–Lambert Law. It follows from (1.3), (1.4), and (1.5) that

$$\alpha(\sigma) = \frac{1}{d}\ln\left(\frac{I_0(\sigma)}{I(\sigma)}\right) - \frac{1}{d}\ln\left(\frac{1}{\tau_R(\sigma, d)}\right) \quad [\text{Np m}^{-1}]. \tag{1.6}$$

The second term represents a correction which must be applied to the combination

$$\frac{1}{d}\ln\left(\frac{I_0(\sigma)}{I(\sigma)}\right) = \alpha^\circ(\sigma) \quad [\text{Np m}^{-1}] \tag{1.7}$$

of the measured quantities in order to arrive at the true absorption coefficient. Unfortunately, it is very common to find that absorption coefficients quoted in the literature are of the erroneous $\alpha^\circ(\sigma)$ kind, and there is seldom sufficient additional information from which the correction can be inferred.

The calculation of $\tau_R(\sigma, d)$ in the general case is very difficult and it is desirable, if possible, to eliminate it from the calculation by a suitable modification of the experimental technique. For a suitably thick specimen, not only will $\tau_R(\sigma, d)$ be much reduced relative to $\tau_A(\sigma)$, but it becomes possible virtually to eliminate the multiple-beam interference effects and hence make $\tau_R(\sigma, d)$ essentially independent of d. If then we have that $\tau_R(\sigma, d) \rightarrow \tau_R(\sigma)$, it follows that, by making measurements of the powers transmitted by two specimens of identical composition but differing thicknesses d_1 and d_2, it is possible to determine $\alpha(\sigma)$ directly since

$$\alpha(\sigma) = \frac{1}{d_2 - d_1}\ln\left(\frac{I_1(\sigma)}{I_2(\sigma)}\right) \quad [\text{Np m}^{-1}]. \tag{1.8}$$

From this equation it will be seen that reliable determination of $\alpha(\sigma)$ depends on accurate measurement of specimen thickness. This becomes difficult when $\alpha(\sigma)$ is large, for then d will have to be very small, and consequently hard to measure accurately, if $\tau_A(\sigma)$ and $\tau(\sigma)$ are not to be too small for meaningful measurement.

The scope of modern spectroscopy has recently been considerably extended in that, in addition to being able to determine the characteristic wavenumbers σ_i, it is now possible to determine the integrated absorption strengths

$$A_i = \int_i \alpha(\sigma)\, d\sigma \quad [\text{Np m}^{-2}]. \tag{1.9}$$

4

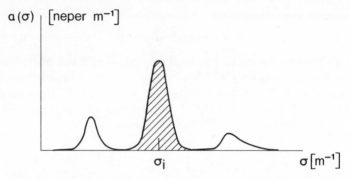

Figure 1.2. Absorption spectrum corresponding to the transmission spectrum of Figure 1.1. The shaded area yields the integrated absorption strength $A_i = \int \alpha(\sigma)\,d\sigma$ of the feature at σ_i

A_i is the area of the ith absorption feature and is a measure of the strength of the oscillator of frequency $\nu_i = c\sigma_i$.

1.1.2. Emission spectrometry

Emission spectra may be continuous or have a discrete structure. While the observer may be concerned with both the shape and the magnitude of the spectrum in either case, he is frequently interested in the absolute power level of the continuous spectra and in the shapes and relative intensities of the features in the line or band spectra. Since the measurements concern the source rather than any passive absorbing system, it is important to know the transmission characteristic of the complete spectrometric system over the range of interest. This knowledge is not easily acquired.

1.1.3. Resolution and resolving power

The amount of detail seen in an experimental spectrum is limited and can be described in terms of the resolution R or the resolving power $\mathcal{R}$. The resolution R is measured in frequency or wavenumber units and represents the smallest spectral interval that can be meaningfully discerned. Hence, the *smaller* the numerical value of R the more detailed the spectrum and the *better* the resolution. Every spectroscopic instrument has a scanning function (or apparatus function) which we can regard as the observed spectral record obtained for a strictly monochromatic input to the spectrometer. The function has a finite width which is related to the resolution limit. The choice of definition of the criterion for the resolution limit is to some extent arbitrary, but more detailed treatment, in Chapters 6 and 8, discusses the various definitions that may be employed.

Sometimes it is more convenient to describe the quality of the resolution in terms of the resolving power $\mathcal{R}$, which is a measure of the ability to separate close spectral lines. $\mathcal{R}$ is a pure (dimensionless) number given by

$$\mathcal{R}(\sigma) = \sigma/R. \tag{1.10}$$

It is obvious that $\mathcal{R}$ is numerically greater the better the detail seen in the spectrum.

1.2. FOURIER TRANSFORM SPECTROMETRY

It is, of course, possible to measure $I(\sigma)$, and hence $\tau(\sigma)$, directly using a number of well-established techniques. There are, however, alternatives, of which an important one is the method of Fourier transform spectroscopy (FTS). This may be defined as the technique whereby a spectrum is determined by the explicit application of a Fourier transformation to the output of an optical† apparatus—generally a two-beam interferometer.

The procedure is, roughly, to divide a beam of radiation from the source into two parts; make provision for the introduction of a known, variable phase delay of one part relative to the other; and then detect the resultant power when the two beams are recombined. This power is recorded as a function of the phase delay and shows fluctuations that are basically periodic. When all the detailed fluctuations of this power record are subject to Fourier transformation we obtain a *fully characterized spectrum*; when only the trends shown by the fluctuations are subject to Fourier transformation we obtain a *partially characterized spectrum*. This latter technique is suitable for application only to spectra having a narrow bandwidth.

1.2.1. Fully characterized spectrometry

In fully characterized spectrometry, the required spectrum, the power distribution of the detected radiation as a function of frequency (or wavenumber), is obtained by Fourier transformation of the variable part of the record of power versus phase delay. As we shall see below, we call the total record the *interference function* and the variable part of it the *interferogram*. The spectrum is, therefore, the Fourier transform of the interferogram. That there is an explicit dependence of the interferogram on the spectrum may be simply seen from the following discussion.

Consider the schematic interferometer shown in Figure 1.3. The partial beam that has traversed the path XP_1Y to the detector has travelled a distance that is greater than that travelled by the other beam along XP_2Y by an amount x ($0 \leqslant x < \infty$) which we call the path difference. By assuming the interferometer to be otherwise symmetrical and evacuated, the optical and geometrical path differences are identical. Because of the path difference x,

† We shall use the term 'optical' to mean an array of lenses, mirrors, etc. when pertaining to apparatus, and to mean some part of $\hat{n}(\sigma)$ when pertaining to constants; the terms 'light' and 'visible' will be applied only to radiation detectable by the eye. Optical path length is defined, as usual, by the product refractive index × path length.

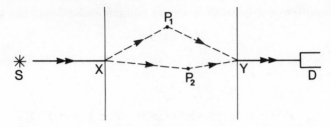

Figure 1.3. Schematic arrangement for a two-beam interferometer. The radiation from the source S is divided, within the interferometer, into two partial beams passing, respectively, via P_1 and P_2. These beams recombine before falling on the detector, which they reach with a relative phase delay determined by the path difference $XP_1Y - XP_2Y$ within the interferometer

the two beams arrive at the detector with a phase delay $2\pi\sigma x$ for any component σ and show interference which is governed by this delay. As we shall see below, the power at the detector is, in fact,

$$I(x) = B(\sigma)\,d\sigma + B(\sigma)\cos 2\pi\sigma x\,d\sigma \qquad [\text{W}] \qquad (1.11)$$

for each spectral component of power $B(\sigma)\,d\sigma$ in the interval σ to $\sigma + d\sigma$. We assume the contribution from each path SP_1D and SP_2D to be the same, $\frac{1}{2}B(\sigma)\,d\sigma$.

When the source is strictly monochromatic and of wavenumber σ_0, the power from each partial beam is б_0 (the Russian symbol б is known as 'buki')

$$I(x) = \text{б}_0(1 + \cos 2\pi\sigma_0 x) \qquad (1.12\text{a})$$

$$= 2\text{б}_0\cos^2 \pi\sigma_0 x \qquad [\text{W}] \qquad (1.12\text{b})$$

and is the interference function. It has the familiar form of cosine fringes which extend to infinite values of x without change of either $I_{\max} = 2\text{б}_0$ or $I_{\min} = 0$ (Figure 1.4). If, however, the source is made more realistic and given a finite, but small, width $\Delta\sigma$ the fringes are still basically cosinusoidal of period σ_0^{-1}; the interference between components from either side of the feature is now constructive at $x = 0$, but becomes increasingly destructive as x increases. This has the effect of modulating the fringes whose $I_{\max}$ falls to zero at about $x = \Delta\sigma^{-1}$.

When the source has a broad bandwidth, such as is required for most spectroscopic measurements, all components are in phase at $x = 0$, but the detected intensity fluctuates rapidly to zero as x is increased and we have the equivalent of white light fringes observed in broad-band visible interferometry. The detected power is given by adding all the components represented singly by equation (1.11):

$$I(x) = \int_0^\infty B(\sigma)\,d\sigma + \int_0^\infty B(\sigma)\cos 2\pi\sigma x\,d\sigma \qquad [\text{W}]. \qquad (1.13)$$

Because

$$I(0) = 2 \int_0^\infty B(\sigma)\, d\sigma \qquad [\text{W}] \qquad\qquad (1.14)$$

when $x = 0$, we may write

$$I(x) = \tfrac{1}{2}I(0) + \int_0^\infty B(\sigma) \cos 2\pi\sigma x\, d\sigma \qquad [\text{W}] \qquad\qquad (1.15)$$

for the interference function, noting that it consists of a part $\tfrac{1}{2}I(0) = I$ invariant with x, and a variable part $I(x) - I = F(x)$. We call this variable part the *interferogram* and rewrite equation (1.15) in terms of it as

$$F(x) = \int_0^\infty B(\sigma) \cos 2\pi\sigma x\, d\sigma \qquad [\text{W}], \qquad\qquad (1.16)$$

which shows that the interferogram depends on the spectrum according to a

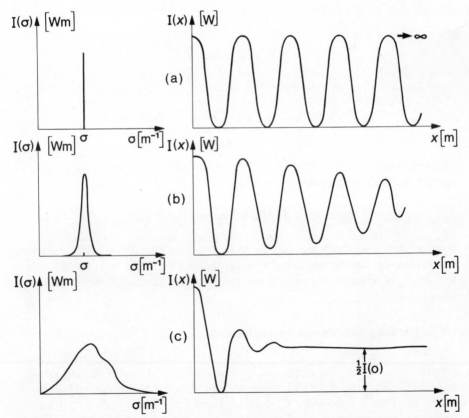

Figure 1.4. Relation between the detected spectrum $I(\sigma)$ and the two-beam interference signal $I(x)$ produced by it: (a) monochromatic spectrum; (b) quasi-monochromatic spectrum; (c) broad-band spectrum

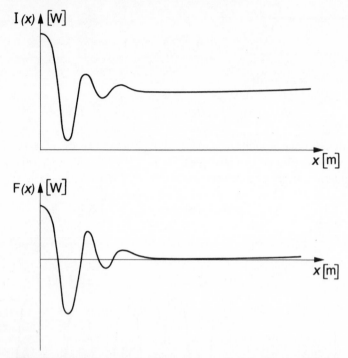

Figure 1.5. Relation between the total interference signal $I(x)$ and the interferogram $F(x)$ which is the difference $I(x) - \frac{1}{2}I(0)$ between the total signal and the constant background signal

cosine Fourier integral. This relation may, by use of the inversion theorem applicable to such integrals, be written as

$$B(\sigma) = 4 \int_0^\infty F(x) \cos 2\pi\sigma x \, dx \qquad [\text{W m}], \qquad (1.17)$$

showing how the spectrum may be calculated by the operation of Fourier transformation on the measured interferogram $F(x)$. This relation, which is the basis of Fourier transform spectroscopy, is discussed more fully in Chapter 4.

1.2.2. Partially characterized spectrometry

We have seen that a source of finite, but small, spectral width gives essentially periodic two-beam interference fringes of slowly falling power and the broader the bandwidth the more rapidly the fringe power falls as x increases. Thus a knowledge of the variation of the fringe power enables us to deduce the spectral bandwidth. This is the basis of partially characterized spectrometry in which we can find not only the bandwidth but also the band shape.

Let the spectral range of the source be very small so that we represent the power at the detector by $b(\sigma - \sigma_0) = b(s)$, where the new variable $s = \sigma - \sigma_0$ occupies only a small range.

From (1.13), we see that the total detected power may be expressed as

$$I(x) = \int_{-\infty}^{\infty} b(s)\, ds + \int_{-\infty}^{\infty} b(s) \cos 2\pi(s + \sigma_0)x\, ds \qquad (1.18a)$$

$$= I + C(x) \cos 2\pi\sigma_0 x - S(x) \sin 2\pi\sigma_0 x \qquad [\text{W}], \qquad (1.18b)$$

where

$$C(x) = \int_{-\infty}^{\infty} b(s) \cos 2\pi s x\, ds \qquad [\text{W}] \qquad (1.19a)$$

and

$$S(x) = \int_{-\infty}^{\infty} b(s) \sin 2\pi s x\, ds \qquad [\text{W}] \qquad (1.19b)$$

are Fourier integrals.

The expression (1.18b) represents the fluctuations in the interference function which occur as x is varied. Since the source is spectrally narrow, these fluctuations extend to large values of x.

The relatively rapidly varying circular functions of σ_0 are slowly modulated by the factors (1.19); thus the interference record shows fringes having roughly constant period but slowly varying power. The difference

$$I_{\max}(x) - I_{\min}(x) \qquad [\text{W}] \qquad (1.20)$$

between the powers $I_{\max}(x)$ and $I_{\min}(x)$ of bright and dark fringes adjacent to some path difference x will slowly vary as x is varied. Moreover, the fringes are more readily discerned the greater this power difference. The

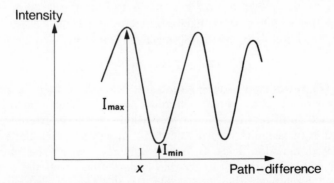

Figure 1.6. The intensities $I_{\max}$ and $I_{\min}$ used to define the visibility $\mathcal{V}(x)$ at a point such as x in an interference pattern

visibility $\mathcal{V}(x)$ of the fringes is defined by the normalized difference

$$\mathcal{V}(x) = \frac{I_{\max}(x) - I_{\min}(x)}{I_{\max}(x) + I_{\min}(x)}. \qquad (1.21)$$

In the present case, we find (see equation (5.17)) that the visibility is

$$\mathcal{V}(x) = \frac{\{[C(x)]^2 + [S(x)]^2\}^{1/2}}{I}. \qquad (1.22)$$

This expression enables the Fourier integrals (1.19) to be explicitly linked to the visibility according to

$$C(x) = I\mathcal{V}(x)\cos\theta(x) \qquad [\text{W}] \qquad (1.23a)$$

and

$$S(x) = -I\mathcal{V}(x)\sin\theta(x) \qquad [\text{W}], \qquad (1.23b)$$

where

$$\theta(x) = 2\pi\sigma_0 x \qquad [\text{rad}] \qquad (1.24)$$

is a phase factor. Equations (1.19) and (1.23) together lead to the Fourier integral relation

$$\frac{b(s)}{6} = \tfrac{1}{2} \int_{-\infty}^{\infty} \mathcal{V}(x)\cos\left[2\pi sx + \theta(x)\right] \mathrm{d}x \qquad [\text{m}] \qquad (1.25)$$

between the normalized spectral distribution $b(s)/6$ and the visibility.

Thus it follows that the power distribution within a narrow spectrum can be found from observations on the (slowly-varying) visibility of the two-beam interference fringes produced from it. In fact, the relative spectral power is the Fourier transform[†]

$$\frac{b(s)}{6} = \tfrac{1}{2} \operatorname{Re} \int_{-\infty}^{\infty} \hat{\mathcal{V}}(x)\exp 2\pi isx \, \mathrm{d}x \qquad [\text{m}] \qquad (1.26)$$

of a complex visibility $\hat{\mathcal{V}}(x)$, having as its modulus the observed visibility function $\mathcal{V}(x)$ and as its phase the term $\theta(x)$, that is,

$$\hat{\mathcal{V}}(x) = \mathcal{V}(x)\exp i\theta(x). \qquad (1.27)$$

Since $\mathcal{V}(x)$ varies only slowly with x, its magnitude may be recorded at widely separated values of path difference, thus enabling a spectral profile to be obtained from relatively few observations on the interference function. Application of visibility measurements is, therefore, particularly important for isolated emission lines at high frequency, where the interferogram function will fluctuate so rapidly that observation of all the detail of the fringes will be difficult if not impossible. For this reason, the Fourier relation

---

† Throughout this work the symbol 'Re' means 'real part of'.

(1.25) has been applied to the measurement of line-shapes in the visible spectral range, while the Fourier integral (1.17) has been applied to spectra lying at longer wavelengths.

1.3. REALIZATION OF FOURIER TRANSFORM SPECTROMETRY

Numerical evaluation of the Fourier integrals demands the use of a high-speed computer or analyser if the technique is to be efficient and viable compared with alternative procedures. The disadvantage this represents is constantly diminishing in importance now that computers are becoming widely available and is, in any case, generally outweighed by the advantages the techniques possess, particularly when applied to the measurement of infrared spectra.

Since high-speed computers were not readily available until the late 1950's, Fourier transform spectroscopy developed only slowly prior to 1960. Since that time however, its growth has been rapid. In order to place this development in context, it is desirable to examine some of the relevant experiments which predate Fourier transform methods as we now know them.

1.3.1. The work of Michelson

The first interferometer to be applied to any form of spectral analysis was the two-beam instrument that Michelson invented[5,6] in 1882 and which now bears his name. The phase delay between the two beams, shown in the schematic instrument of Figure 1.3, is achieved in practice by dividing the incident radiation beam, as shown in Figure 1.7, with a semi-transparent layer and then reflecting the two component partial beams from the plane mirrors M_1 and M_2 back to the layer where they recombine and proceed to the detector. The angle of incidence at the beam divider of the rays from the source is usually 45° and the mirrors M_1 and M_2 are mutually perpendicular as shown. Inspection of the interferometer from D reveals an image M_1', in the beam divider, of the mirror M_1 in the arm containing M_2. These planes M_1' and M_2 constitute a virtual plane-parallel layer of optical thickness x (equal to twice the actual separation of M_1 and M_2). This gives rise to interference between the rays reflected from the two component reflectors. If a stop placed before the detector transmits only the central disc of the circular interference pattern produced in the focal plane of the condensing lens, the detected power, which is the power in that spot, is described by equation (1.11) for each spectral component.

In the last decade of the nineteenth century, Michelson made many classic observations with his interferometer. Among the many applications he studied were some attempts at spectroscopic analysis, although this was not Fourier spectrometry. He realized that the intensity distribution in the

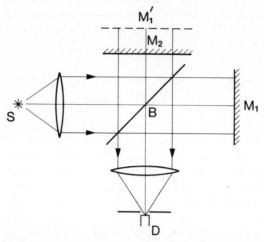

Figure 1.7. The two-beam interferometer invented by Michelson. S = source, M_1 = fixed mirror, M_2 = movable mirror, B = beam divider, and D = detector. M_1' = image of M_1 in M_2 as 'seen' from D

circular interference fringes he observed was dependent on the spectral distribution of the radiation transmitted by the instrument and hence, for a transparent interferometer and linear detector, on that of the radiation emitted by the source. By using a prism predisperser before the interferometer, he was able to select single spectral lines that could be regarded as quasi-monochromatic; this ensured that the interference patterns were reasonably simple. His experiments are described extensively in the literature[5,6] and, less technically, in two of his well-known books.[7,8] The quantity recorded by Michelson was the visibility $\mathscr{V}(x)$. This quantity he observed at the central spot of the fringe pattern as a function of the path difference x introduced in the interferometer by moving one of the mirrors away from the position of zero difference of path. The visibility was recorded at intervals of pathlength of about 1 mm and a visibility curve was plotted. In order to relate numerically the visibility curve to the spectrum he invented a mechanical harmonic synthesizer in which eighty harmonic motions could be superimposed and used to drive a pen that plotted a visibility profile.[7] Each of the eighty elements was built to have an amplitude which could be chosen to lie between zero and some maximum value. At a time when the origins of hyperfine structure in spectral lines were unexplained, he showed the presence of splitting in many apparently simple spectral lines, for example the doublet nature of the red Balmer line of hydrogen and the extremely complex structure of the mercury green line. An additional extremely important finding was that of the very narrow spectral width of the red line of cadmium, radiation from which gave interference over path-differences as

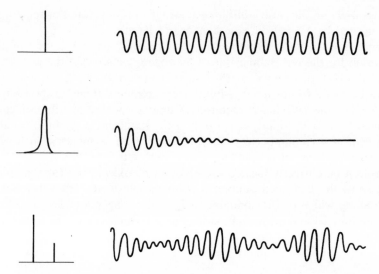

Figure 1.8. Schematic representations of the interference functions which would be produced by a Michelson interferometer illuminated by radiation having the simple spectra shown on the left. The envelope of the interference fringes is a measure of the visibility

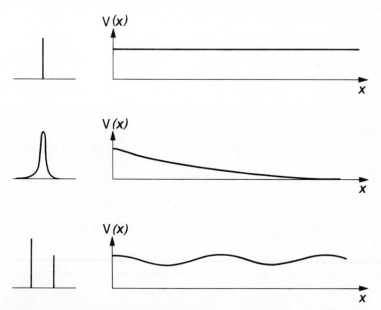

Figure 1.9. Schematic representations of the visibility functions observed with a Michelson interferometer illuminated by the simple spectra shown on the left. The visibility curves are the upper envelopes of the curves shown in Figure 1.8

14

great as 240 mm (i.e. over 400 000 wavelengths). On account of this, the line was adopted by Michelson as a standard of length.

Figure 1.10(b) shows the visibility curve and spectral form found by Michelson for the red Balmer line of hydrogen; we see that the weaker line of the doublet is assigned to the longer wavelength. Such an assignment cannot be made from simple visibility measurements alone as knowledge of the phase of the visibility is required. A unique spectral distribution can be obtained from the visibility curve only if the spectrum is symmetrical about some central wavenumber σ_0. The presence of an asymmetry, such as the weaker component of a doublet, can be detected from the visibility curve but cannot be correctly located spectrally in relation to σ_0: for example, in the case of the hydrogen doublet it is impossible to tell from the visibility curve alone which of the components is at the longer wavelength. Michelson[5] used a trial and error method for the determination of the phase. Lord Rayleigh[9] first pointed out to Michelson that his method did not lead to a unique distribution. He indicated that the phase should be explicitly meas-

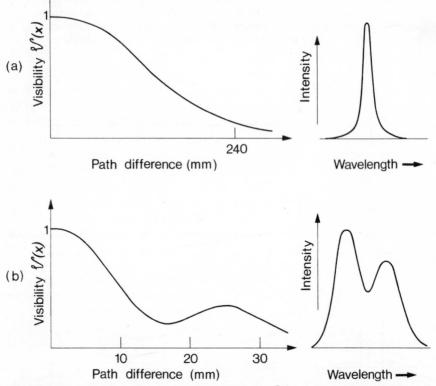

Figure 1.10. Visibility curves obtained by Michelson[7] for radiation from (a) the red line of cadmium at 644 nm and (b) the red Balmer ($n = 3 \rightarrow n = 2$) line of hydrogen at 656 nm. The spectral distributions deduced by Michelson, by trial and error, using his harmonic analyser, are given to the right of the visibility curves

ured. Visibility measurements fell into disuse but have been revived recently by Terrien.[10,11]

1.3.2. The work of Rubens and Wood—fully characterized spectrometry

Michelson did not record interferograms and used only partially characterized spectrometry. His attention was confined solely to emission lines in the visible region of the spectrum. It was left for workers at longer wavelengths to pioneer the use of fully characterized spectrometry. Again, the application was to emission measurements. Rubens and Wood[12] working in the far infrared reported the first experimental observation of a true interferogram in 1911. The interferogram was produced by an interferometer consisting of two thin crystalline quartz plates mounted with their faces parallel to one another. Their separation was variable from 0 to 85 μm and was measured by counting the number of sodium D line fringes formed in the air gap between the plates. They were studying the spectral emission from a Wellsbach mantle and used a focal isolator to separate the far infrared components and a microradiometer to detect the radiant flux. The interferogram was observed by selecting a path difference, taking a microradiometer reading, and then moving to the next path difference and repeating the procedure. The spectra were not obtained by a direct transformation of the interferogram but by a trial and error method. The position of the peak of the transmission curve of the system was estimated from the position in the interferogram of the first minimum of intensity; the termination on the short wavelength side was estimated from the known absorption of quartz, while the long wavelength side of the distribution was estimated by assuming that the source was non-selective and that the power per unit wavelength decreased with the fourth power of the wavelength. The resultant curve was then divided into vertical strips, each of which represented relatively narrow-band radiation. The interference curves (cosine functions) relevant to each of the various strips were then drawn and superimposed with the expectation that the resultant would resemble very closely the curve obtained with the interferometer. Rubens and Wood found that the curves generated in this way were very similar to the experimental result and concluded that the spectrum shown in Figure 1.11 represented the approximate distribution of energy in the radiation which showed explicitly, for the first time, radiation of 150 μm or even 200 μm wavelength. Later the same year, Rubens and Von Baeyer[13] published some further results obtained in this manner, but little more interferometric work was to be done for over forty years.

1.3.3. The work of Fellgett and Jacquinot—advantages of interferometric spectrometers

In spite of the use by Michelson and Rubens of the interferometer as a spectroscopic tool, one of its significant advantages does not appear to have

16

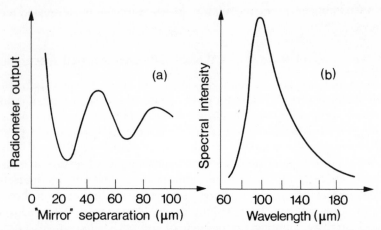

Figure 1.11. One of the first interferograms (a) recorded by Rubens and Wood[12] and the spectral distribution (b) determined by trial and error from the form of the interferogram

been noticed. Jacquinot[14] pointed out in 1954 that an interferometer, being an instrument possessing circular symmetry, has an angular admission advantage over a conventional spectrometer, which employs slits and consequently has no such symmetry. All spectrometers must, of necessity, have a limited angular admission if they are to have non-zero resolving power. Jacquinot showed that when prism, grating, and interference spectrometers were compared at *equal resolving power*, the radiant throughput (or *étendue*) of the interference spectrometer was much higher than that of the grating spectrometer, which had itself a considerably greater *étendue* than that of the prism spectrometer.

The second advantage, pointed out by Fellgett in 1951 in his doctoral thesis,[15] is perhaps, less obvious than the throughput advantage of Jacquinot.† It arises as a direct consequence of the fact than in an interferometer the whole of the spectral band is observed for the whole of the duration of the experiment, whereas in a grating (or any dispersive) spectrometer the spectral elements are observed sequentially for short periods which sum to give the total time of the experiment. Fellgett gave the name *multiplex spectrometry* to that variety in which all the spectral elements are simultaneously observed.

Suppose that it is desired to measure M spectral elements in the range σ_{min} to σ_{max} (that is $M = (\sigma_{max} - \sigma_{min})/R$) in a time T (see Figure 1.12). With a sequential spectrometer each spectral element is observed for a time T/M, while in the multiplex case the observation time is T. If the signal observed integrates in direct proportion to T and the noise integrates in proportion to $T^{1/2}$, as it does in a detector-noise-limited system, then the sensitivity

---

† Jacquinot's advantage is always available in Fourier transform spectrometry, but Fellgett's (see later) is only present under conditions of detector-noise limitation.

(proportional to the ratio of signal to noise) is in direct proportion to $T^{1/2}$ for the multiplex device, while in the sequential case it is in proportion to $(T/M)^{1/2}$. The gain in sensitivity is therefore $M^{1/2}$. This result may be expressed in several ways,[16] in comparing the sensitivity in the multiplex method with that in the sequential method:

(a) the signal to noise ratio can be multiplied by $\sqrt{M}$ if the total duration of the measurement is the same in the two cases;
(b) for equal signal-to-noise ratio, the duration of the experiment is divided by M;
(c) for a given signal-to-noise ratio and measurement time, an increase in resolution can be obtained and thus problems, otherwise inaccessible, can be treated.

This fundamental advantage can be eliminated or even become a disadvantage, depending on the type of noise that perturbs the recorded signal. Broadly speaking, there are three types of noise: detector noise, photon noise, and fluctuation noise. Detector noise is independent of the incident radiation signal and is observed when thermal or photo-conductive detectors are used. In the infrared, where such detectors are used and photon noise is negligible, the multiplex advantage fully applies. At shorter wavelengths, photo-emissive detectors are employed. These are sensitive to statistical fluctuations in the numbers of photons measured. Under these circumstances, the noise due to M spectral elements adds up quadratically, the

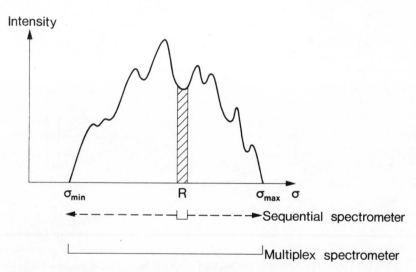

Figure 1.12. The principal of the multiplex advantage. In the multiplex spectrometer the whole of the radiation band $\sigma_{max} - \sigma_{min}$ composed of $(\sigma_{max} - \sigma_{min})/R = M$ elements, is observed for a time T, whereas in the sequential spectrometer the spectrum is scanned by a window of width R which spends a time T/M observing any given part of the spectrum

signal-to-noise ratio is divided by $M^{1/2}$ (if all the spectral elements are of comparable intensity), and the multiplex advantage is completely lost. Fluctuation noise embraces the category of errors that are proportional to the detected signal level (for example, in the observation of astronomical signals). If all the spectral elements are of comparable intensity, the noise level is proportional to M. By comparison with Fellgett's advantage of $M^{1/2}$ this leads to a disadvantage in signal-to-noise ratio of $M^{-1/2}$. The question of noise in Fourier spectrometry and a more rigorous derivation of the multiplex advantage will be given in Chapter 9.

1.3.4. Recent practical application of the principles of Fellgett and Jacquinot

The almost simultaneous publication of these two outstanding advantages of certain two-beam interferometric systems combined with the increasing availability of reliable digital computers capable of computing the necessary Fourier transforms revived interest in the technique. Fellgett[15] published the first numerically transformed interferogram in 1951 and vigorously applied the multiplex method to stellar spectroscopy,[17,18] where the sources are weak and the implicit gain of the method is realized. Initially, a laboratory

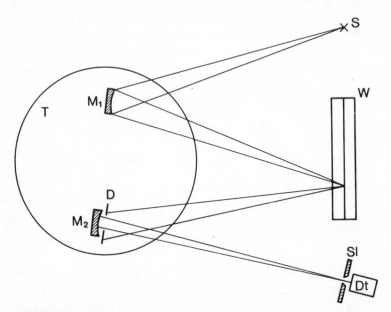

Figure 1.13. The first modern multiplex spectrometer: this was used by Fellgett in 1950 to demonstrate the multiplex advantage. The path difference was introduced by the air wedge W and was varied by rotating the turntable T which carried the two mirrors M_1, M_2. Other symbols are: S = source, D = diaphragm, to limit angular spread of the beam, Sl = slit, and Dt = detector

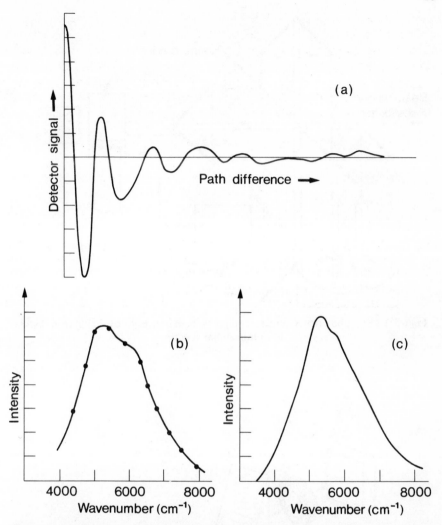

Figure 1.14. Interferogram (a) and corresponding spectrum (b) obtained by Fellgett using his air wedge multiplex spectrometer (Figure 1.13). The spectrum was obtained by numerical transformation using Beevers–Lipson strips and agrees well with the spectrum (c) obtained with a grating spectrometer

interferometer employing an air wedge formed by two sheets of glass was used, as shown in Figure 1.13. The wedge was fixed, but the radiation to be analysed was scanned along the wedge so that the path difference between the beams reflected from each of the plates was increased.[18] The fringe pattern was transformed numerically using Beevers–Lipson strips,[19] and the resultant spectrum of the emission from a mercury lamp was compared with the spectrum obtained using a grating spectrometer (Figure 1.14). This was the first spectrum obtained by a direct 'forward' Fourier transformation of

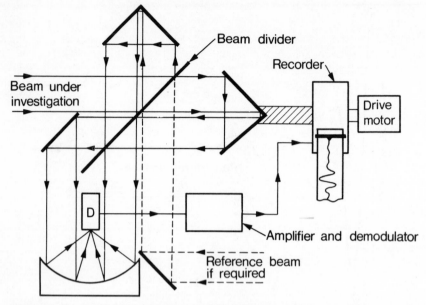

Figure 1.15. Cube-corner interferometer (schematic) used by Fellgett in 1958 to obtain stellar spectra

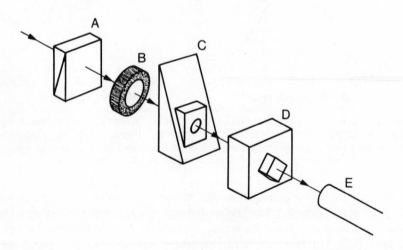

Figure 1.16. The variable retardation plate interferometer of Mertz[20] for use in the visible region: A = Wollaston prism, B = electro-optic retardation plate, C = Bravais compensator, D = polarizer carrying a fixed birefringent plate, E = photomultiplier detector

an interferogram function. Fellgett subsequently constructed a Michelson-type interferometer having cube-corner reflectors; the beam divider was a thin layer of zinc selenide deposited on a calcium fluoride substrate. Cube-corner reflectors have the merit that the beams reflected from them are not displaced if the cube-corner reflectors are rotated. This tends to happen significantly at visible and near visible wavelengths due to imperfect instrumentation and the use of cube-corner reflectors introduces a desirable element of stability. Cube-corner reflectors also have the advantage that the ingoing and outgoing beams are spatially separated. Fellgett also used a double-sided reflecting chopper and two lead sulphide detectors with combined outputs and with this equipment the amount of power transferred from one output beam to the other could be measured. A resolving power of about 60 in the region 4000 to 8000 cm^{-1} was obtained.

Mertz devised an interferometer to operate in the visible region of the spectrum;[20,21] this consisted essentially of a variable retardation plate between polarizers (Figure 1.16). The variable path was achieved with a double wedge of crystalline quartz, the total retardation available being determined by the product of the thickness of the material and its birefringence. In practice the maximum attainable resolution is low. The effects of scintillation noise were minimized by use of an electro-optic shutter

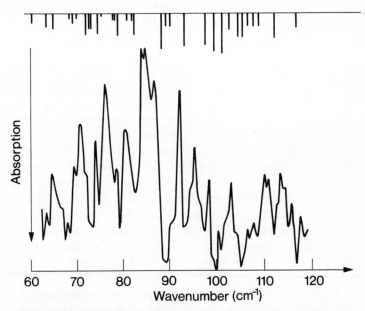

Figure 1.17. The first spectrum obtained by digital transformation of an atmospheric interferogram. The observations were made by Gebbie and Vanasse[23] using the lamellar grating interferometer of Strong and Vanasse.[22,24] The theoretical (line) spectrum of H_2O vapour is shown at the top for comparison

22

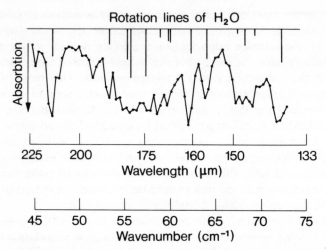

Figure 1.18. The first spectrum, obtained by Gebbie,[25,26] using an interferometer with a dielectric film beam divider

operating at 3000 Hz. The interferograms obtained were not Fourier transformed, the hope being that an empirical classification of star types could be made without such a procedure. In 1954 Mertz also devised a polarization interferometer for the far infrared, but published details of this instrument very much later.[21] Again, however, the interferograms were not transformed.

In the latter half of the 1950's, a number of workers took up interference spectroscopy using a variety of approaches. Strong[22] constructed a lamellar grating interferometer—an instrument intended primarily for use at long wavelengths. The essential element of this interferometer is a plane grating which has variable groove depth. Interference occurs between the beams reflected from the front surface and the recessed surface of the grooves. In 1956, Gebbie and Vanasse[23] published the first digitally computed far infrared spectrum showing water vapour absorption lines that were situated in satisfactory agreement with theory (Figure 1.17). Further results followed with the application of the technique to high-altitude absorption spectroscopy and to the observation of the far infrared solar spectrum. The interferometer, subsequently described in detail,[22,24] was of the lamellar grating variety which is one of the three main types currently used at long wavelengths. The type now most widely used was first described in 1959 by Gebbie, who had built a Michelson-type interferometer having a dielectric film to divide the beams.[25,26] He also showed the first spectra obtained with such a device (Figure 1.18).

Concurrent with these long-wave developments, progress was also being made with astronomical applications in the near infrared. J. Connes[27] gave some preliminary observations of interferograms in 1958 and two years later published, with Gush,[28,29] interferometric spectra of the night sky in the region of $6000\ cm^{-1}$. A two-beam Michelson interferometer was used to give spectra of resolution 1000 (Figure 1.19). As Fellgett has pointed out,[30] these results revealed in a very striking manner one of the aspects of the gain due to the multiplex principle, for prior to the observations of Connes and Gush an effective resolution of only 150 was obtained from an average of up to ten spectra obtained with a grating, whereas the interferometric spectrum obtained from just a single interferogram had a resolution over six times greater.

An early disadvantage of Fourier transform spectroscopy was that, to obtain a given resolution, the number of input observations to the computer increased in direct proportion to the observing frequency. Thus whereas for a resolution of $10\ m^{-1}$ (i.e. $0.1\ cm^{-1}$ or 3 GHz) in the $5000\ m^{-1}$ region (i.e. the far infrared) only 500 input observations would be required, to achieve the same resolution in the $200\,000\ m^{-1}$ region (i.e. 5 μm) would require $20\,000$ input points. The situation became worse if the whole spectrum were

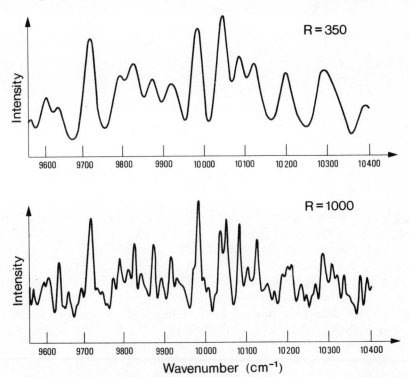

Figure 1.19. Part of the night sky spectrum observed by Connes and Gush.[28,29] This was the first night sky spectrum of such quality to be observed in the near infrared. The structure arises from rotational lines within the second overtone (i.e. $\Delta \nu = 3$) vibrational bands of the OH radical. The improvement in resolution as more of the interferogram is transformed is very obvious

to be computed, for then the number of operations in the computer went up as the *square* of the maximum frequency. The situation has been almost completely relieved by, first, the development of faster computers but, most significantly, by the introduction of the fast Fourier transform (FFT) algorithm developed by Cooley and Tukey.[31] Using this, Connes and his colleagues routinely transform interferograms containing over a million observations.

The current position is that FTS reigns supreme at long wavelengths and is steadily invading the traditional territory of the dispersive instrument in the near and mid-infrared. In future it is likely that all demanding experiments in the infrared will be done by either FTS or else by the use of tunable lasers. Dispersive instruments will tend to be relegated to the routine analysis department.

CHAPTER 2

Mathematical background— the Fourier transformation and some related topics

A full understanding of the principles and practice of Fourier transform spectrometry can only be accomplished with some knowledge of the manipulation of Fourier integrals and some familiarity with the theorems related to them. It is therefore desirable to develop this necessary theory as a self-contained outline, with some emphasis on the more important mathematical devices, before going on to the spectroscopic applications. This branch of mathematics exists, of course, as an abstract topic in its own right, but it has always been associated with practical science and its development owes much to the promptings of the experimental physicists. Fourier transform spectroscopy is but the latest example of a long and venerable line.

The special devices encountered in Fourier spectrometry are Fourier transformations, Fourier series, sampling, and random functions. The first two of these arise, broadly speaking, in deriving the relationships between the recorded (interference) and required (spectral) signals, the third in sampling for numerical computation, and the fourth in the analysis of the noise which must inevitably accompany the signal. An extensive treatment of the Fourier transform is out of place, especially as comprehensive texts already exist, but the more important properties are given. Likewise, no extensive treatment of random variables is given, but an account adequate enough for spectroscopic applications is developed.

The treatment begins with some properties of functions, goes on to introduce the Fourier transform, and then uses this as the basis for all the remaining concepts.

2.1. EVEN AND ODD FUNCTIONS

A general real function $f(k)$, where k is a real variable, is usually neither even nor odd but it may be divided into components that are themselves even and odd. We write

$$f(k) \equiv f_u(k) = f_e(k) + f_o(k), \tag{2.1}$$

where

$$f_e(k) = f_e(-k) \tag{2.2a}$$

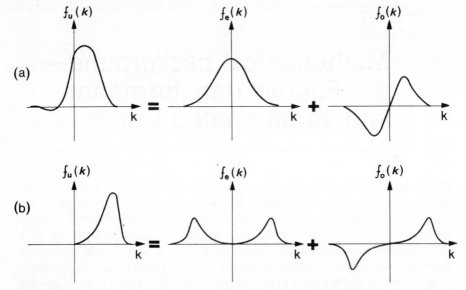

Figure 2.1. The resolution of a general unsymmetrical function (a) into even and odd components. When the function exists only for $k > 0$, (b), the components are simply related

and

$$f_o(k) = -f_o(-k), \qquad (2.2b)$$

while for the function $F(x)$ in the reciprocal domain ($x = 1/k$)—see later—we have the similar expression

$$F(x) \equiv F_u(x) = F_e(x) + F_o(x) \qquad (2.3)$$

where

$$F_e(x) = F_e(-x) \qquad (2.4a)$$

and

$$F_o(x) = -F_o(-x). \qquad (2.4b)$$

The subscripts u, e, and o signify unsymmetrical, even, and odd, respectively.

In the general case, there is no explicit relationship between the even and odd parts, which are merely particular functions which sum to produce the given function. There is, however, a special case that occurs when these parts are numerically the same, that is when

$$|f_e(k)| = |f_o(k)|. \qquad (2.5)$$

When this condition is satisfied,

$$f(k) \equiv f_u(k) = \begin{cases} 2f_e(k), & k > 0, \\ 0, & k < 0; \end{cases} \qquad (2.6)$$

that is, the given function is finite only for $k > 0$. We see, therefore, that a function existing only for $k > 0$ can be represented as the sum of related even and odd parts, the even part, for $k > 0$, being exactly half the given function since, from (2.6),

$$f_e(k) = \tfrac{1}{2} f(k), \qquad k > 0. \tag{2.7}$$

We shall find that in Fourier spectrometry it is extremely useful to represent the spectrum $B(\sigma)$ in terms of its even or odd parts $B_e(\sigma)$ and $B_o(\sigma)$. In particular, because it is easier mathematically to discuss the even component than it is to discuss the spectrum itself, a considerable amount of the analysis uses $B_e(\sigma)$ and then, by virtue of (2.7), yields the required spectrum by considering only positive values of σ, since

$$B(\sigma) = 2B_e(\sigma), \qquad \sigma > 0. \tag{2.8}$$

We have considered no particular form for the function $f(k)$. If it is different from zero only in a narrow region near k', it might be more convenient to discuss the function in terms of a new variable $v = k - k'$ than to use k. We can write

$$f(k) \triangleq h(k - k') = h(v), \tag{2.9}$$

where $\triangleq$ means 'is defined as equal to'. The symmetry behaviour of the function is altered by change of variable since it is with respect to the initial origin that this behaviour is determined. We can split $h(k - k')$ into even and odd components with respect to the origin $k = k'$. We write

$$h(v) \triangleq h_e(v) + h_o(v), \tag{2.10}$$

where, as before,

$$h_e(v) = h_e(-v) \tag{2.11a}$$

and

$$h_o(v) = -h_o(-v). \tag{2.11b}$$

The function representing the reflection of $h(v)$ about the $v = 0$ axis is $h(-v)$, which is given by the difference $h_e(v) - h_o(v)$.

There is an important relation between $h(v)$ and the even part $f_e(k)$ of $f(k)$ which we may derive as follows. We have seen that for a function which only exists for $k > 0$.

$$f(k) = 2f_e(k) \qquad (k > 0). \tag{2.12}$$

The function $f_e(k)$ exists for all k and, in particular, by definition,

$$f_e(k) = f_e(-k).$$

To distinguish between $f_e(k)$ and $f_e(-k)$ analytically, we may use the

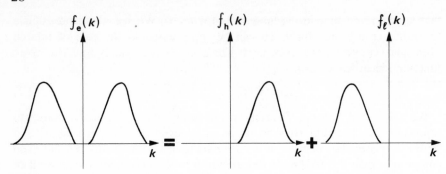

Figure 2.2. Decomposition of an even function $f_e(k)$ into its two components $f_h(k)$ and $f_l(k)$ which exist only for $k>0$ and $k<0$, respectively

1725

Heaviside step function (see section 2.9.5) given by

$$H(k) = \begin{cases} 0, & k<0, \\ \frac{1}{2}, & k=0, \\ 1, & k>0, \end{cases} \tag{2.13}$$

to write

$$f_e(k) = f_e(|k|)H(k) + f_e(|k|)H(-k) \tag{2.14a}$$

$$= f_h(k) + f_l(k), \tag{2.14b}$$

where

$$f_h(k) = f_e(|k|)H(k) \tag{2.15a}$$

$$= f_e(k), \quad k>0, \tag{2.15b}$$

exists only when k is positive and

$$f_l(k) = f_e(|k|)H(-k) \tag{2.16a}$$

$$= f_e(k), \quad k<0, \tag{2.16b}$$

exists only when k is negative. We may describe $f_h(k)$ and $f_l(k)$ as the high and low k-components of $f_e(k)$. Thus (2.12) becomes

$$f(k) = 2f_h(k) \tag{2.17a}$$

or

$$f_h(k) = \tfrac{1}{2}f(k), \tag{2.17b}$$

and so

$$f_h(k) = \tfrac{1}{2}h(v) = \tfrac{1}{2}h_e(v) + \tfrac{1}{2}h_o(v) \tag{2.18}$$

shows that the high k-component of $f_e(k)$ is readily expressed in terms of $h_e(v)$ and $h_o(v)$.

We can also relate the components of $h(v)$ to the low k-component of $f_e(v)$. We change the variable to $u = k + k'$ but note that k is negative and write, therefore, $u = -|k| + k'$, which is related to v by $u = -v$. We define

$$f_l(k) = \tfrac{1}{2}l(u) = \tfrac{1}{2}l(-v) \tag{2.19}$$

and seek to relate $l(-v)$ to $h(u)$. It is clear, from inspection, that $f_l(k)$ has the same form as $h(-v)$; this may be shown by considering the relation between $f_l(k)$ and $f_h(k)$. From (2.15) and (2.16) it is clear that

$$f_l(k) = f_h(-k), \tag{2.20}$$

while, from (2.18),

$$f_h(-k) \equiv \tfrac{1}{2}h(-k - k'), \qquad k < 0, \tag{2.21a}$$
$$= \tfrac{1}{2}h(|k| - k')$$
$$= \tfrac{1}{2}h(v). \tag{2.21b}$$

Thus, comparison of (2.19), (2.20), and (2.21) shows that

$$l(-v) = h(v) \tag{2.22}$$

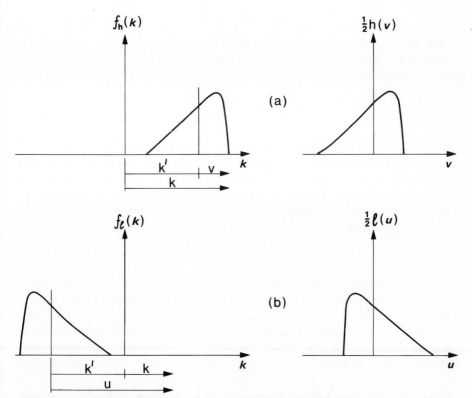

Figure 2.3. Representation of the band-limited functions $f_h(k)$ and $f_l(k)$ by the functions $\tfrac{1}{2}h(v)$ and $\tfrac{1}{2}l(u)$ which are non-zero only in the region of $v = 0$ and $u = 0$, respectively. The changed variables v and u are given for the two cases by $v = |k| - k'$ since $k > 0$ and $u = k' - |k|$ since $k < 0$

or

$$l(u) = h(-u). \tag{2.23}$$

Thus we can represent the low k-component of $f_e(k)$ by

$$f_l(k) = \tfrac{1}{2}h(-v) \tag{2.24a}$$

$$= \tfrac{1}{2}h_e(v) - \tfrac{1}{2}h_o(v). \tag{2.24b}$$

For the moment, the functions $f_h(k)$ and $f_l(k)$ may seem to be rather elaborate analytical ways of describing a situation which pictorially is self-evident, but later, and especially after the concept of the Fourier transform of a product has been introduced, they will be seen to be most useful.

2.2. THE FOURIER INTEGRAL

The function $F(x)$, of the real variable x, is said to be the (full) Fourier transform of the function $f(k)$, of the real variable k, if

$$F(x) = \int_{-\infty}^{\infty} f(k) \exp(2\pi i k x) \, dk. \tag{2.25}$$

(The sign chosen for the exponent in (2.25) is arbitrary; if a negative sign is chosen, the sign in (2.26) is positive.) The right-hand side of equation (2.25) is a Fourier integral and one may therefore apply Fourier's integral theorem, with the result

$$f(k) = \int_{-\infty}^{\infty} F(x) \exp(-2\pi i k x) \, dx. \tag{2.26}$$

One may hence regard $f(k)$ and $F(x)$ as Fourier transforms of each other. The variables k and x have reciprocal dimensions, for example m^{-1} and m or s^{-1} and s. The range of each abscissa variable is said to form a domain and x and k therefore lie in reciprocal domains. The pre-exponential function in a Fourier integral is often referred to as the 'kernel'. The relation (2.25) is only valid if

$$\int_{-\infty}^{\infty} |f(k)| \, dk \tag{2.27}$$

is finite and definite, and if $f(k)$ has only a finite number of finite discontinuities. Similar conditions hold if $F(x)$ is to have a transform. We call $f(k)$ and $F(x)$ a *Fourier transform pair* (Figure 2.4).

The Fourier transforms shown in (2.25) and (2.26) reduce to special forms when $F(x)$ or $f(k)$ satisfy certain conditions. If $F(x)$ is real and symmetrical, that is, if it is an even function $F_e(x)$, then (see Figure 2.5(a)) we see that

$$f(k) = \int_{-\infty}^{\infty} F_e(x) \cos 2\pi k x \, dx - i \int_{-\infty}^{\infty} F_o(x) \sin 2\pi k x \, dx$$

$$= 2 \int_{0}^{\infty} F_e(x) \cos 2\pi k x \, dx \tag{2.28}$$

F(x) f(k)

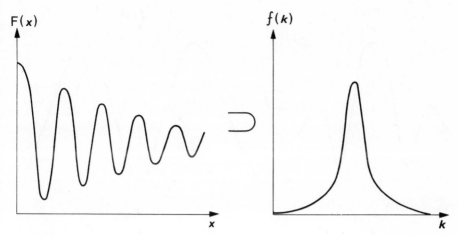

Figure 2.4. A schematic example of a Fourier transform pair. The symbol ⊃ means 'is the Fourier transform of'

is real and is given by the *cosine* Fourier transform of $F_e(x)$. It is obvious, by changing the sign of k in equation (2.28), that $f(k) = f(-k)$ and thus $f(k)$ is also even. The inverse relation to (2.28) is

$$F_e(x) = \int_{-\infty}^{\infty} f(k) \cos 2\pi kx \, dk$$

$$= 2\int_0^{\infty} f_e(k) \cos 2\pi kx \, dk. \tag{2.29}$$

If $F(x)$ is real but odd (see Figure 2.5(b)), that is, $F(x) = F_o(x)$, then

$$f(k) = \int_{-\infty}^{\infty} F_o(x) \cos 2\pi kx \, dx - i \int_{-\infty}^{\infty} F_o(x) \sin 2\pi kx \, dx$$

$$= -2i \int_0^{\infty} F_o(x) \sin 2\pi kx \, dx \tag{2.30}$$

is pure imaginary, odd, and given by the *sine* Fourier transform of $F_o(x)$. By inversion of (2.30) we find that

$$F_o(x) = i \int_{-\infty}^{\infty} f(k) \sin 2\pi kx \, dk$$

$$= 2i \int_0^{\infty} f_o(k) \sin 2\pi kx \, dk. \tag{2.31}$$

More generally, if $F(x)$ is real but neither even nor odd, that is, unsymmetrical, then we have

$$f(k) = \int_{-\infty}^{\infty} F_u(x) \exp(-2\pi ikx) \, dx. \tag{2.32}$$

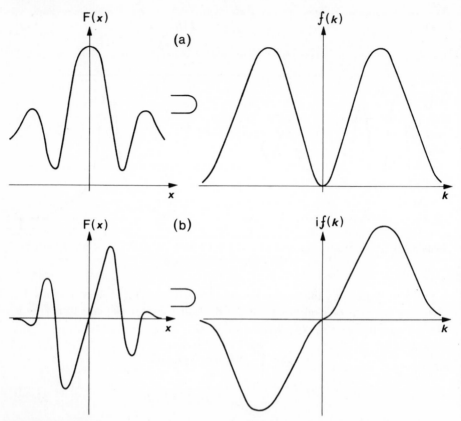

Figure 2.5. The Fourier transform of a real even function (a) is a real even function: the Fourier transform of a real odd function (b) is a pure imaginary odd function. The interferogram $F(x)$ shown in (a) is similar to that obtained in practice with an amplitude-modulated interferometer; that shown in (b) is similar to that obtained with a phase-modulated interferometer

Since we may write an unsymmetrical function as the sum of even and odd parts, it is clear, from (2.28) and (2.30), that

$$\hat{f}(k) = \int_{-\infty}^{\infty} F_e(x) \cos 2\pi kx \, dx - i \int_{-\infty}^{\infty} F_o(x) \sin 2\pi kx \, dx \qquad (2.33)$$

since the cosine transform of an odd function and the sine transform of an even function are zero. $f(k)$ is therefore complex and is written $\hat{f}(k)$ to bring this out explicitly. We may write the cosine transform as $p(k)$ and the sine transform as $q(k)$ so that (2.33) may be expressed as

$$\hat{f}(k) = p(k) - iq(k). \qquad (2.34)$$

From a comparison of (2.33) and (2.34) it follows that $p(k) = p(-k)$ and

$q(k) = -q(-k)$, so that

$$\hat{f}(-k) = p(k) + iq(k) \tag{2.35}$$

and the complex conjugate is

$$\hat{f}^*(-k) = p(k) - iq(k) = \hat{f}(k). \tag{2.36}$$

When the value of a complex number with positive real variable is equal to the complex conjugate of that number with negative real variable that number is said to be Hermitian, that is, $\hat{f}(k) = \hat{f}^*(-k)$ represents a Hermitian function.

It is implied in (2.33) that the Fourier transformation is a linear process. This can readily be proved: one writes

$$F(x) = F_1(x) + F_2(x) + \ldots = \sum_j F_j(x). \tag{2.37}$$

The Fourier transform is

$$f(k) = \int_{-\infty}^{\infty} \left| \sum_j F_j(x) \right| \exp(-2\pi i k x) \, dx$$

$$= \sum_j \int_{-\infty}^{\infty} F_j(x) \exp(-2\pi i k x) \, dx$$

$$= \sum_j f_j(k), \tag{2.38}$$

where

$$f_j(k) = \int_{-\infty}^{\infty} F_j(x) \exp(-2\pi i k x) \, dx, \tag{2.39}$$

thus showing the linearity of the transformation.

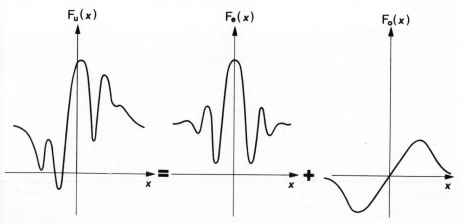

Figure 2.6. A real unsymmetrical interferogram $F_u(x)$ can be resolved into its even and odd components. Its Fourier transform is seen to be a complex Hermitian function with the real part even and the imaginery part odd. Interferograms such as $F_u(x)$ are often obtained in practice when the interferometer is badly misaligned

A convenient short-hand notation for the representation of the Fourier transform is given by Bracewell, although there are alternative forms that could be adopted. We shall use $\mathscr{F}\{\ \}$ to denote 'the Fourier transform of' and shall use $\supset$ to denote 'is the Fourier transform of'; hence

$$\mathscr{F}\{F(x)\} = f(k), \tag{2.40a}$$

$$\mathscr{F}\{f(k)\} = F(x), \tag{2.40b}$$

and

$$F(x) \supset f(k). \tag{2.41}$$

In purely mathematical applications, the variables k and x have no dimensions, but in physical applications they do (m^{-1} and m for instance), and it must be borne in mind that the symbols $\mathscr{F}$ and $\supset$ imply change of dimension. The symbol $\mathscr{F}$ is equivalent to the operation defined in (2.26), i.e. with the exponent negative. The equivalent operation to (2.25), i.e. with the exponent positive, may be denoted $\mathscr{F}^{-1}$ since it is the inverse of $\mathscr{F}$. Both operations have the same effect on an even function but give opposite signs when acting on an odd function.

2.3. THE FOURIER TRANSFORM OF A PRODUCT—THE CONVOLUTION

Let the Fourier transform of the product $F(x)G(x)$ be written

$$\mathscr{C}(k) = \int_{-\infty}^{\infty} F(x)G(x) \exp(-2\pi i k x)\, dx. \tag{2.42}$$

Suppose the Fourier transform of $F(x)$ is†

$$f(h) = \int_{-\infty}^{\infty} F(x) \exp(-2\pi i h x)\, dx, \tag{2.43}$$

then, by inversion,

$$F(x) = \int_{-\infty}^{\infty} f(h) \exp(2\pi i h x)\, dh. \tag{2.44}$$

This may be substituted in (2.42) to give

$$\mathscr{C}(k) = \int_{-\infty}^{\infty} dx \int_{-\infty}^{\infty} dh\, f(h)G(x) \exp[-2\pi i(k-h)x] \tag{2.45a}$$

$$= \int_{-\infty}^{\infty} f(h)g(k-h)\, dh, \tag{2.45b}$$

where

$$g(k-h) = \int_{-\infty}^{\infty} G(x) \exp[-2\pi i(k-h)x]\, dx \tag{2.46}$$

---

† h has the same dimensions and occupies the same range as k—a different symbol is used because k is parametric in (2.42) whereas h is required to be a continuous variable in (2.44).

is the Fourier transform of $G(x)$. We call the right-hand side of (2.45b) a convolution integral and write

$$\mathscr{C}(k) = \int_{-\infty}^{\infty} f(h)g(k-h)\, dh = f(k) * g(k). \qquad (2.47)$$

The operation of convolution is denoted by the asterisk symbol and this symbol implies the dimension of the argument of either f or g. Bracewell[32] examines at some length the physical significance of convolution, an operation which arises in many branches of physics and engineering. It is known by several alternative names such as 'smoothing', 'blurring', 'scanning', and 'smearing'. Convolution is equivalent to first displacing one function, $g(h)$, over the other, $f(h)$, until the new origin of g, namely $h = k$, is reached: g is then reflected about the ordinate $h = k$. The transposed and reflected ordinates of g, at each value of h, are multiplied by the unaffected ordinates of f and the resulting function integrated over the entire range of h. This operation is shown in Figure 2.7.

It should be noted that, since we could just as reasonably have chosen $G(x)$ as the function for the steps (2.43) and (2.44),

$$f(k) * g(k) = g(k) * f(k); \qquad (2.48)$$

in other words the operation of convolution is commutative. By similar arguments it follows that it is associative,

$$f(k) * [g(k) * h(k)] = [f(k) * g(k)] * h(k), \qquad (2.49)$$

and distributive over addition,

$$f(k) * [g(k) + h(k)] = f(k) * g(k) + f(k) * h(k). \qquad (2.50)$$

The asterisk therefore behaves as though it were a multiplication sign in algebra.

2.4. THE SELF-CONVOLUTION AND THE AUTOCORRELATION FUNCTION

The self-convolution of a real function $f(k)$ is

$$f(k) * f(k) = \int_{-\infty}^{\infty} f(h)f(k-h)\, dh, \qquad (2.51)$$

which represents the cumulative effect of scanning $f(k)$ over itself. Suppose, however, we evaluate

$$\int_{-\infty}^{\infty} f(h)f(h-k)\, dh, \qquad (2.52)$$

where the sign of the argument of one of the component factors is reversed relative to the convolution integral (see Figures 2.7 and 2.8). Since $f(h)$ is displaced relative to itself *without reversal* of its shape, it follows by

36

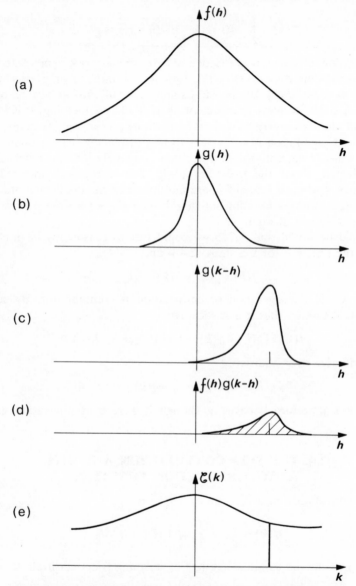

Figure 2.7. Schematic representation of a convolution operation. The general (not necessarily smooth) function $f(h)$ in (a) is to be convolved with $g(h)$ in (b) according to $f(k) * g(k)$

symmetry arguments that the value of the integral will be the same whether k be positive or negative. To prove this result analytically one changes to the variable $u = h - k$ and writes

$$\int_{-\infty}^{\infty} f(h)f(h-k)\,\mathrm{d}h = \int_{-\infty}^{\infty} f(u+k)f(u)\,\mathrm{d}(u+k)$$

$$= \int_{-\infty}^{\infty} f(u+k)f(u)\,\mathrm{d}u.$$

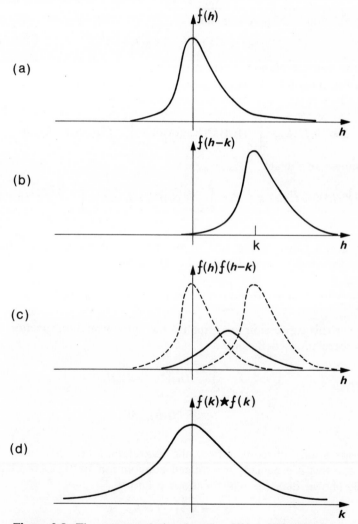

(a)

(b)

(c)

(d)

Figure 2.8. The autocorrelation function $f(k) \star f(k)$. The function $f(h)$ is multiplied by a displaced version $f(h-k)$ of this same function (b)

Now, by changing from u to the equally dummy variable h, it follows that

$$\int_{-\infty}^{+\infty} f(h)f(h-k)\,dh = \int_{-\infty}^{\infty} f(h+k)f(h)\,dh. \qquad (2.53)$$

The right-hand side of equation (2.53) is known as the autocorrelation function of $f(k)$ and, following Bracewell,[32] we write

$$f(k) \star f(k) = \int_{-\infty}^{\infty} f(h+k)f(h)\,dh, \qquad (2.54)$$

using the pentagram $\star$ to distinguish the operation from a convolution $*$. From (2.52) and (2.53) we see that

$$f(k) \star f(k) = f(-k) \star f(-k), \qquad (2.55)$$

that is, the autocorrelation function is even.

When the function $f(k)$ is complex, $\hat{f}(k)$, the autocorrelation function is, from (2.53),

$$\hat{f}(k) \star \hat{f}^*(k) = \int_{-\infty}^{\infty} \hat{f}(h)\hat{f}^*(h-k)\,dh = \int_{-\infty}^{\infty} \hat{f}^*(h)\hat{f}(h+k)\,dh, \qquad (2.56)$$

with complex conjugate

$$[\hat{f}(k) \star \hat{f}^*(k)]^* = \hat{f}^*(k) \star \hat{f}(k) = \int_{-\infty}^{\infty} \hat{f}^*(h)\hat{f}(h-k)\,dh = \int_{-\infty}^{\infty} \hat{f}(h)\hat{f}^*(h+k)\,dh. \qquad (2.57)$$

The maximum value of the real or the complex autocorrelation function occurs at the origin where $k = 0$ and is given by

$$\hat{f}^*(0) \star \hat{f}(0) = \int_{-\infty}^{\infty} \hat{f}^*(h)\hat{f}(h)\,dh = \int_{-\infty}^{\infty} |\hat{f}(h)|^2\,dh. \qquad (2.58)$$

We can use this expression to normalize the autocorrelation function to give the autocorrelation coefficient

$$C_{\hat{f}\hat{f}}(k) = \frac{\displaystyle\int_{-\infty}^{\infty} \hat{f}(h)\hat{f}^*(h-k)\,dh}{\displaystyle\int_{-\infty}^{\infty} |\hat{f}(h)|^2\,dh}. \qquad (2.59)$$

The operations of convolution and correlation are obviously closely related and this is especially true of self-convolution and autocorrelation. It is readily shown that, for real functions g and h,

$$g * h = g(-) \star h, \qquad (2.60a)$$

from which it immediately follows that, for $g = h = f$,

$$f * f(-) = f \star f. \qquad (2.60b)$$

When f is complex, say $\hat{f}(k)$, equivalent expressions can be derived. The self-convolutions take the form

$$\hat{f}(k) * \hat{f}^*(k) = \int_{-\infty}^{+\infty} \hat{f}(h)\hat{f}^*(k-h)\, \mathrm{d}h \qquad (2.61a)$$

and

$$\hat{f}(k) * \hat{f}^*(-k) = \int_{-\infty}^{\infty} \hat{f}(h)\hat{f}^*(h-k)\, \mathrm{d}h. \qquad (2.61b)$$

The latter, by comparison with (2.57), reveals the relation

$$\hat{f}(k) * \hat{f}^*(-k) = \hat{f}(k) \star \hat{f}^*(k), \qquad (2.62)$$

which together with (2.60b) shows that the correlation function is a special case of self-convolution for both real and complex functions.

2.5. FOURIER TRANSFORM OF THE AUTOCORRELATION FUNCTION—THE WIENER–KHINCHIN THEOREM

The Fourier transform of the autocorrelation function (2.56) is

$$\int_{-\infty}^{\infty} [\hat{f}(k) \star \hat{f}^*(k)] \exp 2\pi i k x\, \mathrm{d}k$$

$$= \int_{-\infty}^{\infty} \mathrm{d}k \int_{-\infty}^{\infty} \mathrm{d}h \hat{f}^*(h)\hat{f}(h+k) \exp 2\pi i k x$$

$$= \int_{-\infty}^{\infty} \mathrm{d}k \int_{-\infty}^{\infty} \mathrm{d}h \hat{f}(h)\hat{f}^*(h-k) \exp 2\pi i h x \exp 2\pi i (k-h)x$$

$$= \int_{-\infty}^{\infty} \mathrm{d}h \hat{f}(h) \exp 2\pi i h x \int_{-\infty}^{\infty} \mathrm{d}k \hat{f}^*(h-k) \exp 2\pi i (k-h)x \qquad (2.63a)$$

$$= \hat{F}(x)\hat{F}^*(x) = |F(x)|^2; \qquad (2.63b)$$

thus the Fourier transform of the autocorrelation function $\hat{f}(k) \star \hat{f}^*(k)$ is the modulus squared, $|\hat{F}(x)|^2$, of the Fourier transform of the function $\hat{f}(k)$. If $\hat{f}(t)$ is a time-correlation function, then $|\hat{F}(\nu)|^2$ is defined as the power spectrum of $\hat{f}(t)$—a result known as the Wiener–Khinchin theorem.

2.6. RAYLEIGH'S THEOREM

A result related to the Wiener–Khinchin theorem and first derived by Rayleigh may be obtained as follows. The correlation function

$$\hat{f}(k) \star \hat{f}^*(k) = \int_{-\infty}^{\infty} \hat{f}(h)\hat{f}^*(h-k)\, \mathrm{d}h \qquad (2.64)$$

can, from (2.62), be expressed as the convolution $\hat{f}(k) * \hat{f}^*(-k)$. As a convolution is the result of finding the Fourier transform of a product, we may write

$$\hat{f}(k) * \hat{f}^*(-k) = \int_{-\infty}^{\infty} \hat{F}(x)\hat{F}^*(-x) \exp(-2\pi ikx) \, dx, \qquad (2.65)$$

where

$$\hat{F}(x) = \int_{-\infty}^{\infty} \hat{f}(k) \exp 2\pi ikx \, dk \qquad (2.66)$$

and

$$\hat{F}^*(-x) = \int_{-\infty}^{\infty} \hat{f}^*(-k) \exp 2\pi ikx \, dk. \qquad (2.67)$$

Since (2.64) and (2.65) are different expressions for the same quantity, we have

$$\int_{-\infty}^{\infty} \hat{f}(h)\hat{f}^*(h-k) \, dh = \int_{-\infty}^{\infty} \hat{F}(x)\hat{F}^*(-x) \exp(-2\pi ikx) \, dx, \qquad (2.68)$$

which becomes

$$\int_{-\infty}^{\infty} |\hat{f}(h)|^2 \, dh = \int_{-\infty}^{\infty} |\hat{F}(x)|^2 \, dx \qquad (2.69)$$

when we take the special case of $k = 0$. The statement (2.69) is known as Rayleigh's theorem (see Bracewell[32]). It is analogous to Parseval's theorem, which arises in connection with fourier series. If one expresses a periodic function (even for convenience) in the form

$$f(t) = \sum_{0}^{\infty} a_n \cos 2\pi n\nu t, \qquad (2.70a)$$

then Parseval's theorem states that

$$\nu \int_{0}^{1/\nu} |f(t)|^2 \, dt = a_0^2 + \tfrac{1}{2} \sum_{1}^{\infty} a_n^2. \qquad (2.70b)$$

2.7. THE CROSS-CORRELATION FUNCTION

This is simply defined, by analogy with (2.53) and (2.54), as

$$f(k) \star g(k) = \int_{-\infty}^{\infty} f(h)g(h-k) \, dh \qquad (2.71a)$$

$$= \int_{-\infty}^{\infty} f(h+k)g(h) \, dh \qquad (2.71b)$$

when $f(k)$ and $g(k)$ are real, and is given by

$$\hat{f}(k) \star \hat{g}^*(k) = \int_{-\infty}^{\infty} \hat{f}(h)\hat{g}^*(h-k)\,dh \qquad (2.72a)$$

$$= \int_{-\infty}^{\infty} \hat{f}(h+k)\hat{g}^*(h)\,dh \qquad (2.72b)$$

when $\hat{f}(k)$ and $\hat{g}(k)$ are complex. The cross-correlation function, unlike the convolution, is not commutative.

2.8. AVERAGE CORRELATION FUNCTIONS

It is commonly the case that neither the autocorrelation function nor the cross-correlation function can be properly evaluated owing to the functions concerned not being known from $-\infty$ to $+\infty$. In that case an average is calculated: for autocorrelation (in the general case) we have

$$\gamma_{f\!f}(k) \triangleq \langle \hat{f}(h+k)\hat{f}^*(h)\rangle = \frac{1}{2K}\int_{-K}^{K} \hat{f}(h+k)\hat{f}^*(h)\,dh \qquad (2.73)$$

and for cross-correlation

$$\gamma_{f\!g}(k) \triangleq \langle \hat{f}(h+k)\hat{g}^*(h)\rangle = \frac{1}{2K}\int_{-K}^{K} \hat{f}(h+k)\hat{g}^*(h)\,dh. \qquad (2.74)$$

(In some instances it is necessary to take the limit of these integrals as $K \to \infty$.) These averages may be normalized, in which case we obtain the corresponding correlation coefficients.

2.9. FOURIER TRANSFORMS OF SOME PARTICULAR FUNCTIONS

Some special functions occur frequently in Fourier transform spectrometry; it is therefore helpful to have these functions and their Fourier transforms collected together.

2.9.1. The rectangle function, $\sqcap$

This is an even function of unit height and also unit base if this is measured in units of $x/(2D)$:

$$\sqcap\left(\frac{x}{2D}\right) = \begin{cases} 0, \\ \frac{1}{2}, \\ 1, \end{cases} \quad \left|\frac{x}{2D}\right| \begin{cases} > \frac{1}{2}, \\ = \frac{1}{2}, \\ < \frac{1}{2}. \end{cases} \qquad (2.75)$$

It is useful to describe the truncation of functions outside the range $x = \pm D$ (Figure 2.9(a)). It is often called a *box-car* function. From (2.26) the Fourier transform is

$$f(kD) = \int_{-\infty}^{\infty} \sqcap\left(\frac{x}{2D}\right) \exp\left(-2\pi i k x\right) dx \qquad (2.76a)$$

$$= 2 \int_{0}^{D} \cos 2\pi k x \, dx$$

$$= 2D \operatorname{sinc}(2kD), \qquad (2.76b)$$

where

$$\operatorname{sinc}(2kD) = \frac{\sin 2\pi k D}{2\pi k D}. \qquad (2.77)$$

This function has the form shown in Figure 2.9(a). The fourier transform pair is, therefore,

$$\sqcap\left(\frac{x}{2D}\right) \supset 2D \operatorname{sinc}(2kD). \qquad (2.78)$$

This pairing illustrates that sharp cutoff in one domain implies damped oscillatory behaviour in the other. We note that $\sqcap(x/2D)$ is dimensionless, but $f(kD)$ has the dimensions of D.

2.9.2. The sinc function

Sharp cutoff is not confined to the x (distance, time) domain, it commonly happens that k (wavenumber, frequency) has a sharp upper bound, say $\pm k_M$. In this case, either by the transform theorem or more simply by changing variables in (2.78), one has

$$2k_M \operatorname{sinc}(2xk_M) \supset \sqcap(k/2k_M). \qquad (2.79)$$

In practice $\sqcap(k/2k_M)$ might be a model for a band-limited 'white' spectrum or the characteristic of an ideal band-pass filter.

If some function $F(x)$ is convolved in the x-domain with $\operatorname{sinc}(2xk_M)$, the properties of the convolution integral show us that the equivalent operation in the k-domain is a multiplication

$$\mathcal{F}\{F(x)\}\mathcal{F}\{\operatorname{sinc}(2xk_M)\} = f(k) \times \frac{1}{2k_M} \sqcap\left(\frac{k}{2k_M}\right) \qquad (2.80)$$

in which the Fourier transform $f(k) = \mathcal{F}\{F(x)\}$ is filtered by the ideal low pass filter $\sqcap(k/2k_M)$; all components at $k > k_M$ are removed, all components at $k < k_M$ are passed unaltered.

2.9.2.1. The sine integral, Si (x)

Because of the frequent appearance of the sinc function in Fourier transform theory in general and Fourier spectrometry in particular it will be

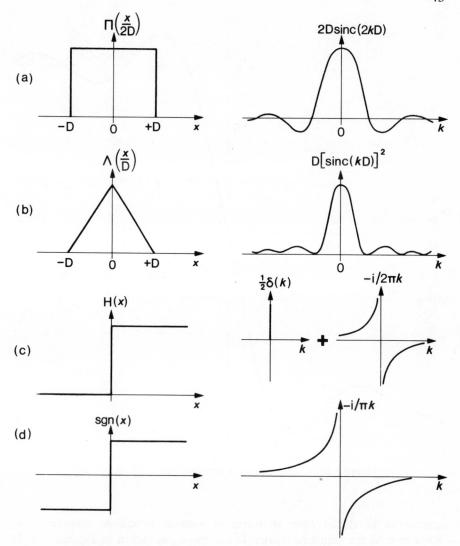

Figure 2.9. Some special functions and their Fourier transforms

useful to consider its integral. We introduce the integral defined by

$$\text{Si}(x) = \int_0^x \frac{\sin x'}{x'} \, dx', \qquad (2.81)$$

note that $\text{sinc}\, x' = (\sin \pi x')/\pi x'$ and find that

$$\int_0^x \text{sinc}\, x' \, dx' = \frac{1}{\pi} \text{Si}(\pi x). \qquad (2.82)$$

The sine integral $\text{Si}(\pi x)$ is a transcendental function which cannot be

44

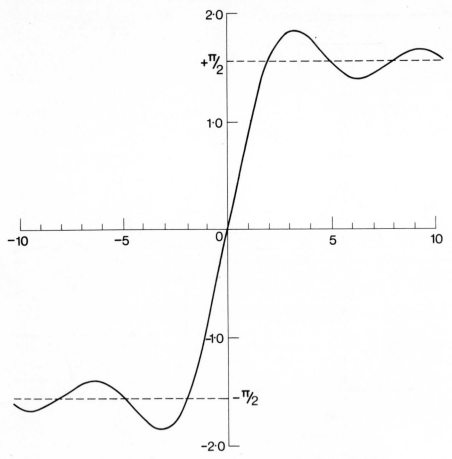

Figure 2.10. The sine integral function Si $(x) = \int_0^x \sin x'/x' \, dx'$

expressed in closed form in terms of simpler functions: however it is tabulated in the standard works. It has the form shown in Figure 2.10. It oscillates about $-\pi/2$ for large negative x, the amplitude of the oscillations increasing near $x = 0$. At $x = 0$ the function is zero, but as x increases positively the function oscillates about $+\pi/2$ with decreasing amplitude. Since Si $(-\infty) = -\pi/2$, we see, from (2.82), that

$$\int_{-\infty}^{0} \text{sinc } x' \, dx' = -\int_{0}^{-\infty} \text{sinc } x' \, dx' = \tfrac{1}{2}, \tag{2.83}$$

thus enabling us to write

$$\int_{-\infty}^{x} \text{sinc } x' \, dx' = \frac{1}{\pi}\left\{\frac{\pi}{2} + \text{Si}\,(\pi x)\right\}. \tag{2.84}$$

2.9.3. The triangle function, $\bigwedge$

This is, by definition, the even function

$$\bigwedge\left(\frac{x}{D}\right) = \left\{ \begin{array}{l} 0, \\ 1 - |x/D|, \end{array} \right. \quad |x| \left\{ \begin{array}{l} > D, \\ < D, \end{array} \right. \tag{2.85}$$

which has the transform

$$f(kD) = \int_{-\infty}^{\infty} \bigwedge\left(\frac{x}{D}\right) \exp\left(-2\pi i k x\right) dx \tag{2.86a}$$

$$= 2 \int_{0}^{D} \left(1 - \frac{x}{D}\right) \cos 2\pi k x \, dx$$

$$= D[\text{sinc}\,(kD)]^2, \tag{2.86b}$$

which is shown in Figure 2.9(b); thus

$$\bigwedge(x/D) \supset D[\text{sinc}\,(kD)]^2. \tag{2.87}$$

2.9.4. The impulse or delta function, δ

Physically, the delta function can be thought of as a unit-area impulse of great intensity and such brevity that the system to which the pulse is subjected cannot distinguish between it and an even shorter pulse; we shall see that this function enables us to calculate the properties of many physical systems and, in particular, the finite resolution that any spectrometer is bound to have. Mathematically the function is defined by the relations

$$\delta(x) = 0, \qquad x \neq 0 \tag{2.88}$$

and

$$\int_{-\infty}^{\infty} \delta(x) \, dx = 1. \tag{2.89}$$

$\delta(x)$ is not therefore a function in the normal sense. It has been called a generalized function by Lighthill and an improper function by Dirac, who interpreted the integral (2.89) in a limiting way as

$$\lim_{\chi \to 0} \int_{-\infty}^{\infty} \frac{1}{\chi} \prod\left(\frac{x}{\chi}\right) dx. \tag{2.90}$$

The integrand in (2.90) always has unit area, but as $\chi \to 0$ the function becomes confined to the region near the origin and as its width shrinks its height becomes infinite to compensate (Figure 2.11).

The Fourier transform of $\delta(x-m)$ is

$$f(k) = \int_{-\infty}^{\infty} \delta(x-m) \exp\left(-2\pi i k x\right) dx \tag{2.91a}$$

$$= \exp\left(-2\pi i k m\right) \tag{2.91b}$$

$$= \cos 2\pi k m - i \sin 2\pi k m, \tag{2.91c}$$

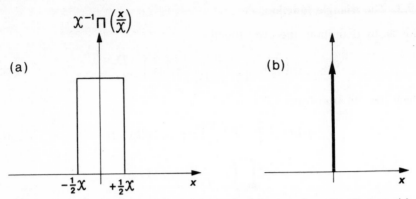

Figure 2.11. The Dirac delta function. A narrow pulse of unit area (a), becomes a delta pulse of infinitesimal width and infinite height (b) as χ tends to zero

and the transform of $\delta(x)$ itself is therefore simply unity. It is obvious from (2.91c) that

$$\mathcal{F}\{\delta(x-m)+\delta(x+m)\}=2\cos 2\pi km \qquad (2.92a)$$

and

$$\mathcal{F}\{\delta(x-m)-\delta(x+m)\}=-2i\sin 2\pi km \qquad (2.92b)$$

showing that, in the reverse form,

$$\mathcal{F}\{\cos 2\pi km\}=\tfrac{1}{2}\delta(x+m)+\tfrac{1}{2}\delta(x-m), \qquad (2.93a)$$
$$\mathcal{F}\{\sin 2\pi km\}=\tfrac{1}{2}\delta(x+m)-\tfrac{1}{2}\delta(x-m). \qquad (2.93b)$$

We can also consider the reverse transform of $\delta(k)$, but, since this will formally be defined as

$$F(x)=\int_{-\infty}^{+\infty}\delta(k)\exp 2\pi ikx\,dk=1, \qquad (2.94)$$

the equivalents of (2.92) and (2.93) will involve sign changes in the terms involving $\sin 2\pi mx$.

If the scale of x is diminished by a factor a, the height of the pulse is modified to ensure that the areal property (2.89) is satisfied:

$$\int_{-\infty}^{\infty}\delta(ax)\,dx=\frac{1}{a}\int_{-\infty}^{\infty}\delta(x')\,dx'=\frac{1}{a}; \qquad (2.95)$$

hence we may write

$$\delta(ax)=\frac{1}{a}\,\delta(x) \qquad (2.96)$$

and

$$\delta(ax-m)=\frac{1}{a}\,\delta\!\left(x-\frac{m}{a}\right) \qquad (2.97)$$

There are two properties of the delta function that we shall find to be of considerable relevance in Fourier spectrometry: its effect when used as (a) a multiplying factor and (b) a factor in convolution. In multiplication it behaves as a sampling operator, yielding only the value of the function at the exact value at which the delta pulse is itself non-zero, that is,

$$F(x) \times \delta(x - m) = F(m). \tag{2.98}$$

In convolution it acts as a shifting operator, for

$$F(x) * \delta(x - m) = \int_{-\infty}^{\infty} F(x') \, \delta(x - x' - m) \, \mathrm{d}x'$$
$$= F(x - m); \tag{2.99}$$

that is, the function $F(x)$ is shifted by convolution with $\delta(x - m)$ to $F(x - m)$.

2.9.5. The Heaviside step function $H(x)$

This step function (see Figure 2.9(c)), often named after Heaviside, is extremely useful in the representation of finite discontinuities and functions

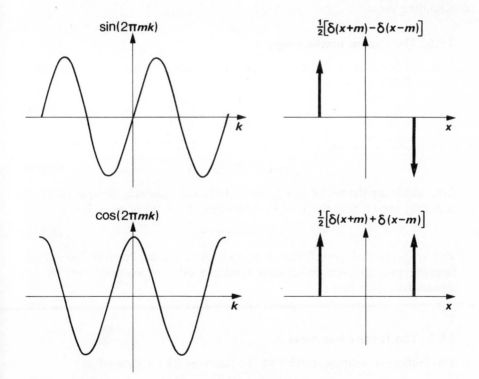

Figure 2.12. The sine and cosine functions and their Fourier transforms. In the upper inset, if either function is real the other is pure imaginary

existing over only a single half-axis:

$$H(x) = \begin{cases} 0, \\ \frac{1}{2}, \\ 1, \end{cases} \quad x \begin{cases} <0, \\ =0, \\ >0. \end{cases} \quad (2.100)$$

It has the Fourier transform

$$f(k) = \int_{-\infty}^{\infty} H(x) \exp(-2\pi i k x)\, dx \quad (2.101a)$$

$$= \int_{0}^{\infty} (\cos 2\pi k x - i \sin 2\pi k x)\, dx$$

$$= \tfrac{1}{2}\delta(k) - i \frac{1}{2\pi k}; \quad (2.101b)$$

that is,

$$H(x) \supset \tfrac{1}{2}\delta(k) - i \frac{1}{2\pi k}. \quad (2.102)$$

This is illustrated in Figure 2.9(c). The Heaviside step function does not, of course, satisfy (2.27), and equation (2.102) has therefore to be interpreted in a limiting sense.

2.9.6. The signum function, sgn (x)

This odd function is equal to $+1$ or -1 according to the sign of x; that is (Figure 2.9(d)),

$$\text{sgn}(x) = \begin{cases} -1, \\ +1, \end{cases} \quad x \begin{cases} <0, \\ >0. \end{cases} \quad (2.103)$$

From the definition of the Heaviside function $H(x)$ it is clear that

$$\text{sgn}(x) = 2H(x) - 1, \quad (2.104)$$

from which, by the use of (2.93) and (2.101) and assuming linearity (2.39), it is readily shown that the Fourier transform of sgn (x) is given by

$$\text{sgn}(x) \supset -i/\pi k. \quad (2.105)$$

This is shown in Figure 2.9(d). It is apparent from the form of the signum function that an even function is rendered odd by multiplication by the signum function; that is,

$$\text{sgn}(x) \cdot F_e(x) = F_o(x). \quad (2.106)$$

2.9.7. The Hilbert transform $\mathcal{H}$

The Hilbert transform $\mathcal{H}\{f(k)\}$ of the function $f(k)$ is defined as

$$\mathcal{H}\{f(k)\} = \frac{P}{\pi} \int_{-\infty}^{\infty} \frac{f(h)}{h-k}\, dh, \quad (2.107)$$

where P means 'Cauchy principal value of'. It will readily be seen that, from the definition of convolution, the Hilbert transform can also be written

$$\mathcal{H}\{f(k)\} = -\frac{1}{\pi k} * f(k). \tag{2.108}$$

The great power of the Hilbert transform in the analysis of physical systems stems from a remarkable theorem, the convolution theorem (see section 2.3), which states that if

$$h(x) = f(x) * g(x) \tag{2.109a}$$

and if $H(k)$, $F(k)$, and $G(k)$ are the Fourier transforms of $h(x)$, $f(x)$, and $g(x)$, then

$$H(k) = F(k) \cdot G(k). \tag{2.109b}$$

In words, the theorem states that the Fourier transform of a convolution is equal to the product of the Fourier transforms of the functions being convolved. Applying this result to equation (2.108) at once gives

$$\mathcal{F}\{\mathcal{H}\{\hat{f}(k)\}\} = i \operatorname{sgn}(x) \cdot \hat{F}(x)$$
$$\equiv \exp\left[\tfrac{1}{2}\pi i\right] \operatorname{sgn}(x) \cdot \hat{F}(x). \tag{2.110}$$

The interpretation of this relation is that the phase of $\hat{F}(x)$ is shifted by $\pi/2$ for x positive and by $-\pi/2$ for x negative when $f(k)$ undergoes Hilbert transformation. From this result, it follows that, if the Hilbert transformation is carried out twice, the phases will be shifted by $-\pi$ for x negative and $+\pi$ for x positive—that is, the function will be reversed in sign. The same result can be derived less heuristically by noting that

$$\mathcal{H}\{\mathcal{H}\{\hat{f}(k)\}\} = \frac{1}{\pi k} * \frac{1}{\pi k} * \hat{f}(k) \tag{2.111}$$

and hence

$$\mathcal{F}\{\mathcal{H}\{\mathcal{H}\{\hat{f}(k)\}\}\} = \mathcal{F}\left\{\frac{1}{\pi k} * \frac{1}{\pi k} * \hat{f}(k)\right\} = [-i \operatorname{sgn}(x)]^2 \hat{F}(x)$$
$$= -\hat{F}(x) = \exp(i\pi)\hat{F}(x) \tag{2.112}$$

and therefore, by inversion,

$$\mathcal{H}\{\mathcal{H}\{\hat{f}(k)\}\} = -\hat{f}(k). \tag{2.113}$$

The relation (2.108) shows that Hilbert transformation does not change a real function into one that is imaginary or complex; an even function is, however, transformed into an odd one and vice versa.

2.9.7.1. Causality and the transfer function

The principle of causality is believed to be a universal law of physics; it states baldly that events cannot precede their causes. Therefore when the variable is t and we are considering causes that do not operate before the time $t = 0$, the response function will necessarily be confined to the positive half axis. In other words, if the response of a physical system to an impulse applied at $t = 0$ is $R(t)$, then

$$R(t) = H(t)F(t), \tag{2.114}$$

where $F(t)$ is some function that is identical to $R(t)$ when $t \geqslant 0$. The spectrum of the temporal response, which we call the transfer function, is

$$\hat{T}(\nu) = \int_0^\infty R(t) \exp(-2\pi i\nu t)\, dt \tag{2.115a}$$

$$= \int_{-\infty}^\infty H(t)F(t) \exp(-2\pi i\nu t)\, dt. \tag{2.115b}$$

To evaluate this integral, we express $R(t)$ in terms of its even and odd parts and then use the signum function to give

$$R(t) = [1 + \mathrm{sgn}\,(t)]\,\mathrm{Re}\,(t). \tag{2.116}$$

Taking the Fourier transform of both sides gives

$$\hat{T}(\nu) = \phi(\nu) + \mathscr{F}\{\mathrm{sgn}\,(t) \cdot \mathrm{Re}\,(t)\}, \tag{2.117}$$

where $\phi(\nu)$ is the (real) Fourier transform of $\mathrm{Re}\,(t)$. Now, by the inverse of (2.110), one may write

$$\mathscr{H}\{\phi(\nu)\} = i\mathscr{F}^{-1}\{\mathrm{sgn}\,(t) \cdot \mathrm{Re}\,(t)\}$$

$$= -i\mathscr{F}\{\mathrm{sgn}\,(t) \cdot \mathrm{Re}\,(t)\}. \tag{2.118}$$

One can substitute for the second term in the right-hand side of (2.117) with the result

$$\hat{T}(\nu) = \phi(\nu) + i\mathscr{H}\{\phi(\nu)\}. \tag{2.119}$$

This remarkable result is the mathematical formulation of causality: in words, it states that a causal transfer function must be such that its real and imaginary parts are Hilbert transforms of one another.

Another aspect of the same theory emerges by considering a stimulus $E_1(t)$ applied to a physical system; this stimulus has the spectrum

$$\hat{S}_1(\nu) = \mathscr{F}\{E_1(t)\}. \tag{2.120}$$

The output spectrum must be given by

$$\hat{S}_2(\nu) = \hat{S}_1(\nu) \cdot \hat{T}(\nu), \tag{2.121}$$

from which it follows that $E_2(t) = \mathscr{F}\{\hat{S}_2(\nu)\}$ is given by

$$E_2(t) = E_1(t) * R(t). \tag{2.122}$$

Thus we see that the output from the system is smoothed by convolution of the input signal with the response function. This result can be checked by invoking the delta function introduced earlier. If $E_1(t) = \delta(t)$, then $\hat{S}_1(\nu) = 1$: $\hat{S}_2(\nu)$ therefore equals $\hat{T}(\nu)$ and the output signal is $\hat{R}(t)$, as expected.

The squared modulus of the transfer function $|\hat{T}(\nu)|^2$ also appears in certain aspects of Fourier spectrometry. By Rayleigh's theorem it follows directly that

$$|\hat{S}_2(\nu)|^2 = |S_1(\nu)|^2 \, |\hat{T}(\nu)|^2. \tag{2.123}$$

2.9.8. The Gaussian distribution Erf (x)

This function, illustrated in Figure 2.13, is given by

$$\mathrm{Erf}\,(ax) = \exp\,(-\pi a^2 x^2); \tag{2.124}$$

it has unit height, and its width at half height is given by

$$\Delta x_{1/2} = 2\pi^{-1/2} a^{-1} (\log_e 2)^{1/2} = 0.9396 a^{-1}. \tag{2.125}$$

The total area under the curve is

$$\int_{-\infty}^{+\infty} \exp\,(-\pi a^2 x^2)\,\mathrm{d}x = a^{-1}. \tag{2.126}$$

The Fourier transform of the Gaussian distribution function is

$$\begin{aligned}
f(k) &= \int_{-\infty}^{+\infty} \exp\,(-\pi a^2 x^2) \exp\,(-2\pi i k x)\,\mathrm{d}x \\
&= a^{-1} \exp\,(-\pi k^2 / a^2)
\end{aligned} \tag{2.127}$$

and is hence also a Gaussian distribution function.

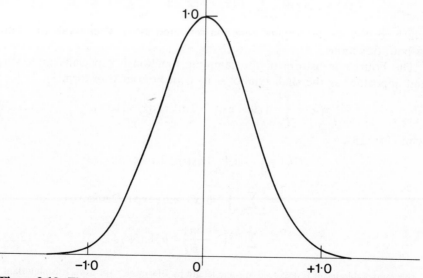

Figure 2.13. The error or Gaussian Function $\exp\,(-\pi a^2 x^2)$ shown for the special case $a = 1$

2.9.9. The sampling function—'shah'—$\sqcup\!\sqcup(x)$

A widespread, and in fact nearly essential, part of the method for carrying out the numerical Fourier transformation of the interferograms which are produced by Fourier spectrometry consists of sampling the interferogram at equal intervals of path difference. The reason for this practice, in a nutshell, is that the interferogram is a continuous but not a simple analytic function and, since the Fourier transformation has to be carried out in a finite time, only a finite number of samples of the interferogram can be processed and the integration must be replaced by a summation. The concept of an infinite one-dimensional array of delta functions is extremely useful for the representation of this sampling. We follow Bracewell's proposal[32] for the use of the descriptive symbol 'shah' for the sampling function and write

$$\sqcup\!\sqcup(x) = \sum_{m=-\infty}^{\infty} \delta(x-m) \qquad (2.128)$$

or, more generally,

$$\sqcup\!\sqcup\left(\frac{x}{b}\right) = \sum_{m=-\infty}^{\infty} b\delta(x-mb). \qquad (2.129)$$

Multiplication of a function $F(x)$ by $\sqcup\!\sqcup(x)$ effectively samples it at equal intervals:

$$F(x)\sqcup\!\sqcup(x) = \sum_{m=-\infty}^{\infty} F(m) \qquad (2.130)$$

and retains information about $F(x)$ only at the values $x = m$. The convolution of a function $F(x)$ with the sampling comb gives a function

$$F(x) * \sqcup\!\sqcup(x) = \sum_{m=-\infty}^{\infty} F(x) * \delta(x-m) = \sum_{m=-\infty}^{\infty} F(x-m), \qquad (2.131)$$

which consists of the original function repeated at equal intervals to infinity in both directions.

The Fourier transform of the sampling function is especially interesting and important as the shah symbol is its own Fourier transform:†

$$f(k) = \int_{-\infty}^{\infty} \sqcup\!\sqcup(x) \exp(-2\pi ikx)\, dx \equiv \sqcup\!\sqcup(k), \qquad (2.132)$$

while for $\sqcup\!\sqcup(x/b)$:

$$f(k) = \int_{-\infty}^{\infty} \sqcup\!\sqcup\left(\frac{x}{b}\right) \exp(-2\pi ikx)\, dx$$

$$= \sum_{m=-\infty}^{\infty} \int_{-\infty}^{\infty} \delta(x-mb) \exp(-2\pi ikx)\, dx$$

$$= \sum_{m=-\infty}^{\infty} \delta\left(k-\frac{m}{b}\right) = b\sqcup\!\sqcup(bk); \qquad (2.133)$$

† This remarkable result is discussed at some length by Bracewell[32] (p. 214). It is important to realize that shah, like delta, must be considered as the limit of an otherwise proper function as some defining parameter is allowed to tend to zero.

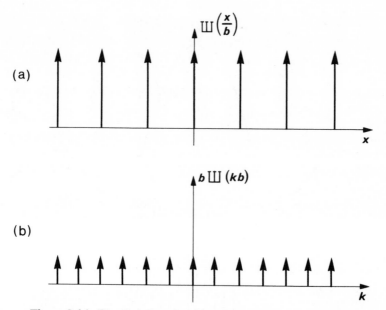

Figure 2.14. The shah function (a) and its Fourier transform (b)

thus

$$\text{Ш}(x) \supset \text{Ш}(k) \qquad (2.134)$$

and

$$\text{Ш}(x/b) \supset b\,\text{Ш}(bk). \qquad (2.135)$$

These relations are only true when the array of Dirac delta functions, or the proper functions used to define the limit, are truly infinite in extent.

2.10. THE SAMPLING THEOREM

When a continuous function $F(x)$ is sampled, according to (2.130), to give a series of discrete values, it is obvious that the extent to which the series represents the original function is determined to a considerable degree by the size chosen for the sampling intervals (Figure 2.15). If the sampling interval is infinitesimal, very little information is discarded; if it is large, a lot is lost and, under certain circumstances, it is conceivable that the sampled versions of two completely different functions may be identical. The sampling theorem tells us the size of the interval that is necessary to ensure a unique and meaningful sampled version of a given function; in fact, it is possible to recover the intervening values between the sampled values without any loss of accuracy.

A simple consideration of a few of the properties of sampled Fourier integrals will soon show how the sampling theorem arises. Consider the

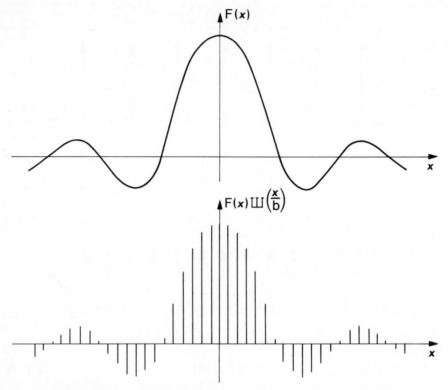

Figure 2.15. An analogue function and its sampled form $F(x) \sqcup \! \sqcup (x/b)$

sampled function

$$F(x) \sqcup \! \sqcup \left(\frac{x}{b} \right) = b \sum_{m=-\infty}^{\infty} F(x) \delta(x - mb). \tag{2.136}$$

If the Fourier transform of $F(x)$ is $f(k)$, then the Fourier transform of (2.136) is

$$b \sqcup \! \sqcup (bk) * f(k), \tag{2.137}$$

This relation shows that the spectral function is replicated at intervals $1/b$. If we specify $f(x)$ to be band-limited, so that $f(k) \equiv 0$ for $k > k_c$, where k_c is the cutoff value of k, then we may represent the true function $f(k)$ in terms of (2.137) by writing

$$[b \sqcup \! \sqcup (bk) * f(k)] \sqcap (k/2k_c), \tag{2.138}$$

which is identical to

$$bf(k), \qquad |k| \leqslant k_c. \tag{2.139}$$

This means that an analogue band-limited function $f(k)$ has been obtained from a *sampled* function $F(x) \sqcup \! \sqcup (x/b)$; moreover, the analogue record is unique if $k_c < (2b)^{-1}$ and therefore, since it will yield a unique Fourier transform, it follows that $F(x)$ itself can be uniquely reconstructed knowing only $F(x) \sqcup \sqcup (x/b)$.

When $k_c \nless (2b)^{-1}$, the replicated versions of $f(k)$ which are $(b)^{-1}$ apart will overlap since the range of $f(k)$ is now greater than the separation. This situation is shown in Figure 2.16. It will not now be possible to recover $F(x)$ uniquely from the transformed values of $F(x) \sqcup \sqcup (x/b)$ because the calculated values $[bf(k) * \sqcup \sqcup (kb)] \sqcap (k/2k_c)$ differ from $bf(k)$ in the overlap regions. When the span exactly equals b^{-1}, the uniqueness of the transform is just preserved. Thus it is evident that the sampling interval for $F(x) \sqcup \sqcup (x/b)$ that will just ensure the unique reconstruction of $F(x)$ is given by $b^{-1} = 2k_c$, that is, by $b = (2k_c)^{-1}$. We can, therefore, state the sampling theorem. A function whose Fourier transform is zero for $|k| > k_c$ is fully specified by sampled values placed at equal intervals not exceeding $(2k_c)^{-1}$. (A corollary is generally added by remarking that this statement is true except for any harmonic term with zeros at the sampling points.)

2.11. THE FOURIER SERIES

If one has a well-behaved periodic function $F(x) = F(x + T_0)$, then this function may be expanded in a (usually infinite) linear series of cosine and sine terms, thus

$$F(x) = \tfrac{1}{2}a_0 + \sum_{n=1}^{n=\infty} [a_n \cos(2\pi nx/T_0) + b_n \sin(2\pi nx/T_0)]. \qquad (2.140)$$

This series in the cosines and sines of harmonics of the fundamental frequency is called the Fourier series. The coefficients in (2.140) are given by the finite Fourier transforms

$$a_n = \frac{1}{T_0} \int_{-T_0/2}^{+T_0/2} F(x) \cos(2\pi nx/T_0) \, dx \qquad (2.141a)$$

and

$$b_n = \frac{1}{T_0} \int_{-T_0/2}^{+T_0/2} F(x) \sin(2\pi nx/T_0) \, dx. \qquad (2.141b)$$

If $F(x)$ is even, $b_n \equiv 0$; if $F(x)$ is odd, $a_n \equiv 0$. Under these circumstances the function is given by a Fourier cosine series or by a Fourier sine series, respectively.

The Fourier series arose originally in connection with harmonic analysis in the real world and, of course, there only positive frequencies occur. Nevertheless, it is very useful to generalize the idea of a Fourier series by letting the running integer n take on all values from $-\infty$ to $+\infty$. When this is done, it is clear from the symmetry properties of the integrals in (2.141) that $a_n = a_{-n}$ and $b_n = -b_{-n}$. Having done this, one can then go on to write

56

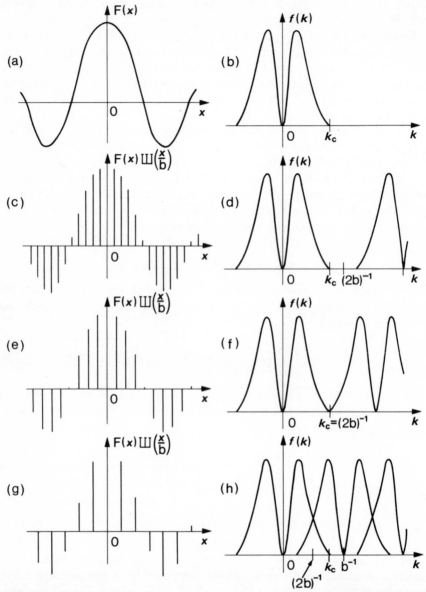

Figure 2.16. The sampling theorem. At the left are shown a function $F(x)$ and three sampled versions of it: to the right are shown the corresponding Fourier transforms. The pairs (c)–(d), (e)–(f), and (g)–(h) correspond to values of the sampling parameter b which are subcritical, critical, and supercritical, respectively. The maximum possible value of b if $F(x)$ is to be uniquely recovered from its sampled form is given by $2b = k_c$

(2.140) in the compact form

$$F(x) = \tfrac{1}{2} \sum_{-\infty}^{+\infty} (a_n + ib_n) \exp(-2\pi inx/T_0).$$ (2.142)

The complex pre-exponential coefficient can then be found from

$$a_n + ib_n = \frac{1}{T_0} \int_{-T_0/2}^{+T_0/2} F(x) \exp(2\pi inx/T_0)\, dx.$$ (2.143)

The series (2.140) and (2.142) are definite only if the function $F(x)$ is such that the integrals (2.141) and (2.143) exist. The prime condition for this is that the integral of $|F(x)|$ from $-T_0/2$ to $+T_0/2$ be finite and that $F(x)$ has no more than a finite number of discontinuities in a finite interval of x.

2.11.1. The finite Fourier series

In practice it is usually the case that functions have to be calculated when only a finite number of their Fourier components can be observed. Naturally the sum of the series is then only an approximation to the true function $F(x)$. Let the sum over N terms be

$$F_N(x) = \tfrac{1}{2} a_0 + \sum_{n=1}^{n=N} (a_n \cos 2\pi nx/T_0 + b_n \sin 2\pi nx/T_0).$$ (2.144)

Introducing the dummy variable y (since x is parametric in (2.144)), one may use equations (2.141) to substitute for a_0, a_n, and b_n with the result

$$F_N(x) = \frac{2}{T_0} \int_{-T_0/2}^{+T_0/2} F(y) \left\{ \tfrac{1}{2} + \sum_{n=1}^{n=N} \cos(2\pi n(y-x)/T_0) \right\} dy,$$ (2.145a)

which may also be written

$$F_N(x) = \frac{2}{T_0} \int_{-T_0/2}^{+T_0/2} F(y) \cdot \left\{ \frac{\sin(2N+1)\pi(y-x)/T_0}{\sin \pi(y-x)/T_0} \right\} dy.$$ (2.145b)

This integral is discussed in standard texts on Fourier series: the only point that needs to be made here is that it only has a significant value in the region where $y - x \sim 0$.

2.12. THE GIBBS' PHENOMENON

The sine and cosine functions are both smooth and slowly varying and therefore any finite number of them added together must produce a resultant which has the same properties. From this it follows that the finite Fourier series approximation can agree closely with the true function in regions where this is smooth and slowly varying but will depart significantly in regions where there are discontinuities. A good example is provided by the square wave odd function which is shown in Figure 2.17. The full

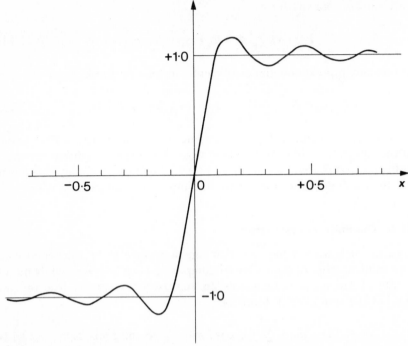

Figure 2.17. The square-wave function and its representation near the origin by a finite Fourier series. The ten-term series clearly demonstrates the Gibbs' phenomenon near the discontinuity at $x = 0$

Fourier series for the square wave is given by

$$F(x) = \frac{4}{\pi}\left[\sin x + \frac{1}{3}\sin 3x + \frac{1}{5}\sin 5x + \text{etc.}\right]. \tag{2.146}$$

The ten-term approximation $F_{10}(x)$ is also shown for comparison in Figure 2.17. It will be seen that, in the regions away from the discontinuities, the ten-term truncation gives a very good approximation to the true function, but near the discontinuities there is 'overshoot' (similar to 'ringing' in electronic circuits) and the function has a finite slope at the discontinuities of the true function. This overshoot and the attendant oscillations form the Gibbs' phenomenon, which distorts any discontinuity in a periodic function if that function is represented by only a finite number of terms in its Fourier series.

To analyse the Gibbs' phenomenon for this particular case in more detail, one makes use of the theory developed in the preceding section. The function over the repeat unit can be written $\text{sgn}\,(x)\,\sqcap(x/2\pi)$ and therefore, from (2.145b),

$$F_N(x) = \frac{1}{2\pi}\int_0^\pi \frac{\sin\,(N+\tfrac{1}{2})(y-x)}{\sin\tfrac{1}{2}(y-x)}\,\mathrm{d}y - \frac{1}{2\pi}\int_{-\pi}^0 \frac{\sin\,(N+\tfrac{1}{2})(y'-x)}{\sin\tfrac{1}{2}(y'-x)}\,\mathrm{d}y'. \tag{2.147}$$

By changing the variable in each integral, we obtain

$$F_N(x) = \frac{1}{2\pi} \int_{-x}^{\pi-x} \frac{\sin(N+\frac{1}{2})v}{\sin\frac{1}{2}v} \, dv + \frac{1}{2\pi} \int_{\pi+x}^{x} \frac{\sin(N+\frac{1}{2})v}{\sin\frac{1}{2}v} \, dv$$

$$= \frac{1}{2\pi} \int_{-x}^{x} \frac{\sin(N+\frac{1}{2})v}{\sin\frac{1}{2}v} \, dv + \frac{1}{2\pi} \int_{\pi+x}^{\pi-x} \frac{\sin(N+\frac{1}{2})v}{\sin\frac{1}{2}v} \, dv. \quad (2.148)$$

Since the integrand is significant only in the region of $v = 0$, the second integral (not including $v = 0$ in its range) is negligible by comparison with the first. Near $v = 0$, $\sin\frac{1}{2}v \sim \frac{1}{2}v$ and, since

$$\text{Si}\,[(N+\tfrac{1}{2})x] = \tfrac{1}{2} \int_{-x}^{+x} \frac{\sin(N+\frac{1}{2})v}{v} \, dv, \quad (2.149)$$

it follows that

$$F_N(x) = \frac{2}{\pi} \text{Si}\,[(N+\tfrac{1}{2})x] \quad (2.150)$$

within the range of x where $F(x)$ is identical with $\text{sgn}\,(x)$. Now, as N gets very large, the range of x where there is sensible oscillation gets smaller and smaller and, in the limit, the function can be set equal to $\text{sgn}\,(x)$ at arbitrarily small values of x. However, there does remain the mathematical curiosity that the *magnitude* of the overshoot does not decrease as N increases; all that happens is that the oscillatory behaviour becomes more and more confined.

2.12.1. The Fejer series

This is a series which better represents, over a partial sum of N terms, the function $F(x)$. We write

$$F_N(x) = \tfrac{1}{2}a_0 + \sum_{n=1}^{n=N} \left(1 - \frac{n}{N}\right)(a_n \cos(2\pi nx/T_0) + b_n \sin(2\pi nx/T_0)). \quad (2.151)$$

The Gibbsian behaviour of the finite Fourier series approximation at a discontinuity is due to the lack of uniform convergence of that series in the neighbourhood of the discontinuity. The Fejer series damps out the spurious oscillations and there is no 'ringing'.

2.13. RANDOM PROCESSES—AVERAGE VALUES

Noise is a random process; that is to say that the noise function is indeterminate in any particular single record. Large collections of records, however, show strong statistical relationships and provided sufficient are available the random process may be fully characterized. We define a random or stochastic process as an ensemble or set of time functions such that the ensemble can be characterized by statistical properties.

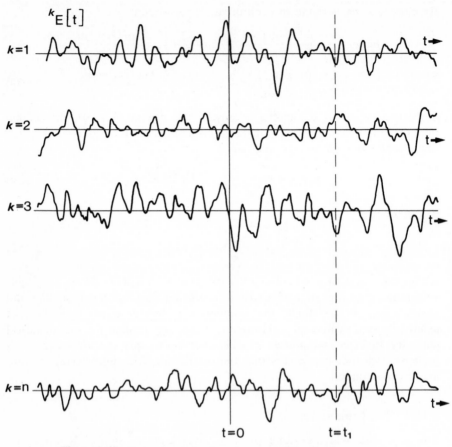

Figure 2.18. Some examples from an ensemble of noise records

Let us consider a particular set $^kE(t)$ of random functions. The first random record for $-\infty < t < \infty$ is different from the second, and so on. To examine the properties of the ensemble we need to select a fixed time, say t_1, and compare the magnitudes of $^kE(t_1)$ for the various k. The average value is

$$\widetilde{E(t_1)} = \lim_{N \to \infty} \frac{1}{N} \sum_{k=1}^{N} {}^kE(t_1), \qquad (2.152)$$

where $\overset{\sim}{}$ denotes average over the ensemble. In the most general case, we find that at different times t_1 and t_2 the ensemble or statistical averages $\widetilde{E(t_1)}$ and $\widetilde{E(t_2)}$ are not the same, that is,

$$\widetilde{E(t_1)} \neq \widetilde{E(t_2)}. \qquad (2.153)$$

Processes such as this are common in physics and are known as non-stationary random processes, since the statistical properties need to be determined from ensemble averages and depend on the time t_1 of the observation.

There is, however, an important class of random functions that does not satisfy (2.153); averages over the ensemble of records at different times are not different in magnitude, that is,

$$\widetilde{E(t_1)} = \widetilde{E(t_2)} = \ldots = \widetilde{E(t_m)}. \tag{2.154}$$

We conclude, therefore, that the statistical properties are invariant with respect to time translations, that is,

$$\widetilde{E(t_1)} = \widetilde{E(t_1 + t)}. \tag{2.155}$$

Stochastic processes having time invariance of statistical properties are known as stationary random processes. Ensemble averages are still required in making statistical measurements, but a great saving in labour is effected by comparison with the non-stationary case for only one time need be selected for the study.

The study of the random process may be further simplified by applying another restriction. We shall assume each record of the ensemble to be statistically equivalent to every other record. As a consequence, ensemble averages over a large number of records at a fixed time may be replaced by corresponding time averages on a single representative record of the ensemble. Stationary random processes such as this are said to be ergodic.

We can easily see the equivalence of the ensemble average and the time average when the process is ergodic. Suppose one record in an ensemble is infinite in extent, then this can be divided into sections of length $2T$, where T is large. There will be an infinite number of such sections and these can be stacked as if they were themselves records within an ensemble. The ensemble average is clearly the same as the time average.

The time average is represented by

$$\overline{{}^k E} = \lim_{T \to \infty} \frac{1}{2T} \int_{-T}^{T} {}^k E(t)\, dt, \tag{2.156}$$

which is independent of t but is apparently a function of the particular record used. In fact, in the light of the preceding, it appears that the time average is the same for all records, that is,

$$\overline{{}^1 E} = \overline{{}^2 E} = \ldots = \overline{{}^m E}. \tag{2.157}$$

So we can drop the set indicator and write $\bar{E}$.

It often happens that the mean value of a random function is zero, being, on average, equally positive and negative. In such cases and, in fact, in some others, it is necessary to turn to a higher power of the function and find the average of these higher powers in a particular record.

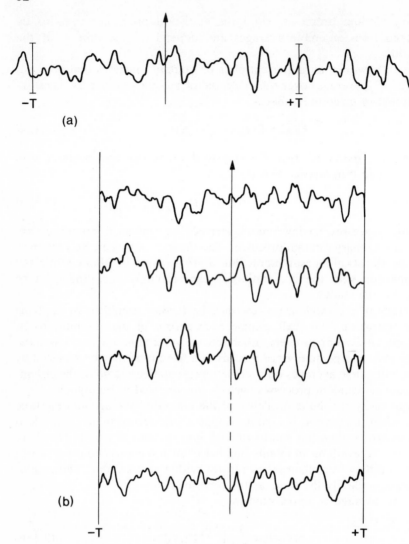

Figure 2.19. Infinite stationary ergodic record (a) divided into an infinite number of sections of length $2T$. These are then assembled (b) into a time-limited ensemble

2.14. AUTOCORRELATION FUNCTIONS OF RANDOM VARIABLES

2.14.1. Autocorrelation

When we are dealing with random functions we are frequently concerned with the extent to which they show similarity of behaviour, for example the extent to which the behaviour at two times t_1 and t_2 is correlated from set to set. If we consider the values $^kE(t_1)$ and $^kE(t_2)$ at those times, we can use the average

$$\gamma_{EE}^k(t_1, t_2) = \overline{^kE(t_1)^kE(t_2)} \tag{2.158}$$

of their product over all ensembles to represent a measure of the correlation. We call $\gamma_{EE}^k(t_1, t_2)$ the ensemble (or statistical) autocorrelation function. In general, its value depends on the times t_1 and t_2 chosen.

When the function is stationary the value of (2.158) will not change if temporal translation occurs, therefore

$$\gamma_{EE}^k(t_1, t_2) = \overline{^kE(t_1+t)^kE(t_2+t)} = \overline{^kE(t)^kE(t_2-t_1+t)} \tag{2.159}$$

and, in particular,

$$\gamma_{EE}^k(t_1, t_2) = \overline{^kE(0)^kE(t_2-t_1)}. \tag{2.160}$$

This shows that, in the special case of stationarity, $\gamma_{EE}^k(t_1, t_2)$ depends only on the difference $\tau = t_2 - t_1$. The last member of (2.159) may be written as

$$\gamma_{EE}^k(t_1, t_2) \triangleq \gamma_{EE}(\tau) = \overline{^kE(t)^kE(\tau+t)}. \tag{2.161}$$

Remembering that ergodicity enables us to replace ensemble averages by temporal averages, we may write

$$\gamma_{EE}(\tau) = \langle {}^kE(t)^kE(\tau+t)\rangle = \lim_{T\to\infty} \frac{1}{2T} \int_{-T}^{T} {}^kE(t)^kE(\tau+t)\, dt. \tag{2.162}$$

The right-hand side of this equation is an autocorrelation integral and, in alternative notation, we have

$$\gamma_{EE}(\tau) = {}^kE(\tau) \star {}^kE(\tau). \tag{2.163}$$

2.14.2. Autocovariance

A particular form of the correlation function arises when it is the variation about the mean that is to be correlated. The quantities concerned are

$$^kE(\tau+t) - \overline{^kE} \quad \text{and} \quad {}^kE(t) - \overline{^kE}$$

and the correlation is expressed in the form of the average:

$$C_{EE}(\tau) = \langle[{}^kE(t) - \overline{^kE}][{}^kE(\tau+t) - \overline{^kE}]\rangle. \tag{2.164}$$

We call this the autocovariance. The autocovariance is identical with the autocorrelation when the mean is zero.

Finally, it should be noted that τ can, in principle, assume any value. There is a particular value worthy of special consideration, namely $\tau = 0$. We then have

$$\gamma_{EE}(0) = \langle {}^kE(t)^kE(t)\rangle = {}^kE(0) \star {}^kE(0) \tag{2.165}$$

for the autocorrelation function at zero lag and

$$C_{EE}(0) = \langle [{}^k E(t) - \overline{{}^k E}][{}^k E(t) - \overline{{}^k E}] \rangle \tag{2.166}$$

for the autocovariance. For a random function $E(t)$, therefore, $C_{EE}(0)$ is simply the variance. We shall see that these quantities have an important relationship with the spectral density of a temporal random variable (section 2.16 below).

2.14.3. Cross-correlation and covariance

The above discussion can be easily extended to the question of correlation between two different random variables, ${}^k E(t)$ and ${}^k F(t)$ at times t_1 and t_2. The statistical average

$$\gamma_{EF}(t_1, t_2) = {}^k E(t_1)^k F(t_2) \tag{2.167}$$

is called the ensemble cross-correlation function. This is identical to the temporal cross-correlation function

$$\gamma_{EF}(\tau) = \langle {}^k E(t)^k F(\tau + t) \rangle \tag{2.168}$$

when ${}^k E(t)$ and ${}^k F(t)$ are stationary ergodic functions. When the means are subtracted, we have the temporal covariance:

$$C_{EF}(\tau) = \langle [{}^k E(t) - \overline{{}^k E}][{}^k F(t) - \overline{{}^k F}] \rangle. \tag{2.169}$$

The covariance and the cross-correlation are identical when the means $\overline{{}^k E}$ and $\overline{{}^k F}$ are both zero.

2.15. RANDOM FUNCTIONS AND PROBABILITY

2.15.1. Probability

In discussing random functions, we have seen that we are frequently concerned with the value of the function at some instant. Because they are random, we cannot specify the actual value, but have to be satisfied with the ascription of a probable value.

Suppose $R(u)$ is a random variable, a real-valued function of u, where u is an arbitrary element of a random array. We suppose that $R(u)$ may have any value: $-\infty < R(u) < \infty$. The probability that $R(u)$ may have a value within the total infinite range is unity; the probability that the value be within some finite portion of the infinite range is less than unity. We can write the probability of the value being less than or equal to E as

$$D(E) = \text{Prob} \,|R(u) \leqslant E|, \tag{2.170}$$

where we call $D(E)$ the distribution function. Clearly,

$$\left. \begin{array}{l} D(\infty) = \text{Prob}\,[R(u) \leqslant \quad \infty] = 1 \\ D(-\infty) = \text{Prob}\,[R(u) \leqslant -\infty] = 0 \end{array} \right\} \tag{2.171}$$

and

$$D(a) < D(b) \quad \text{when} \quad a < b. \tag{2.172}$$

$D(E)$ is therefore a monotonically increasing function.

2.15.2. Probability density function

Let us suppose that $D(E)$ is differentiable, that is,

$$f(E) = \frac{d}{dE} D(E) \tag{2.173}$$

exists. This derivative, which is called the probability density function, is always positive because of the monotonic increase of $D(E)$. The probability that a random variable has a precise value is, of course, zero, but one may define a narrow range of the variable dE so that the probability that the variable lies within this range will be finite. From (2.173) it is natural to define the chance that the variable $R(u)$ lies within the range $E_i - (E_i + dE)$ as $f(E_i)\, dE$. By integration, the chance that the variable lies between a and b is given by

$$\int_a^b f(E)\, dE = D(b) - D(a) = \text{Prob}\,[a < R(u) < b]. \tag{2.174}$$

2.15.3. Expected values

The expected value of $R(u)$ is the limit of the sum of all the possible values multiplied by the appropriate probability of occurrence, that is,

$$\langle R(u) \rangle = \lim_{N \to \infty} \sum_{i=1}^{N} E_i \, \text{Prob}\,[R(u) = E_i]$$

$$\equiv \int_{-\infty}^{\infty} E f(E)\, dE. \tag{2.175}$$

This is equivalent to the mean value of $R(u)$, so

$$\overline{R(u)} = \langle R(u) \rangle = \int_{-\infty}^{\infty} E f(E)\, dE = \alpha_1. \tag{2.176}$$

The expected value of $R^2(u)$ is, by analogy,

$$\langle R^2(u) \rangle = \int_{-\infty}^{\infty} E^2 f(E)\, dE = \alpha_2 \tag{2.177}$$

and so on for higher powers:

$$\langle R^n(u) \rangle = \int_{-\infty}^{\infty} E^n f(E)\, dE = \alpha_n \tag{2.178}$$

These integrals are known as the moments of the distribution, the nth moment being denoted by α_n.

In many cases it is more useful to deal with moments in relation to the mean. These are known as central moments and are denoted by

$$\mu_n = \langle \{R(u) - \overline{R(u)}\}^n \rangle = \int_{-\infty}^{\infty} (E - \bar{R}(u))^n f(E) \, \mathrm{d}E. \tag{2.179}$$

In particular, the very significant second moment is given by

$$\mu_2 = \langle \{R(u) - \overline{R(u)}\}^2 \rangle = \int_{-\infty}^{\infty} (E - \overline{R(u)})^2 f(E) \, \mathrm{d}E. \tag{2.180}$$

It is immediately evident that the first central moment is zero. The second central moment is equivalent to the variance, and its positive root is the standard deviation, that is,

$$\sigma = +\sqrt{\mu_2} = +\sqrt{\overline{\{R(u) - R(u)\}^2}} \tag{2.181}$$

A consideration of the distribution of the probability density enables us, therefore, to evaluate the mean and the variance of a random variable. When the variable is a function of time we have, in effect, the temporal mean (which is generally zero) and the temporal variance (which is generally the autocorrelation function), thus

$$\langle R(u) \rangle = \alpha_1 (= 0) \quad \text{and} \quad \langle R^2(u) \rangle = \alpha_2 = \gamma(0). \tag{2.182}$$

2.15.4. The normal distribution

A random variable x is said to be normally distributed if its probability density function has the Gaussian form

$$f(x) = \frac{1}{\sqrt{2\pi}\sigma_n} \exp\left\{-\frac{(x-m)^2}{2\sigma_n^2}\right\}. \tag{2.183}$$

The multiplying factor is chosen so that the integral of $f(x)$ from $-\infty$ to $+\infty$ shall be unity. The parameters m and σ_n are the mean and the standard deviation, respectively.

2.15.5. The central limit theorem

If a random variable x is composed from a sum of mutually independent random variables $x_1, x_2, \ldots, x_N$, the mean of x will be the sum of the means of each component and the variance of x will be the sum of the individual variances. The central limit theorem states that provided no individual variable x_i dominates the sum, the random variable x will become normally distributed as $N \to \infty$.

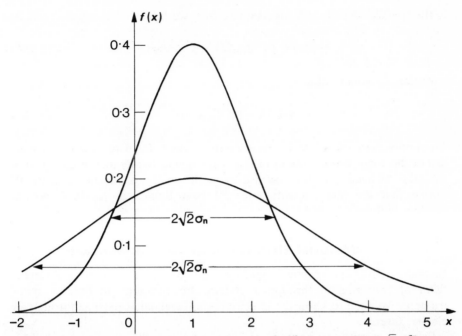

Figure 2.20. The Gaussian function $f(x) = (2\pi\sigma^2)^{-1/2} \exp\left[-\{(x-m)/\sqrt{2}\sigma\}^2\right]$ shown for $m = 1$ and the two values of σ, 1.0 and 2.0. As σ increases the peak height falls but the width increases to compensate

2.16. RANDOM SIGNALS AND SPECTRAL DENSITY

The random variables with which one deals may take one of a variety of functional forms of which the most common is a temporal dependence. It will, therefore, be the temporal statistical properties that we deal with. We can regard the temporal noise signal as being composed of a large range of sinusoids of differing frequency and random amplitude and phase. Our understanding of the nature and properties of the noise signal will be enhanced by some knowledge of the frequency distribution of the sinusoids.

Fortunately, we can readily find this distribution when the random signal is stationary and ergodic. In that case the autocorrelation function is

$$\gamma_{EE}(\tau) = \langle E(t)E(\tau+t)\rangle. \tag{2.184}$$

From section 2.5 it follows that the spectral density of the noise is simply the Fourier transform of (2.184), that is,

$$\mathfrak{P}_{EE}(\nu) = \int_{-\infty}^{\infty} \gamma_{EE}(\tau) \exp(-2\pi i\nu\tau)\,\mathrm{d}\tau \tag{2.185}$$

is the spectral density. By inversion of this, we have

$$\gamma_{EE}(\tau) = \int_{-\infty}^{\infty} \mathfrak{P}_{EE}(\nu) \exp(2\pi i \nu \tau) \, d\nu, \qquad (2.186)$$

with the particular value

$$\gamma_{EE}(0) = \int_{-\infty}^{\infty} \mathfrak{P}_{EE}(\nu) \, d\nu. \qquad (2.187)$$

The importance of $\gamma_{EE}(0)$ is immediately evident, for it represents a measure of the total power content of the noise signal. In any measurement, it is desirable to keep $\gamma_{EE}(0)$ as small as possible. In this connection it is worth noting that the noise density *must* be band-limited if $\gamma_{EE}(0)$ is not to become infinite.

2.17. RANDOM SIGNALS AND LINEAR SYSTEMS

In dealing with random noise signals we are concerned with the effect upon them of transmission through a system; for example, in Fourier spectrometry we need to know the effect on a noise signal of passage through an electric filter.

Suppose a random signal $^kE_1(t)$ is passed into a filter having an impulse response function $R(t)$. The output signal, which is smoothed in relation to the input signal, is the convolution

$$^kE_2(t) = {}^kE_1(t) * R(t) = \int_{-\infty}^{\infty} {}^kE_1(t-t')R(t') \, dt'. \qquad (2.188)$$

The type of distribution to which $^kE_2(t)$ belongs is the same as that for $^kE_1(t)$ but the spectral density differs as we can show. The squared value of the output, that is,

$$^kE_2(t)^kE_2(t) = \int_{-\infty}^{\infty} dt' \int_{-\infty}^{\infty} dt'' {}^kE_1(t-t')^kE_1(t-t'')R(t')R(t''), \quad (2.189)$$

can be averaged over the ensemble to give the statistical autocorrelation function

$$^kE_2(t)^kE_2(t) = \int_{-\infty}^{\infty} dt' \int_{-\infty}^{\infty} dt'' \overline{{}^kE_1(t-t')^kE_1(t-t'')}R(t')R(t''). \quad (2.190)$$

The statistical average over the entire right member reduces to an average over the random terms since it is only for these that the average is meaningful.

This general relation simplifies considerably, as usual, when the random process is stationary and ergodic for then the averages can be replaced by their temporal counterparts, the ensemble label can be omitted, and the average squared value of the output noise is expressible as

$$\gamma_{E_2 E_2}(0) = \int_{-\infty}^{\infty} dt' \int_{-\infty}^{\infty} dt'' \gamma_{E_1 E_1}(t' - t'') R(t') R(t''), \tag{2.191}$$

where

$$\gamma_{E_2 E_2}(0) = \lim_{T \to \infty} \frac{1}{2T} \int_{-\infty}^{\infty} E_2(t) E_2(t) \, dt = E_2(0) \star E_2(0) \tag{2.192}$$

is the mean power of the output noise and

$$\gamma_{E_1 E_1}(t' - t'') = \lim_{T \to \infty} \frac{1}{2T} \int_{-T}^{T} E_1(t + t'' - t') E_1(t) \, dt \tag{2.193}$$

or

$$\gamma_{E_1 E_1}(\tau) = E_1(\tau) \star E_1(\tau), \tag{2.194}$$

where

$$\tau = t' - t'', \tag{2.195}$$

is the (temporal) autocorrelation function of the input noise.

We thus see how a formulation of the mean power of the noise leaving a linear system, such as a filter, necessarily involves an autocorrelation function—that of the filter.

2.18. SPECTRAL DENSITY OF FILTERED RANDOM SIGNAL

The spectral density of the output noise is, by analogy with (2,185), the Fourier transform

$$\mathfrak{P}_{E_2 E_2}(\nu) = \int_{-\infty}^{\infty} \gamma_{E_2 E_2}(\tau') \exp(-2\pi i \nu \tau') \, d\tau' \tag{2.196}$$

of the autocorrelation function of the output signal. We need, therefore, to find this function. For a stationary ergodic output we have

$$\gamma_{E_2 E_2}(\tau') = \lim_{T \to \infty} \frac{1}{2T} \int_{-\infty}^{\infty} E_2(t + \tau') E_2(t) \, dt, \tag{2.197}$$

which can be written as

$$\gamma_{E_2 E_2}(\tau') = \lim_{T \to \infty} \frac{1}{2T} \int_{-T}^{T} dt \int_{-\infty}^{\infty} dt' \int_{-\infty}^{\infty} dt'' E_1(t + \tau' - t') E_1(t - t'') R(t') R(t'') \tag{2.198}$$

by using (2.193). The expression

$$\lim_{T \to \infty} \frac{1}{2T} \int_{-T}^{T} E_1(t + \tau' - t') E_1(t - t'') \, dt \tag{2.199}$$

can be written as

$$\gamma_{E_1 E_1}(\tau' + t'' - t')$$

owing to its invariance to time shift, thus (2.198) becomes

$$\gamma_{E_2 E_2}(\tau') = \int_{-\infty}^{\infty} dt' \int_{-\infty}^{\infty} dt'' \gamma_{E_1 E_1}(\tau' + t'' - t') R(t') R(t''); \qquad (2.200)$$

an expression that obviously reduces to (2.191) when $\tau' = 0$. The spectral density of the output noise is, therefore,

$$
\begin{aligned}
\mathfrak{P}_{E_2 E_2}(\nu) &= \int_{-\infty}^{\infty} d\tau' \int_{-\infty}^{\infty} dt' \int_{-\infty}^{\infty} dt'' \gamma_{E_1 E_1}(\tau' + t'' - t') R(t') R(t'') \exp(-2\pi i \nu \tau') \\
&= \int_{-\infty}^{\infty} du \int_{-\infty}^{\infty} dt' \int_{-\infty}^{\infty} dt'' \gamma_{E_1 E_1}(u) \exp(-2\pi i \nu u) R(t') \\
&\quad \times \exp(-2\pi i \nu t') R(t'') \exp(2\pi i \nu t'') \\
&= \mathfrak{P}_{E_1 E_1}(\nu) T(\nu) T(-\nu), \qquad (2.201)
\end{aligned}
$$

where we have introduced, for convenience, the dummy variable $u = \tau' + t'' - t'$. $T(\nu)$ is the transfer function, the Fourier transform $\mathscr{F}\{R(t)\}$ of the impulse response function. Since $T(\nu)$ is Hermitian,

$$T(\nu) T(-\nu) = |T(\nu)|^2 = \mathscr{G}(\nu)$$

and

$$\mathfrak{P}_{E_2 E_2}(\nu) = \mathfrak{P}_{E_1 E_1}(\nu) \mathscr{G}(\nu) \qquad (2.202)$$

shows how the output spectral density is related to that of the input via the filter function $\mathscr{G}(\nu)$, which is positive and real. Moreover, since the mean power of the noise is given by the integral of $\mathfrak{P}_{E_2 E_2}(\nu)$, we have

$$\gamma_{E_2 E_2}(0) = \int_{-\infty}^{\infty} \mathscr{G}(\nu) \mathfrak{P}_{E_1 E_1}(\nu) \, d\nu \qquad (2.203)$$

for the mean output power.

When the input spectrum is 'white', that is, when the spectral density is constant, at say $\mathfrak{P}_0$, then

$$\gamma_{E_2 E_2}(0) = \mathfrak{P}_0 \int_{-\infty}^{\infty} \mathscr{G}(\nu) \, d\nu \qquad (2.204)$$

shows that the output spectral density depends directly on the 'area' $\int_{-\infty}^{\infty} \mathscr{G}(\nu) \, d\nu$ of the filter function.

CHAPTER 3

Mathematical description of electromagnetic radiation

3.1. THE ELECTROMAGNETIC WAVE-FIELD

Before deriving the dependence of the signal at the output of a two-beam interferometer on the phase delay within it, it will prove useful to summarize some of the properties of the electromagnetic field and the mathematical descriptions that we shall require.

In a strictly monochromatic (ideal) wave-field, the amplitude $E(\nu_0)$ of the electric vector of the electromagnetic field at a given point in the field is constant in time while the phase $\phi(\nu_0) - 2\pi\nu_0 t$ varies linearly with time. The periodic field fluctuation emitted by an ideal elementary radiator consists of a wave-train of infinite extent. However, a real elementary radiator emits only for finite (short) times at random intervals and so the signal emitted consists of wave-trains of finite length. A large number of these pass the observer at random time intervals during the time taken to make an observation. Moreover a real 'point' source is not just a single elementary radiator, but consists of a large number of such radiators. The radiators, each of which actually emits polarized radiation, have random orientations and the resultant field emitted by the ensemble of radiators varies randomly in amplitude phase and plane of polarization. If this field could be observed by an apparatus having an exceedingly short response time, there would appear to be a brief time ΔT during which no perceptible variation of the amplitude or phase occurred. This time ΔT is called the *coherence time* and the corresponding distance travelled is called the *coherence length*. The coherence length may alternatively be regarded as the length of the component wave-trains from which the resultant signal is composed.

According to Fourier's theorem, any finite wave-train can be expressed as the sum of strictly monochromatic (i.e. infinitely long) wave-trains. The infinitely long wave-trains may be regarded as superimposing in such a way as to give zero disturbance outside the time interval of the actual wave-train. This description is strictly equivalent to attributing a finite frequency range $\Delta\nu$ to the Fourier components of the source radiation. The Fourier spectrum of real spectral lines always has a finite width, that is, even the sharpest line is quasi-monochromatic rather than truly monochromatic. We deduce the same width whether we integrate the instantaneous effect of an ensemble of oscillators or integrate, over repeated measurements, the effect produced by a single oscillator. It follows that the emission of the wave-trains is a random process that is stationary and ergodic. The individual fluctuations of optical signals occur, of course, too frequently to be detected and the random variations which are observed represent merely an integrated effect.

Interference can occur when two or more similar wave-trains are superimposed. The wave-trains required for the superposition can be produced either by division of the amplitude of the incident beam or by division of its wavefront. The Michelson interferometer operates by the former method, the lamellar grating interferometer by the latter. Either case may be taken as a particular example of the general situation obtaining when radiation from two secondary sources is combined. The relation between the radiation fields produced by the two sources can be determined by considering the radiation fields at two points irradiated by a primary source and then assuming that these two points act as secondary sources.

The fluctuations occurring at two points chosen at similar distances from a source are *on average* identical but are quite independent and different *at any particular instant* unless the points are close together. If the two points P_1, P_2 lie nearby in a plane irradiated by the source and placed some distance from it, the fluctuations of amplitude and phase as the wave-trains pass are not independent provided the points are only a few wavelengths apart and are both many wavelengths from the source. The signals at the two points are derived from the same wave-train and differ by only a relative phase delay due to the difference $(SP_1 - SP_2)$ between the distances SP_1, SP_2 of the points from the source S (Figure 3.2). If P_1 and P_2 are kept fixed whilst the source becomes finite, the signals become less similar but remain correlated to some degree provided $(SP_1 - SP_2)$ does not exceed the coherence length for all points S in the source plane. There is, therefore, a region of coherence round any point in a wave-field, the extent of which is governed by the coherence length, source size, and distance from the source.

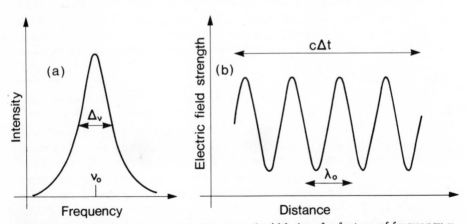

Figure 3.1. Interdependence of (a) the spectral width $\Delta\nu$ of a feature of frequency ν_0 and (b) the length of the corresponding wave-train of wavelength $\lambda_0 = c/\nu_0$ and time duration $\Delta t = 1/\Delta\nu$

Figure 3.2. Relation between two points P_1 and P_2 in an interference plane and an infinitesimal area dS on the source S. The relative magnitudes of S and SP determine the size of a region of coherence around P and the extent of the overlap of the regions centred on P_1, P_2 determines the degree of correlation in the wave-fields at these points

When the source includes ensembles of oscillators of widely differing mean frequencies it is said to be *polychromatic* (or *broad-band* in the limiting case of a wide continuous distribution). The coherence time is shorter and the region of coherence is reduced. The wave-field can be adequately described by introducing a measure of the correlation that exists between the fluctuations at different points such as P_1, P_2. When the correlation is high, P_1 and P_2 are said to be *coherent points* and the radiation field that would result on combining the radiations from the two points is an *interference field*, the intensity in which is found using the principle of amplitude superposition. When correlation is absent, P_1 and P_2 are incoherent points, no observable interference is possible, and the resultant intensity is found by adding the (independent) intensities detected when the wave-trains from P_1 and P_2 are observed separately. The most general case lies between the two extremes, when we describe P_1 and P_2 as *partially coherent points*.

The electric and magnetic field fluctuations are, of course, perpendicular to the direction of propagation of the wave and may be referred to as vector quantities. However, unless we are dealing with plane-polarized radiation, the directions of the electric and magnetic vectors vary randomly in the plane that contains them and the wave is called a scalar wave. Physically, the electric and magnetic fields are real quantities, but it is convenient to represent them in complex form when discussing the wave-field mathematically and then select the real parts as appropriate.

3.2. THE MONOCHROMATIC WAVE-FIELD

3.2.1. The electric field

If we have a monochromatic wave propagating (*in vacuo*) along the z direction, then the electric field vector $\mathbf{E}$, in terms of the usual $\mathbf{i}$, $\mathbf{j}$, $\mathbf{k}$ unit vectors, may be written

$$\mathbf{E}'(z, t) = \mathbf{i}[E_0 \cos (2\pi\nu_0 z/c - 2\pi\nu_0 t)] \qquad [\text{V m}^{-1}] \qquad (3.1\text{a})$$

$$= \mathbf{i}[E_0 \cos (\phi_0 - 2\pi\nu_0 t)] \qquad [\text{V m}^{-1}], \qquad (3.1\text{b})$$

where E_0 is the scalar amplitude and ν_0 is the frequency. Now, as mentioned above, it is convenient to introduce complex representations of periodic phenomena and therefore one writes

$$\hat{\mathbf{E}}(z, t) = \mathbf{i}[E_0 \exp i(2\pi\nu_0 z/c - 2\pi\nu_0 t)] \qquad [\text{V m}^{-1}] \qquad (3.2\text{a})$$

$$= \mathbf{E}'(z, t) - i\mathbf{E}''(z, t) \qquad [\text{V m}^{-1}]. \qquad (3.2\text{b})$$

The field vector (3.1a) can be regarded, therefore, as the real part of the complex amplitude (3.2a). Equation (3.2a) can also be written

$$\hat{E}(z, t) = \mathbf{i}[E_0 \exp (i\phi_0)][\exp (-i2\pi\nu_0 t)] = \mathbf{i}\hat{E}_0 \exp (-i2\pi\nu_0 t) \qquad [\text{V m}^{-1}], \qquad (3.3)$$

It is important to note that (3.2) does not contain anything additional to (3.1); both $\mathbf{E}''(z, t)$ and $\hat{\mathbf{E}}(z, t)$ can be immediately deduced once $\mathbf{E}'(z, t)$ is known.

3.2.2. The magnetic field

The magnetic field vector is in phase with the electric vector but directed perpendicularly to both it and the direction of propagation; one therefore has

$$\mathbf{H}'(z, t) = \mathbf{j}[H_0 \cos (2\pi\nu_0 z/c - 2\pi\nu_0 t)] \qquad [\text{A m}^{-1}] \qquad (3.4\text{a})$$

$$= \mathbf{j}[H_0 \cos (\phi_0 - 2\pi\nu_0 t)] \qquad [\text{A m}^{-1}]. \qquad (3.4\text{b})$$

As before, one can also write

$$\hat{\mathbf{H}}(z, t) = \mathbf{j}\hat{H}_0 \exp (-i2\pi\nu_0 t) \qquad [\text{A m}^{-1}] \qquad (3.5)$$

$$= \mathbf{H}'(z, t) - i\mathbf{H}''(z, t).$$

The amplitudes of the magnetic and electric fields are related by the equation

$$H_0 = \sqrt{\frac{\varepsilon_r \varepsilon_0}{\mu_r \mu_0}} \, E_0 \qquad [\text{A m}^{-1}], \qquad (3.6)$$

where ε_r and μ_r are the relative permittivity and permeability of the medium in which the propagation is taking place and ε_0 and μ_0 are the permittivity and the permeability of free space, $8.854 \times 10^{-12} \, \text{F m}^{-1}$ and $1.257 \times 10^{-6} \, \text{H m}^{-1}$, respectively. ε_r and μ_r are pure (dimensionless) numbers and can be taken as unity for air or vacuum. In this case, equation (3.6) simplifies to

$$H_0 = \varepsilon_0 c E_0 \qquad [\text{A m}^{-1}] \qquad (3.7)$$

since ε_0 and μ_0 are related by

$$\varepsilon_0\mu_0 = c^{-2} \quad [\text{m}^{-2}\,\text{s}^2]. \tag{3.8}$$

This result leads to the expression $\sqrt{\mu_0/\varepsilon_0} = 376.7\ \Omega$, the so-called free-space impedance.

3.2.3. The energy flux

The instantaneous flux of energy transported by the monochromatic wave (3.1), (3.4) is given by the magnitude of the Poynting vector $\mathbf{S}(t)$, which is the vector product

$$\mathbf{S}(t) = \mathbf{E}'(t) \times \mathbf{H}'(t) = E_0 H_0 \cos^2(\phi_0 - 2\pi\nu_0 t)\mathbf{i} \times \mathbf{j} \quad [\text{J m}^{-2}\,\text{s}^{-1}]. \tag{3.9}$$

The direction of this vector is therefore coincident with that of propagation since $\mathbf{i} \times \mathbf{j} = \mathbf{k}$. If one now writes

$$\mathbf{S}(t) = \mathbf{k}S(t) \quad [\text{J m}^{-2}\,\text{s}^{-1}], \tag{3.10}$$

then the average value of the Poynting vector magnitude over the cycle time $T_0 = 1/\nu_0$ is

$$\bar{S} = \frac{1}{T_0}\int_0^{T_0} S(t)\,\mathrm{d}t = \frac{\varepsilon_0 c}{T_0}\int_0^{T_0} E_0^2 \cos^2(\phi_0 - 2\pi\nu_0 t)\,\mathrm{d}t = \tfrac{1}{2}\varepsilon_0 c E_0^2 \quad [\text{J m}^{-2}\,\text{s}^{-1}]. \tag{3.11}$$

The time-average of the energy flux flowing along the direction of propagation of the monochromatic wave is proportional to the square of the amplitude of the electric field component of the electromagnetic field. Consequently, we need deal only with the electric field component in our discussions of energy and power. The time-averaged value of the energy flux, the energy passing along the direction of propagation through unit area in unit time (power passing along direction of propagation through unit area), is the *intensity* of the radiation represented by the wave. It follows therefore that intensity is a measure of the square of the electric field strength and vice versa.

Although the result (3.11) is derived via a non-linear (in fact squaring) process, it can also be arrived at using the complex representation since

$$\hat{\mathbf{E}}(t) \times \hat{\mathbf{H}}^*(t) = E_0 H_0 = \varepsilon_0 c E_0^2 \quad [\text{J m}^{-2}\,\text{s}^{-1}], \tag{3.12}$$

and therefore, from (3.11),

$$\bar{S} = \tfrac{1}{2}\varepsilon_0 c E_0^2 = \overline{\tfrac{1}{2}\hat{\mathbf{E}}(t) \times \hat{\mathbf{H}}^*(t)} \quad [\text{J m}^{-2}\,\text{s}^{-1}]. \tag{3.13}$$

3.3. THE POLYCHROMATIC WAVE-FIELD

3.3.1. The real signal

A perfectly monochromatic wave-field, as mentioned earlier, is an impossible abstraction, but, via Fourier's integral theorem, the concept of a monochromatic field is still useful for the analysis of the quasi-monochromatic and polychromatic fields which are encountered in real life. One therefore writes the field amplitude in the form

$$E'(t) = \int_0^\infty a(\nu) \cos\{\phi(\nu) - 2\pi\nu t\} \, d\nu \qquad [\text{V m}^{-1}] \qquad (3.14)$$

and notes that, since a real field only has positive frequency components, the integral in (3.14) covers only the positive half range. This is inconvenient mathematically however, and one therefore makes use of the relations developed in Chapter 2, which permit a function which is confined to positive-only values of its argument to be resolved into simply related components. One writes

$$a(\nu) = a_e(\nu) + a_0(\nu) \qquad [\text{V m}^{-1}\text{s}] \qquad (3.15)$$

and, therefore,

$$E'(t) = \int_{-\infty}^{+\infty} a_e(\nu) \cos\{\phi(\nu) - 2\pi\nu t\} \, d\nu \qquad [\text{V m}^{-1}] \qquad (3.16)$$

since, of course, $\phi(\nu) = -\phi(-\nu)$. This equation can also be written in the form

$$E'(t) = \int_{-\infty}^{+\infty} a_e(\nu) \exp[i\phi(\nu)] \exp(-2\pi i\nu t) \, d\nu \qquad [\text{V m}^{-1}] \qquad (3.17)$$

since $\exp(i\phi)$ is Hermitian and therefore $E'(t)$ can be written formally as a Fourier integral. It must be stressed, however, that although we evaluate $a_e(\nu) \exp(i\phi)$ for all ν, it can only be interpreted physically in the half range $0 \leq \nu \leq \infty$.

3.3.2. The analytic signal

We have just shown that a real signal can be written in complex form, but we find it very convenient, for mathematical purposes, to go one step further and introduce a complex analogue of the real polychromatic signal. We seek, that is, a relation equivalent to (3.2) for use with polychromatic fields. The first move is to introduce an imaginary field,

$$iE''(t) = i\int_0^\infty a(\nu) \sin\{\phi(\nu) - 2\pi\nu t\} \, d\nu \qquad [\text{V m}^{-1}], \qquad (3.18)$$

which, analogously, can be written

$$iE''(t) = \int_{-\infty}^\infty a_0(\nu) \exp[i\phi(\nu)] \exp(-2\pi i\nu t) \, d\nu \qquad [\text{V m}^{-1}]. \qquad (3.19)$$

The complex quantity

$$\hat{E}(t) = E'(t) + iE''(t) \qquad [\text{V m}^{-1}] \qquad (3.20)$$

then takes on the form

$$\hat{E}(t) = \int_0^{+\infty} \hat{a}(\nu) \exp{-(2\pi i \nu t)} \, d\nu \qquad [\text{V m}^{-1}], \qquad (3.21)$$

where

$$\hat{a}(\nu) = a(\nu) \exp{[i\phi(\nu)]} \qquad [\text{V m}^{-1} \text{s}] \qquad (3.22a)$$

$$= 2a_e(\nu) \exp{[i\phi(\nu)]}, \qquad \nu \geqslant 0. \qquad (3.22b)$$

The complex function $\hat{E}(t)$ is known as the *analytic signal*. It is given the epithet because, if t is allowed to take complex values, $\hat{E}(t)$ has no poles in the lower half of the complex plane. This analyticity is just another aspect of causality—an asymmetry (all the poles in one half-plane) in the behaviour of one variable is a consequence of the asymmetry ($a(\nu)$ not defined for negative ν) in that of its conjugate. From the arguments given in Chapter 2 it will be seen that $E'(t)$ and $E''(t)$ are necessarily Hilbert transforms of one another—a general property of an analytic function. That is,

$$E'(t) = -\mathscr{H}\{E''(t)\} \qquad [\text{V m}^{-1}] \qquad (3.23a)$$

and

$$E''(t) = \mathscr{H}\{E'(t)\} \qquad [\text{V m}^{-1}]. \qquad (3.23b)$$

The analytic signal is a widely used concept, especially in communications theory. In optics it provides a means of access to the 'envelope' of a signal, which is often observable even though the individual oscillations defining it are taking place too fast to be observed. When the real signal is quasi-monochromatic, its envelope is given simply by the modulus of the analytic signal. The square of this modulus is, by Rayleigh's theorem, simply related to that of $\hat{a}(\nu)$ and to the squares of its individual components. Explicitly,

$$\int_{-\infty}^{+\infty} |\hat{E}(t)|^2 \, dt = \int_0^{\infty} |\hat{a}(\nu)|^2 \, d\nu = 2\int_{-\infty}^{+\infty} [E'(t)]^2 \, dt = 2\int_{-\infty}^{+\infty} [E''(t)]^2 \, dt$$

$$[\text{V}^2 \text{m}^{-2} \text{s}] \qquad (3.24a)$$

and thus, from (3.20), we finally have

$$\int_{-\infty}^{+\infty} |\hat{E}(t)|^2 \, dt = 2\int_0^{\infty} [a(\nu)]^2 \, d\nu \qquad [\text{V}^2 \text{m}^{-2} \text{s}]. \qquad (3.24b)$$

The analytic signal is readily computed, given $E'(t)$, from the inversion of (3.17) followed by the use of (3.22) and the application of the half-range transform (3.21).

3.3.3. Time-limited signal

In practice, of course, $E'(t)$ can be measured only for a finite time, in which case (3.24) has to be replaced by finite integrals of the form

$$\int_{-T}^{+T} [E'(t)]^2 \, dt \qquad [\text{V}^2 \text{m}^{-2} \text{s}]. \qquad (3.25a)$$

If this expression is divided by $2T$, we obtain the average

$$\overline{E^2} = \frac{1}{2T} \int_{-T}^{+T} [E'(t)]^2 \, dt \qquad [\text{V}^2 \, \text{m}^{-2}], \qquad (3.25b)$$

which is very similar in form to the expression (3.11) representing the magnitude of the Poynting vector of a monochromatic wave. To gain some understanding of what this average represents, we replace $E'(t)$ by the truncated signal

$$E'_T(t) = E'(t) \sqcap \left(\frac{t}{2T}\right) \qquad [\text{V} \, \text{m}^{-1}], \qquad (3.26)$$

which is assumed to be square integrable and can therefore be written as a Fourier integral:

$$E'_T(t) = \int_0^\infty a_T(\nu) \cos \{\phi(\nu) - 2\pi\nu t\} \, d\nu \qquad [\text{V} \, \text{m}^{-1}]. \qquad (3.27)$$

We now proceed analogously to introduce a truncated analytic signal

$$\hat{E}_T(t) = E'_T(t) + iE''_T(t)$$

$$= \int_0^\infty \hat{a}_T(\nu) \exp(-2\pi i\nu t) \, d\nu \qquad [\text{V} \, \text{m}^{-1}], \qquad (3.28)$$

noting however that neither $\hat{E}_T(t)$ nor $E''_T(t)$ is necessarily restricted to the range $-T \leq t \leq +T$. The relations (3.24) then hold with all the quantities replaced by their truncated equivalents. A set of corresponding averages is now formed by dividing each integral by the total duration, $2T$, of the temporal signal: one therefore has

$$\frac{1}{2T} \int_{-\infty}^{+\infty} \{E'_T(t)\}^2 \, dt = \frac{1}{2T} \int_{-\infty}^{+\infty} \{E''_T(t)\}^2 \, dt = \frac{1}{4T} \int_{-\infty}^{+\infty} |\hat{E}_T(t)|^2 \, dt = \frac{1}{2} \int_0^\infty \frac{|\hat{a}_T(\nu)|^2}{2T} \, d\nu$$

$$[\text{V}^2 \, \text{m}^{-2}]. \qquad (3.29)$$

It is now desired to proceed to the limit as $T \to \infty$, but unfortunately, as discussed by Born and Wolf,[33] it is frequently found that the integrals in (3.29) do not tend to limiting values as $T \to \infty$ but merely fluctuate. This difficulty may be overcome, in practice, by remembering that optical signals are made up from the stationary and ergodic signals from a large number of independent radiators and it is in fact the averages

$$2 \int_{-\infty}^{+\infty} \overline{\frac{\{E'_T(t)\}^2 \, dt}{2T}} = 2 \int_{-\infty}^{+\infty} \overline{\frac{\{E''_T(t)\}^2 \, dt}{2T}} = \int_{-\infty}^{+\infty} \overline{\frac{|\hat{E}_T(t)|^2 \, dt}{2T}} = \int_0^{+\infty} \overline{\left(\frac{|\hat{a}_T(\nu)|^2}{2T}\right)} \, d\nu$$

$$[\text{V}^2 \, \text{m}^{-2}] \qquad (3.30)$$

of an ensemble of signals that are detected. It can be shown (see Born and Wolf[33] for references) that the last term of (3.30) tends to a limiting value when $T \to \infty$, and therefore one can write

$$2\langle\{E'(t)\}^2\rangle = 2\langle\{E''(t)\}^2\rangle = \langle|\hat{E}(t)|^2\rangle = \lim_{T \to \infty} \int_0^{+\infty} \left(\frac{|\hat{a}_\mathrm{T}(\nu)|^2}{2T}\right) \mathrm{d}\nu \qquad [\mathrm{V}^2\,\mathrm{m}^{-2}], \tag{3.31}$$

where the angular brackets denote the time average formally defined by

$$\langle E(t)\rangle = \lim_{T \to \infty} \int_{-\infty}^{+\infty} \frac{E_\mathrm{T}(t)}{2T}\,\mathrm{d}t. \tag{3.32}$$

This is identical to $\overline{\langle E(t)\rangle}$ when the ensemble is stationary and ergodic. If we now introduce the factor $\varepsilon_0 c\,[\Omega^{-1}]$, we obtain values for the recorded *intensity*

$$\mathscr{I} = \tfrac{1}{2}\varepsilon_0 c\langle|\hat{E}(t)|^2\rangle \qquad [\mathrm{W}\,\mathrm{m}^{-2}]. \tag{3.33}$$

This is related to the (smoothed) spectral density† $G(\nu)$ by the equation

$$\mathscr{I} = \int_{-\infty}^{+\infty} G(\nu)\,\mathrm{d}\nu = 2\int_0^\infty G_\mathrm{e}(\nu)\,\mathrm{d}\nu \qquad [\mathrm{W}\,\mathrm{m}^{-2}]. \tag{3.34}$$

The modulus of an infinitely long (temporally) randomly oscillating electric field is therefore related to the intensity of radiation observed and to the spectral density of this radiation. $G(\nu)$ is formally defined as the contribution to the mean detected intensity from components lying in the frequency range ν to $\nu+\mathrm{d}\nu$. The quantity $a_\mathrm{T}(\nu)$ does not have a simple physical interpretation, its dimensions being those of action per unit area.

3.4. COHERENCE

We shall now consider the radiation signals at two neighbouring points that are irradiated by a polychromatic primary source. The degree of correlation between these two signals can be assessed from a calculation of the resultant signal that is observed at a third point assumed to be irradiated by the two points acting as secondary sources.

3.4.1. Point source—temporal coherence

3.4.1.1. Mutual coherence

Radiation from the polychromatic point-source $\mathrm{d}S$ (Figure 3.3) falls on the two points P_1, P_2. The two points are similar distances from $\mathrm{d}S$. We assume that $\mathrm{d}SP_1$, $\mathrm{d}SP_2$ are many times $\bar{\lambda}$ but that P_1 and P_2 are only a few times $\bar{\lambda}$ apart, $\bar{\lambda}$ being the mean effective wavelength. Radiation from P_1 and P_2

---

† $G(\nu)$ is also sometimes called the (smoothed) power spectrum, but since $G(\nu)$ does not have the dimensions of power per unit bandwidth but rather has those of power per unit area per unit bandwidth, this practice seems undesirable.

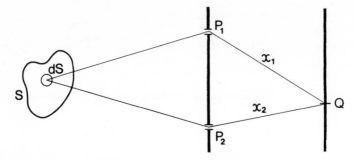

Figure 3.3. Irradiation of a point Q by two secondary sources P_1 and P_2. The radiation is originally derived from an infinitesimal element dS of the extended source S

proceeds along the paths P_1Q (length x_1) and P_2Q (length x_2) to recombine at Q. The instantaneous analytic signal at Q is the sum of contributions from P_1 and P_2, that is,

$$\hat{E}(Q;t) = \hat{K}_1\hat{E}(P_1;t-t_1) + \hat{K}_2\hat{E}(P_2;t-t_2) \qquad [\text{V m}^{-1}], \qquad (3.35)$$

where $\hat{K}_1$, $\hat{K}_2$ are complex factors describing the frequency-independent attenuation and phase shift (if any) that may occur in travelling from P_1P_2 to Q and t_1 and t_2 are the respective transit times. The intensity at Q may now be found from (3.33), with the result†

$$\mathscr{I}(Q) = \tfrac{1}{2}\varepsilon_0 c \langle \hat{E}(Q;t)\hat{E}^*(Q;t)\rangle \qquad [\text{W m}^{-2}], \qquad (3.36)$$

i.e.

$$
\begin{aligned}
2\mathscr{I}(Q) = {}& \varepsilon_0 c\,|\hat{K}_1|^2\,\langle|\hat{E}(P_1;t-t_1)|^2\rangle + \varepsilon_0 c\,|\hat{K}_2|^2\,\langle|\hat{E}(P_2;t-t_2)|^2\rangle \\
& + \varepsilon_0 c\hat{K}_1\hat{K}_2^*\langle\hat{E}(P_1;t-t_1)\hat{E}^*(P_2;t-t_2)\rangle \\
& + \varepsilon_0 c\hat{K}_1^*\hat{K}_2\langle\hat{E}^*(P_1;t-t_1)\hat{E}(P_2;t-t_2)\rangle.
\end{aligned}
\qquad (3.37)
$$

Now, as the field fluctuations are stationary and ergodic, the time origin may be shifted in all terms (see section 2.13) and (3.37) reduces to

$$\mathscr{I}(Q) = |\hat{K}_1|^2\,\mathscr{I}_1 + |\hat{K}_2|^2\,\mathscr{I}_2 + \hat{K}_1\hat{K}_2^*\hat{\Gamma}_{12}(t_2-t_1) + \hat{K}_1^*\hat{K}_2\hat{\Gamma}_{12}^*(t_2-t_1), \qquad (3.38)$$

where $\mathscr{I}_1$ and $\mathscr{I}_2$ are the intensities at P_1 and P_2 and

$$\hat{\Gamma}_{12}(t_2-t_1) = \tfrac{1}{2}\varepsilon_0 c\langle\hat{E}(P_1;t-t_1)\hat{E}^*(P_2;t-t_2)\rangle \qquad [\text{W m}^{-2}] \qquad (3.39)$$

is defined as the *mutual coherence* of the field fluctuations reaching Q from S via P_1 and P_2. If we let $t_2-t_1=\tau$, then it will be seen that $\hat{\Gamma}_{12}(\tau)$ is the

---

† The discussion given here, in common with that in most textbooks, appears to violate the principle of conservation of energy. A detailed treatment of *all* the processes involved resolves the paradox, but since it is purely the *interference* effects which are under consideration here, this treatment is omitted.

average *cross-correlation* function of $\hat{E}(P_1 t)$ and $\hat{E}(P_2 t)$. An average *cross-correlation coefficient* $\hat{\gamma}_{12}(\tau)$ may be defined in the usual way and then (3.38) takes on the form

$$\mathcal{I}(Q) = \mathcal{I}_1(Q) + \mathcal{I}_2(Q) + \sqrt{\mathcal{I}_1(Q)\mathcal{I}_2(Q)}[\hat{\gamma}_{12}(\tau) + \hat{\gamma}_{12}^*(\tau)] \qquad [\text{W m}^{-2}] \quad (3.40)$$

for the usual cases where $\hat{K}_1$ and $\hat{K}_2$ are pure imaginary.

When the points P_1 and P_2 coincide, (3.39) reduces to

$$\hat{\Gamma}_{11}(\tau) = \tfrac{1}{2}\varepsilon_0 c \langle \hat{E}(P; t+\tau)\hat{E}^*(P; t) \rangle \qquad [\text{W m}^{-2}], \quad (3.41)$$

which is called the *self-coherence*: this is equal to the average autocorrelation function of $\hat{E}(P; t)$. The self-coherence reduces to ordinary intensity when $\tau = 0$, that is, when the fluctuations have no relative delay. It follows therefore that

$$\Gamma_{11}(0) = \mathcal{I}_1 \quad \text{and} \quad \Gamma_{22}(0) = \mathcal{I}_2 \qquad [\text{W m}^{-2}]. \quad (3.42)$$

These quantities can be used to normalize the mutual coherence, thus

$$\hat{\gamma}_{12}(\tau) = \hat{\Gamma}_{12}(\tau)/\sqrt{\Gamma_{11}(0)\Gamma_{22}(0)} \quad (3.43a)$$

$$= \hat{\Gamma}_{12}(\tau)/\sqrt{\mathcal{I}_1\mathcal{I}_2}. \quad (3.43b)$$

The quantity $\hat{\gamma}_{12}(\tau)$ has, of course, a modulus and an argument. The value of the argument determines the quality of the interference at Q, but it is the modulus which determines its magnitude. Thus if $|\hat{\gamma}_{12}(\tau)|$ is unity, the interference ranges from perfect constructive to perfect destructive and the points P_1 and P_2 are said to be coherent points. If $|\hat{\gamma}_{12}(\tau)|$ is zero, there is no interference and the beams from P_1 and P_2 simply add their intensities to one another; P_1 and P_2 are then said to be incoherent points. In general, $0 < |\hat{\gamma}_{12}(\tau)| < 1$ and the points are said to be partially coherent points. $\Gamma_{12}(0)$ is called the mutual intensity of the two points. The complex degree of coherence $\hat{\gamma}_{12}(\tau)$ can be experimentally determined from (3.40), and it is therefore an important physical parameter, unlike $\hat{E}(P; t)$ itself which is not usually accessible because of the high frequencies of optical fields.

3.4.1.2. *Mutual spectral density*

The mutual coherence, as noted above, is the cross-correlation of the field fluctuations $\hat{E}(P_1; t+\tau)$ and $\hat{E}^*(P_2; t)$. In general, of course, only the truncated forms $\hat{E}_T(P_1; t+\tau)$ and $\hat{E}_T^*(P_2; t)$ will be available, and then we will be interested in integrals of the form

$$\tfrac{1}{2}\varepsilon_0 c \int_{-\infty}^{\infty} \hat{E}_T(P_1; t+\tau)\hat{E}_T^*(P_2; t)\,\mathrm{d}t. \quad (3.44)$$

It readily follows from (3.28) and its corresponding inverse that the integral (3.44) can be written in the form

$$\tfrac{1}{2}\varepsilon_0 c \int_0^{\infty} \hat{a}_T(P_1; \nu)\hat{a}_T^*(P_2; \nu) \exp(-2\pi i \nu \tau)\,\mathrm{d}\nu. \quad (3.45)$$

The identity of (3.44) and (3.45) is a result akin to Rayleigh's theorem and, in fact, this identity reduces to (3.24b) when $P_1 = P_2$ and $\tau = 0$. If we now divide throughout by $2T$, take the ensemble average, and allow T to tend to infinity, (3.44) becomes the mutual coherence and we can write

$$\hat{\Gamma}_{12}(\tau) = \int_0^\infty G_{12}(\nu) \exp(-2\pi i \nu \tau)\, d\nu \qquad [\text{W m}^{-2}], \qquad (3.46)$$

where

$$G_{12}(\nu) = \tfrac{1}{2}\varepsilon_0 c \lim_{T \to \infty} \left(\frac{\overline{|\hat{a}_T(P_1;\nu)\hat{a}_T^*(P_2;\nu)|^2}}{2T} \right) \qquad [\text{W m}^{-2}\,\text{Hz}^{-1}]. \quad (3.47)$$

$G_{12}(\nu)$ is the *mutual spectral density* of the fluctuations at P_1 and P_2. It is, from (3.46), formally the spectrum of the mutual coherence, and for this reason it is also sometimes called the *cross power spectrum*, but this term is undesirable because it does not have the dimensions of power per unit bandwidth. When $P_1 = P_2$ and $\tau = 0$, the mutual coherence becomes the intensity at P and therefore

$$\Gamma_{11}(0) = \mathscr{I}_1 = \tfrac{1}{2}\varepsilon_0 c \int_0^\infty \lim_{T \to \infty} \left(\frac{\overline{|\hat{a}_T(P;\nu)|^2}}{2T} \right) d\nu \qquad [\text{W m}^{-2}]. \quad (3.48)$$

This shows that the intensity at a point P depends on the integral over all frequencies of the squared modulus of the spectral amplitude of the field fluctuations at the point. Many authors quote this result in a form which is not dimensionally correct, either because constants, which have dimensions, are omitted or else because $\hat{a}_T(P;\nu)$ is not defined properly. The limit in the integral in (3.48) of the ensemble average is equal to the spectral density $G_{11}(\nu)$ at P_1 (or P_2). It is useful to introduce the hypothetical complex conjugates $\hat{U}_1(\nu)$ and $\hat{U}_1^*(\nu)$, with dimensions $[\text{W m}^{-2}\,\text{Hz}^{-1}]^{1/2}$, which satisfy the equation

$$\hat{U}_1(\nu)\hat{U}_1^*(\nu) \triangleq G_{11}(\nu) \qquad [\text{W m}^{-2}\,\text{Hz}^{-1}]. \quad (3.49)$$

Usually, $\hat{U}_1(\nu)$ is a real quantity and then $\hat{U}_1(\nu) = \hat{U}_1^*(\nu) = [G_{11}(\nu)]^{1/2}$. In terms of these quantities, equation (3.48) becomes

$$\Gamma_{11}(0) = \mathscr{I}_1 = \int_0^\infty |\hat{U}_1(\nu)|^2\, d\nu \qquad [\text{W m}^{-1}]. \quad (3.50)$$

$\hat{U}(\nu)$ is called the *complex spectral amplitude of the spectral density of the field fluctuations*.

Since (3.38) contains the real part of the mutual coherence, it is important to know the dependence of this on the mutual spectral density. As usual, we may write

$$\hat{\Gamma}_{12}(\tau) = \Gamma'_{12}(\tau) + i\Gamma''_{12}(\tau) \qquad [\text{W m}^{-2}], \quad (3.51)$$

noting that, since $\hat{\Gamma}_{12}(\tau)$ contains no spectral components belonging to negative frequencies, the two components of $\hat{\Gamma}_{12}(\tau)$ are Hilbert transforms of one another. $\Gamma'_{12}(\tau)$ is an even function and may therefore be written

$$\Gamma'_{12}(\tau) = \int_0^\infty G_{12}(\nu) \cos 2\pi\nu\tau \, d\nu = \int_{-\infty}^{+\infty} G_{(e)12}(\nu) \exp(-2\pi i\nu\tau) \, d\nu \qquad [\text{W m}^{-2}].$$

(3.52)

This result can also be written in terms of the correlation of the real disturbances, thus

$$\Gamma'_{12}(\tau) = \tfrac{1}{2}\varepsilon_0 c \langle E'(P_1; t+\tau)E'(P_2; t)\rangle \qquad [\text{W m}^{-2}]. \qquad (3.53)$$

3.4.2. Extended source—spatial coherence

The argument can now be generalized to include the case of an extended source. The source is imagined to be divided into small elements dS_1, dS_2, etc., centred about the points S_1, S_2, etc., and having linear dimensions that are small by comparison with the wavelengths radiated. At P_1 and P_2 the field fluctuations are the sums

$$\hat{E}_T(P_1; t) = \sum_m \hat{E}_{mT}(P_1; t), \qquad \hat{E}_T(P_2; t) = \sum_m \hat{E}_{mT}(P_2; t). \qquad (3.54)$$

The mutual coherence function then becomes

$$\hat{\Gamma}_{12}(\tau) = \tfrac{1}{2}\varepsilon_0 c \sum_m \langle \hat{E}_{mT}(P_1; t+\tau)\hat{E}^*_{mT}(P_2; t)\rangle$$

$$+ \tfrac{1}{2}\varepsilon_0 c \sum_{m \neq n} \langle \hat{E}_{mT}(P_1; t+\tau)\hat{E}^*_{nT}(P_2; t)\rangle, \quad (3.55)$$

but, since the contributions from the different source elements are assumed to be mutually incoherent, the second summation is identically zero and only the first need be considered. One therefore has

$$\hat{\Gamma}_{12}(\tau) = \tfrac{1}{2}\varepsilon_0 c \lim_{T\to\infty} \sum_m \int_0^\infty \left(\frac{\hat{a}_m(P_1; \nu)\hat{a}^*_m(P_2; \nu)}{2T}\right) \exp(-2\pi i\nu t) \, d\nu \quad (3.56)$$

and, in terms of the complex attenuation factors K, this becomes

$$\hat{\Gamma}_{12}(\tau) = \tfrac{1}{2}\varepsilon_0 c \lim_{T\to\infty} \sum_m \int_0^\infty \left(\frac{|a_m(S_m; \nu)|^2}{2T}\right) \hat{K}_m(P_1; \nu)\hat{K}^*_m(P_2; \nu) \exp(-2\pi i\nu\tau) \, d\nu$$

$$[\text{W m}^{-2}] \quad (3.57\text{a})$$

$$= \sum_m \int_0^\infty G_m(\nu)\hat{K}_m(P_1; \nu)\hat{K}^*_m(P_2; \nu) \exp(-2\pi i\nu\tau) \, d\nu$$

$$[\text{W m}^{-2}], \quad (3.57\text{b})$$

where $G_m(\nu)$ is the spectral density of the mth element of the source. This can be expressed, via surface integration, in terms of the *spectral density per unit area* $L_m(\nu) [(\text{W m}^{-2} \text{ Hz}^{-1}) \text{ m}^{-2}]$, and when the elements become so small and close that they cannot be distinguished from a continuous distribution the summations in the above equations can be replaced by double integrals. One therefore has

$$\hat{\Gamma}_{12}(\tau) = \int_0^\infty d\nu \iint_S dS L_s(\nu) \hat{K}_s(P_1; \nu) \hat{K}_s^*(P_2; \nu) \exp(-2\pi i \nu \tau) \quad [\text{W m}^{-2}],$$

$$(3.58)$$

which can be written

$$\hat{\Gamma}_{12}(\tau) = \int_0^\infty \hat{M}_{12}(\nu) \exp(-2\pi i \nu \tau) \, d\nu \quad [\text{W m}^{-2}], \quad (3.59)$$

where $\hat{M}_{12}(\nu)$ is the mutual spatial coherence. By comparison of (3.59) and (3.46) it will be seen that as the points P_1 and P_2 coalesce the imaginary part of $\hat{M}_{12}(\nu)$ vanishes and as the source size shrinks $\hat{M}_{12}(\nu)$ tends to $G_{12}(\nu)$.

Hopkins[34] wrote the mutual spatial coherence in the form

$$\hat{M}_{12}(\nu) = \iint_S \hat{U}_s(P_1; \nu) \hat{U}_s^*(P_2; \nu) \, dS \quad (3.60)$$

in terms of the complex amplitude of the spectral density per unit area given by

$$\hat{U}_s(P; \nu) = \sqrt{L_s(\nu)} \hat{K}_s(P; \nu) \quad [\text{W m}^{-2} \text{ Hz}^{-1} \text{ m}^{-2}]^{1/2}. \quad (3.61)$$

It is important to note that although this amplitude is of the same type as that introduced in (3.49) its dimensions are different. By analogy with the temporal case, a *complex degree of spatial coherence*

$$\hat{\mu}_{12}(\nu) = \hat{M}_{12}(\nu)/\sqrt{M_{11}(\nu)M_{22}(\nu)} \quad (3.62)$$

can be introduced which is a normalized form of (3.60).

With these concepts we can now proceed to the most general treatment of the two-slit problem. The complex (areal) spectral amplitudes of the disturbances at P_1 and P_2 can be written in the general form

$$\hat{f}(\nu)\hat{U}_s(P; \nu) = \hat{f}(\nu)\sqrt{L_s(\nu)} \hat{K}_s(P; \nu), \quad (3.63)$$

where $\hat{f}(\nu)$ is a factor introduced to allow for the medium of propagation not being a vacuum. The resulting complex spectral amplitude at Q arising from the source element ds is therefore

$$\hat{U}_s(P_1; \nu)\hat{f}_1(\nu) + \hat{U}_s(P_2; \nu)\hat{f}_2(\nu). \quad (3.64)$$

The intensity at Q arising from that element is therefore

$$d\mathscr{I}_s(Q) = \int_0^\infty |\hat{U}_s(P_1; \nu)\hat{f}_1(\nu) + \hat{U}_s(P_2; \nu)\hat{f}_2(\nu)|^2 \, d\nu \quad (3.65)$$

and the total intensity is found by integrating this over the whole source.

The result is

$$\mathscr{I}(Q) = \int_0^\infty M_{11}(\nu)\,|\hat{f}_1(\nu)|^2\,\mathrm{d}\nu + \int_0^\infty M_{22}(\nu)\,|\hat{f}_2(\nu)|^2\,\mathrm{d}\nu$$

$$+ \int_0^\infty [\hat{M}_{12}(\nu)\hat{f}_1(\nu)\hat{f}_2^*(\nu) + \hat{M}_{12}^*\hat{f}_1^*(\nu)\hat{f}_2(\nu)]\,\mathrm{d}\nu. \quad (3.66)$$

The first term represents the contribution from P_1, the second that from P_2, and the remaining two represent the contributions arising from the interactions of the fields from P_1 and P_2. In the case of a small source, this becomes

$$\mathscr{I}(Q) = \int_0^\infty G_{11}(\nu)\,|\hat{f}_1(\nu)|^2\,\mathrm{d}\nu + \int_0^\infty G_{22}(\nu)\,|\hat{f}_2(\nu)|^2\,\mathrm{d}\nu$$

$$+ \int_0^\infty [G_{12}(\nu)\hat{f}_1(\nu)\hat{f}_2^*(\nu) + G_{12}^*(\nu)\hat{f}_1^*(\nu)\hat{f}_2(\nu)]\,\mathrm{d}\nu, \quad (3.67)$$

while if $\hat{f}(\nu) \rightarrow \hat{f} \exp(-2\pi i \nu \tau)$,

$$\mathscr{I}(Q) = |\hat{f}_1|^2\,\Gamma_{11}(0) + |\hat{f}_2|^2\,\Gamma_{22}(0) + [\hat{f}_1\hat{f}_2^*\hat{\Gamma}_{12}(\tau) + \hat{f}_1^*\hat{f}_2\hat{\Gamma}_{12}^*(\tau)], \quad (3.68)$$

which is identical to (3.38).

3.5. POLARIZATION

3.5.1. Types of polarization—the coherency matrix

The mutually perpendicular electric and magnetic vectors, $\mathbf{E}'(t)$ and $\mathbf{H}'(t)$ of a monochromatic wave (see section 3.2) define a plane which is perpendicular to the direction of propagation and, once the direction of one vector has been fixed and the direction of propagation is known, the direction of the other can be inferred at once. The electromagnetic field is therefore a vector field and, although this can be ignored in many situations, there are many others where the state of polarization of the wave must be explicitly considered.

A wave is *polarized* when the direction of the electric vector is not randomly changing as the wave propagates along a fixed direction. Historically, the plane of polarization was that defined by the direction of propagation and the *magnetic* vector (see Born and Wolf[33] for details), but since the majority of electromagnetic interactions that are studied involve the *electric* vector it seems more sensible to redefine the plane of polarization in terms of it. There are several types of polarization: if the plane remains fixed in space, the wave is said to be *linearly* polarized; if it rotates, the wave is either *circularly* or *elliptically* polarized; and if there is no defined plane existing for the time of observation, the wave is *unpolarized*. In the latter case the direction of the electric vector is randomly changing with time.

These cases, of course, represent extremes, and waves encountered in practice will be in general *partially* polarized.

Any vector can be resolved into two coplanar and mutually perpendicular vectors, and the degree of polarization can be related to the intensities carried by these two components and to the correlation (or coherence) which exists between them.[33] To see how this is so, we shall consider two quasi-monochromatic waves travelling in the same z direction. The respective analytic signals may be denoted

$$\hat{E}_x(t) = E_x^\circ(t) \exp i[\phi_x(t) - 2\pi\bar{\nu}t],$$
$$\hat{E}_y(t) = E_y^\circ(t) \exp i[\phi_y(t) - 2\pi\bar{\nu}t]. \tag{3.69}$$

In these equations, $\bar{\nu}$ is the mean frequency and $E_x^\circ(t)$ and $E_y^\circ(t)$ are (slow) functions of time because the wave is only quasi-monochromatic rather than truly monochromatic. A similar remark applies to $\phi_x(t)$ and $\phi_y(t)$. Suppose now that the y component is given a retardation δ relative to the x component, then the component of the analytic signal in a direction oriented at an angle θ to the x direction will be

$$\hat{E}(t; \theta; \delta) = \hat{E}_x(t) \cos \theta + \hat{E}_y(t) \exp (i\delta) \sin \theta. \tag{3.70}$$

Now, following Born and Wolf,[33] the intensity in the θ direction may be written

$$\begin{aligned}
\mathcal{I}(\theta; \delta) &= \tfrac{1}{2}\varepsilon_0 c \langle \hat{E}(t; \theta; \delta)\hat{E}^*(t; \theta; \delta)\rangle \\
&= \tfrac{1}{2}\varepsilon_0 c [J_{xx} \cos^2 \theta + J_{yy} \sin^2 \theta \\
&\quad + (\hat{J}_{xy} \exp (-i\delta) + \hat{J}_{yx} \exp (i\delta)) \sin \theta \cos \theta],
\end{aligned} \tag{3.71}$$

where the quantities J_{xx}, etc., are of the form

$$J_{ij} = \langle E_i E_j^* \rangle.$$

The relation (3.71), in common with all quadratic forms, can be written in matrix notation:

$$\frac{2\mathcal{I}(\theta; \delta)}{\varepsilon_0 c} = \mathbf{R} \mathbf{J} \tilde{\mathbf{R}}^* = [\cos \theta, \sin \theta \exp (i\delta)] \begin{bmatrix} J_{xx} & \hat{J}_{xy} \\ \hat{J}_{yx} & J_{yy} \end{bmatrix} \begin{bmatrix} \cos \theta \\ \sin \theta \exp (-i\delta) \end{bmatrix}, \tag{3.72}$$

where the tilde ($\tilde{\ }$) denotes transposition. The matrix $\mathbf{J}$ is Hermitian, that is, $J_{ij} = J_{ji}^*$ and its diagonal elements are therefore real: it is known as the *coherency matrix*. The trace of the matrix represents, as usual, an invariant quantity—in this case the total intensity, i.e.

$$\mathcal{I} = J_{xx} + J_{yy}. \tag{3.73}$$

The trace may, therefore, be used to normalize the coherency matrix to give

$$\boldsymbol{\mu} = \mathbf{J}/(J_{xx} + J_{yy}), \tag{3.74}$$

which can alternatively be expressed in terms of its elements as

$$\hat{\mu}_{ij} = |\hat{\mu}_{ij}| \exp i\Phi_{ij}, \tag{3.75}$$

where $\Phi_{ij} = 0$ if $i = j$ and $\Phi_{ij} = -\Phi_{ji}$. In terms of these quantities, (3.71) can be rewritten

$$\frac{2\mathscr{I}(\theta; \delta)}{\varepsilon_0 c} = J_{xx} \cos^2 \theta + J_{yy} \sin^2 \theta + 2\sqrt{J_{xx}J_{yy}} \cos \theta \sin \theta \, |\hat{\mu}_{xy}| \cos (\Phi_{xy} - \delta),$$

(3.76)

which is identical in form with the equation (3.38) describing interference in quasi-monochromatic wave-fields. The magnitudes of the off-diagonal elements $\hat{\mu}_{xy}$ are a measure of the form of polarization assumed by the radiation. When $\hat{\mu}_{xy} = \hat{\mu}_{yx} = 0$ there is no correlation and when $\mu_{xx} = \mu_{yy}$ the intensities carried by the two components are the same. If both conditions apply, the radiation is unpolarized, that is, the intensities of the two incoherent component into which we imagine it split are equal and independent of the relative retardation existing between them. The normalized coherency matrix for unpolarized radiation is therefore

$$\boldsymbol{\mu} = \begin{bmatrix} \frac{1}{2} & 0 \\ 0 & \frac{1}{2} \end{bmatrix}.$$

(3.77)

Thus a beam of unpolarized (or *natural*) radiation travelling along the z direction can be resolved into two mutually perpendicular and mutually incoherent components whose orientation is arbitrary and whose intensities are equal.

3.5.2. Polarized and unpolarized radiation

When the radiation is fully polarized, the correlation between the $0x$ and $0y$ components is complete. For simplicity, it will be assumed, at the moment, that the radiation is monochromatic and, since E_x°, E_y°, ϕ_x, and ϕ_y now no longer depend on time, the elements of the coherency matrix are of the form

$$\hat{J}_{ij} = E_i^{\circ} E_j^{\circ} \exp i(\phi_i - \phi_j).$$

(3.78)

Thus the coherency matrix is of unit rank (i.e. it has a zero determinant) and the normalized complex mutual coherence becomes

$$\hat{\mu}_{xy} = \exp i(\phi_x - \phi_y).$$

(3.79)

When $\phi_x - \phi_y = 2m\pi$, the components are in phase, but when $\phi_x - \phi_y = (2m+1)\pi$ the components are out of phase. In both cases, simple vector addition applies and the resultant is a fluctuation which lies in a plane oriented to $0x$ at an angle θ given by

$$\tan \theta = (-1)^n E_y^{\circ}/E_x^{\circ},$$

(3.80)

where n is the number of multiples of π by which the x vibration leads the y vibration. The coherency matrix becomes

$$\mathbf{J} = \begin{bmatrix} (E_x^{\circ})^2 & E_x^{\circ} E_y^{\circ}(-1)^n \\ E_x^{\circ} E_y^{\circ}(-1)^n & (E_y^{\circ})^2 \end{bmatrix}.$$

(3.81)

This describes the case of complete linear polarization in which the vector representing the radiation vibration is confined to a single *plane of polarization*. The orientation of the plane is determined by the relative intensities of the component waves and, since the complex mutual coherence is either ± 1, the component waves are fully coherent. If the directions of the axes $0x$, $0y$ have been chosen so that either E_x° or E_y° is zero, then the normalized coherency matrices take on the particularly simple forms

$$\boldsymbol{\mu} = \begin{bmatrix} 0 & 0 \\ 0 & 1 \end{bmatrix} \quad \text{or} \quad \boldsymbol{\mu} = \begin{bmatrix} 1 & 0 \\ 0 & 0 \end{bmatrix}. \tag{3.82}$$

When the phase difference betweeen the two components is an odd multiple of $\pi/2$, the two are in quadrature. The complex mutual coherence is

$$\hat{\mu}_{xy} = \exp i(\phi_x - \phi_y) = \pm i. \tag{3.83}$$

The normalized coherency matrix then becomes

$$\boldsymbol{\mu} = \begin{bmatrix} \frac{1}{2} & \pm i/2 \\ \mp i/2 & \frac{1}{2} \end{bmatrix} \tag{3.84}$$

for the case when the two components are of equal intensity. This matrix occurs in other areas of physics; it is, for example, related to the spinor matrices used to discuss electron spin, and is always associated with rotational motion. In this instance, the end-point of the vector describing the electric field oscillations describes a circular path and the radiation is said to be *circularly polarized*. When the observer faces the oncoming wave and the vector rotates clockwise, the polarization is said to be right-handed circularly polarized and the positive sign is to be taken in the right upper element of $\boldsymbol{\mu}$. When the rotation is counter-clockwise, it is said to be left-handed circularly polarized and the negative sign is to be taken.

In the more general case, E_x° and E_y° will not be equal and the end-point of the resultant vector traces an ellipse and the radiation is then said to be *elliptically polarized*. In the still more general case, when the phase delay is not a simple multiple of $\pi/2$, the real parts of equation (3.69) have to be solved with the elimination of explicit reference to the time. Introducing the reduced variables $x = E_x'(t)/E_x^\circ$, $y = E_y'(t)/E_y^\circ$ and then making use of the trigonometric identities for $\cos A \pm \cos B$ and $\sin^2 A = 1$ leads to the equation

$$x^2 + 2xy[1 - 2c^2] + y^2 = 4c^2(1 - c^2), \tag{3.85}$$

where $c = \cos (\phi_x - \phi_y)/2$. it will be seen on substitution of the respective values that this equation contains both the limiting cases, i.e. linearly and circularly polarized radiation, but in general it is the equation of an ellipse. Reverting now to the original notation and using the matrix technique to express a quadratic form, one has

$$\begin{bmatrix} E_x' & E_y' \end{bmatrix} \begin{bmatrix} (E_x^\circ)^{-2} & (1 - 2c^2)(E_x^\circ E_y^\circ)^{-1} \\ (1 - 2c^2)(E_x^\circ E_y^\circ)^{-1} & (E_y^\circ)^{-2} \end{bmatrix} \begin{bmatrix} E_x' \\ E_y' \end{bmatrix} = 4c^2(1 - c^2). \tag{3.86}$$

Now it is a readily proved theorem in matrix algebra that the rotation matrix (with respect to the x axes) which diagonalizes a given symmetric 2×2 matrix has $\tan 2\theta = 2M_{xy}/(M_{yy} - M_{xx})$, and, in this particular case,

$$\tan 2\theta = \frac{2E_x^{\circ} E_y^{\circ}}{(E_x^{\circ})^2 - (E_y^{\circ})^2} \cos (\phi_x - \phi_y). \tag{3.87}$$

The value of θ gives the angle at which one of the natural axes of the ellipse is inclined to the x axis. Equations (3.85), (3.86), (3.87) are the most general description of polarization, covering all cases from circular via elliptical to linear. By the further use of some trigonometrical identities, equation (3,87) can be written

$$\tan 2\theta = \tan \left[2 \tan^{-1} \left(\frac{E_y^{\circ}}{E_x^{\circ}} \right) \right] \cos (\phi_x - \phi_y). \tag{3.88}$$

From the above it will be seen that the theory involves only the ratios of amplitudes and the differences of phases, so provided *these* are not functions of time (even if the individual components are) the theory will apply. This means that several cases of quasi-monochromatic radiation can be treated via the above theory, which was, of course, derived for strictly monochromatic radiation.

3.5.3. Partially polarized radiation

Partially polarized radiation, which is, by definition, neither fully polarized nor fully unpolarized, is commonly found and requires, therefore, a description. As any wave may be regarded as the superposition of suitable independent waves, a partially polarized wave may be represented as the sum of a completely polarized and a completely unpolarized wave which are independent of each other. The coherency matrix $\hat{\mathbf{J}}$ may then be represented as the sum of two component matrices representing the component waves:

$$\hat{\mathbf{J}} = \hat{\mathbf{J}}_{\text{unpol}} + \hat{\mathbf{J}}_{\text{pol}} \quad [\text{W m}^{-2}], \tag{3.89}$$

where (from 3.77)

$$\hat{\mathbf{J}}_{\text{unpol}} = \begin{pmatrix} \mathcal{I}_0 & 0 \\ 0 & \mathcal{I}_0 \end{pmatrix} \quad [\text{W m}^{-2}] \tag{3.90}$$

and (from 3.84)

$$\hat{\mathbf{J}}_{\text{pol}} = \begin{pmatrix} \mathcal{I}_1 & \hat{\mathbf{J}} \\ \hat{\mathbf{J}}^* & \mathcal{I}_2 \end{pmatrix} \quad [\text{W m}^{-2}] \tag{3.91}$$

with

$$\mathcal{I}_1 \mathcal{I}_2 = \hat{\mathbf{J}} \hat{\mathbf{J}}^* \quad [\text{W}^2 \, \text{m}^{-4}]. \tag{3.92}$$

It is clear that

$$\hat{\mathbf{J}} \equiv \begin{pmatrix} J_{xx} & J_{xy} \\ J_{yx} & J_{yy} \end{pmatrix} = \begin{pmatrix} \mathcal{I}_0 + \mathcal{I}_1 & \hat{\mathbf{J}} \\ \hat{\mathbf{J}}^* & \mathcal{I}_0 + \mathcal{I}_2 \end{pmatrix} \quad [\text{W m}^{-2}] \tag{3.93}$$

and that the total intensity of the partially polarized wave is

$$\mathscr{I}_{tot} = \mathrm{Tr}\,\hat{\mathbf{J}} = 2\mathscr{I}_0 + \mathscr{I}_1 + \mathscr{I}_2 \quad [\mathrm{W\,m^{-2}}]. \tag{3.94}$$

Born and Wolf[33] show that the total intensity of the polarized part,

$$\mathscr{I}_{pol} = \mathrm{Tr}\,\hat{\mathbf{J}}_{pol} = \mathscr{I}_1 + \mathscr{I}_2 \quad [\mathrm{W\,m^{-2}}], \tag{3.95a}$$

is

$$\mathscr{I}_{pol} = \sqrt{(J_{xx} + J_{yy})^2 - 4\det\hat{\mathbf{J}}} \quad [\mathrm{W\,m^{-2}}], \tag{3.95b}$$

so that the *degree of polarization*, defined as the ratio

$$\rho = \frac{\mathscr{I}_{pol}}{\mathscr{I}_{tot}}, \tag{3.96}$$

is given by

$$\rho = \sqrt{1 - \frac{4\det\hat{\mathbf{J}}}{(J_{xx} + J_{yy})^2}}. \tag{3.97}$$

ρ takes values varying between 0 and 1 for partially polarized radiation, with $\rho = 0$ for unpolarized and $\rho = 1$ for fully polarized. There is a particularly important case which occurs when the two components are mutually incoherent, for then

$$\det\hat{\mathbf{J}} = J_{xx}J_{yy} \quad [\mathrm{W^2\,m^{-4}}] \tag{3.98}$$

and

$$\rho = \left| \frac{J_{xx} - J_{yy}}{J_{xx} + J_{yy}} \right|. \tag{3.99}$$

This relation is applicable to the instances where there is some measure of linear polarization present, such as when radiation is reflected from a mirror or propagated through a beam divider.

3.5.4. The Jones matrix calculus

Having described some of the properties of polarized radiation, a short account will be given of an algebraic system which is based on matrices and which permits a concise description of the propagation of radiation through polarization-sensitive components. The system is known as the Jones calculus.

The electric field fluctuations are represented by a column matrix

$$\mathbf{E} = \begin{pmatrix} E_x \\ E_y \end{pmatrix}, \tag{3.100}$$

$$\begin{pmatrix} E_x \\ 0 \end{pmatrix} \quad \text{or} \quad \begin{pmatrix} 0 \\ E_y \end{pmatrix}, \tag{3.101}$$

respectively.

We now consider a second set of orthogonal axes u and v in the directions **r** and **s**, respectively. The new axes lie at an angle ϕ with respect to the x, y axes. The two sets of axes are related by the rotation matrix $\mathbf{R}(\phi)$ thus:

$$[u \quad v] = \begin{bmatrix} \cos\phi & \sin\phi \\ -\sin\phi & \cos\phi \end{bmatrix}\begin{bmatrix} x \\ y \end{bmatrix}, \tag{3.102}$$

and therefore the vector **E** referred to the new axis system is given by

$$\mathbf{E}^{u/v} = \mathbf{R}(\phi)\mathbf{E}^{x/y}. \tag{3.103}$$

Transformation from the u/v system to the x/y system involves the inverse of $\mathbf{R}(\phi)$, but since $\mathbf{R}(\phi)$ is orthonormal its inverse is its transpose and therefore

$$\mathbf{R}(\phi)^{-1} = \mathbf{R}(-\phi). \tag{3.104}$$

The calculus may now be extended (as described by Jones[35]) to cover reflection, retardation, and polarization. Some examples are

$$\text{reflection, } \hat{\mathbf{r}} = \begin{pmatrix} r_x & 0 \\ 0 & r_y \end{pmatrix}; \quad \text{retardation, } \hat{\mathbf{\Delta}} = \begin{pmatrix} \exp i\delta & 0 \\ 0 & \exp(-i\delta) \end{pmatrix}; \tag{3.105}$$

linear polarization,

$$x \text{ axis} \begin{pmatrix} 1 & 0 \\ 0 & 0 \end{pmatrix} \quad y \text{ axis} \begin{pmatrix} 0 & 0 \\ 0 & 1 \end{pmatrix} \tag{3.106}$$

$$u \text{ axis } \phi = 45° \begin{pmatrix} \frac{1}{2} & \frac{1}{2} \\ \frac{1}{2} & \frac{1}{2} \end{pmatrix} \quad v \text{ axis } \phi = -45° \begin{pmatrix} \frac{1}{2} & -\frac{1}{2} \\ -\frac{1}{2} & \frac{1}{2} \end{pmatrix}; \tag{3.107}$$

circular polarization,

$$\text{right-circular} \begin{pmatrix} \frac{1}{2} & -i/2 \\ i/2 & \frac{1}{2} \end{pmatrix} \quad \text{left-circular} \begin{pmatrix} \frac{1}{2} & i/2 \\ -i/2 & \frac{1}{2} \end{pmatrix}. \tag{3.108}$$

The Jones calculus is applicable only if the incident beam is already polarized and only if the components are of the non-depolarizing type. Scattering devices cannot be described by this calculus.

When the radiation is not strictly monochromatic, the foregoing Jones calculus outlined above can still be used provided the field matrix (3.100) is replaced by

$$\hat{\mathbf{a}}(\sigma) = \begin{pmatrix} \hat{a}_x(\sigma) \\ \hat{a}_y(\sigma) \end{pmatrix}, \tag{3.109}$$

where $\hat{a}_x(\sigma)$ and $\hat{a}_y(\sigma)$ are the complex spectral amplitudes in the x and y directions corresponding to the real fluctuations

$$E_x'(t) = \int_{-\infty}^{\infty} \hat{a}_x(\sigma)\exp(-2\pi i\sigma ct)\, d\sigma \tag{3.110a}$$

and

$$E_y'(t) = \int_{-\infty}^{\infty} \hat{a}_y(\sigma) \exp\left(-2\pi i \sigma c t\right) d\sigma. \qquad (3.110b)$$

The intensity corresponding to (3.110) is

$$\lim_{T \to \infty} \frac{1}{2T} \overline{|\hat{a}_x(\sigma)|^2} + \overline{|\hat{a}_y(\sigma)|^2}, \qquad (3.111)$$

but if either of the polarization components can be isolated by means of a linear polarizer (3.105a) or (3.105b), the intensity of that component is given by just one of the terms in (3.111).

3.6. TERMS USED IN RADIOMETRY

Before we derive, in the next chapter, the expressions for the signals detected with an interferometer, it is worthwhile to define some terms and to give their dimensions. There is in the literature much careless mention of intensity, energy, power, etc., as though these are interchangeable descriptions of a given variable. They are not, although they are closely related.

We shall be concerned both with the total quantity of radiation and, since we are determining spectra, with the quantity of radiation lying in a given spectral interval. The total quantity of radiation in a signal is the *radiant flux* measured in power units, W, or in terms of rate of energy flow, $J\,s^{-1}$. The flux of a particular frequency (and lying in the small interval ν to $\nu + d\nu$ is called the *spectral flux* and is measured in $W\,Hz^{-1} = W\,s$ or, if wavenumber is the variable, the units are $W\,(m^{-1})^{-1} = W\,m$. The *radiant flux density* is the amount of flux flowing through unit area placed perpendicular to the direction of flow and is measured in $W\,m^{-2}$. The *spectral radiant flux density* is measured in $W\,m^{-2}\,s$ or $W\,m^{-1}$ (for unit frequency or unit wavenumber, respectively). The *intensity* of a beam of radiation is the (time-averaged value of the) radiant flux density (measured in $W\,m^{-2}$). The *intensity* of a source is the amount of radiant flux emitted by the whole surface of the source into unit solid angle (measured in $W\,sr^{-1}$). It is not to be confused with the intensity of a beam. The *radiance* of a source is the amount of flux emitted by the unit area of the source into unit solid angle ($W\,m^{-2}\,sr^{-1}$). The *spectral radiance* of a source is the amount of flux emitted per unit bandwidth by unit area of the source into unit solid angle ($W\,m^{-2}\,sr^{-1}\,s$ or $W\,m^{-1}\,sr^{-1}$).

In a measurement of signal it is power that is recorded (not intensity or energy) and from this we determine the distribution of the radiated power over the observed frequency range, that is, we determine the spectral flux or the *power spectrum*.

CHAPTER 4

The two-beam interferometer

4.1. THE IDEAL TWO-BEAM INTERFEROMETER

In an ideal two-beam interferometer, radiation from a point source is collimated by perfect loss-free optics and is divided into two equal beams by a loss-free beam divider. The twin beams proceed to two mirrors from whence they retrace their paths back to the beam divider, at which beam division occurs once more. One of the resulting recombined beams proceeds back to the source, but the other travels through perfect focusing optics to a detector whose aperture is situated in the focal plane. An interference pattern is developed in this plane and this pattern is a function of the phase delay (i.e. path difference) which one of the twin beams has experienced relative to the other. If the detector aperture is small enough, only the central fringe will be detected and the variation of detected power with phase difference—i.e. the interferogram—contains all spectral information about the radiation. The observation of the interferogram and its analysis is the subject matter of Fourier spectrometry. The theory which leads to an expression relating the power at the pinhole stop in the exit focal plane to the phase delay may be developed from one of a number of points of view, each in its own way rewarding the reader with some insight into the mechanism of the interferometer.

Perhaps the most straightforward approach is to find the power from the algebraic sum of the instantaneous magnitudes of the recombined interfering waves. This is a direct application of the principle of coherent superposition. An alternative procedure, requiring more general (but more complicated) algebraic arguments, is to find the correlation between the two interfering beams, that is, to establish the mutual coherence. This is an important approach when the interferometer contains a dispersive path difference. The third point of view we shall adopt has something in common with some aspects of communication theory. We determine the response of the interferometric system to a temporal impulse of radiation (that is, an infinite 'white' spectrum) and use this to find the interference function for an arbitrary input. This analysis leads to an understanding of the instrument in terms of causality—an aspect that is also important when dispersive path differences are present.

4.1.1. Étendue

The ideal interferometer would, unfortunately, not work in practice since a point source is not capable of delivering finite radiant energy. A more

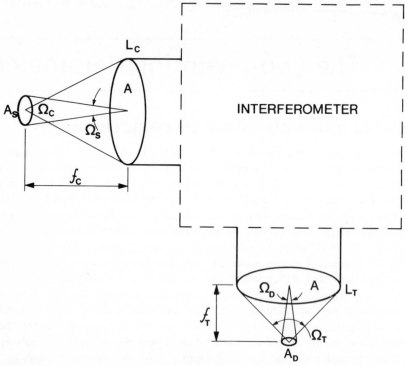

Figure 4.1. Schematic diagram of the optical system to show the quantities used in discussions of *étendue*

realistic interferometer must contain a source of finite size. However, by constraining this to be small and by assuming all the remaining components to be ideal, an idealized system can be envisaged which is free from most departures from 'perfect' performance.

Consider the schematic arrangement of Figure 4.1, in which the collimating and focusing optics are shown. The source (aperture) of area A_s subtends a solid angle Ω_s at the collimating lens L_c of focal length f_c. This lens has area A and subtends a solid angle Ω_c at the centre of the source. The focusing optics are represented by a lens L_T of focal length f_T subtending a solid angle Ω_T at the detector window. This window, of area A_D, subtends an angle Ω_D at the centre of L_T. The source aperture and the detector window are, ideally, exact images of one another and these are the stops that give the image of the source a sharp edge. It is easy to show that Ω_s is effectively equal to Ω_D whatever the values of f_c or f_T. If the images of the source aperture and the window are not equal in size, the smaller provides the limiting or *field stop*, and it is frequently the case in practice that this limitation is provided by the detector window. The amount of radiation that reaches the detector is governed by the *limiting aperture* or *aperture stop*, which frequently coincides with the perimeter of the collimating lens.

From the definitions of solid angle, it follows that for collimating and focusing lenses of equal diameter

$$A = \Omega_c f_c^2 = \Omega_T f_T^2 \qquad (m^2) \tag{4.1}$$

and (since $\Omega_s = \Omega_D$)

$$f_c^2 = \frac{A_s}{A_D} f_T^2, \qquad (m^2) \tag{4.2}$$

so

$$E = \Omega_c A_S = \Omega_T A_D \qquad [m^2 \, sr], \tag{4.3}$$

where E is a measure of the capacity of the system to transmit energy. The products E given in (4.3) are called the *étendue* or *throughput* of the system. The throughput is invariant in a given optical system and is a measure of the radiation collecting-power.

The power of the radiation incident on the interferometer is equal to the power of the radiation collected by the collimator lens if the lens is lossless (otherwise there is a small frequency-dependent reduction of power). When the source area A_s is finite, but very small, the spectral power radiated into unit solid angle is $l(\nu)A_S$, if we assume the spectral radiance $l(\nu)$ of the source element to be isotropic. The spectral power collected by the collimator is, therefore,

$$L(\nu) = l(\nu)A_S\Omega_c = l(\nu)E \qquad [W \, Hz^{-1}], \tag{4.4}$$

which is proportional to the *étendue* E as well as to the spectral radiance $l(\nu)$.

For a loss-free system therefore, the detected power is also given by

$$B_D(\nu) = l(\nu)E \qquad [W \, Hz^{-1}] \tag{4.5}$$

and the detected spectral density (averaged over the exit stop) is

$$G_D(\nu) = \frac{l(\nu)E}{A_D} = \Omega_T l(\nu) \qquad [W \, m^{-2} \, Hz^{-1}]. \tag{4.6}$$

4.1.2. The basic interferometer

The interferometers used for Fourier spectrometry can assume a variety of forms, most of them variants of the simple Michelson-type instrument. The true Michelson interferometer as invented and used by Michelson employed an extended source without collimation; the same configuration, but with a point source and collimation, is strictly, a Twyman–Green interferometer, although it is rarely referred to as such except when used for optical testing. In keeping with common usage we shall loosely call both types of instrument Michelson interferometers.

To establish the basic relations of Fourier spectrometry it will be sufficient to consider the idealized instrument and then modify the results as necessary for application to particular practical arrangements. The instrument has been shown earlier (Figure 1.7). One of the mirrors is mounted so that it can be moved along the optical axis but keeping its reflecting surface accurately parallel to its original plane. In this way a path difference Δx is introduced between the two split beams. The interferometer is therefore a four-arm (or in microwave jargon 'four-port') device: the two passive arms are from the source to beam divider and from the beam divider to the detector, the two active arms are the fixed mirror arm and the moving, or dynamic, mirror arm. Details of how beam dividers operate will be given later: basically there are two types, dielectric film and coated substrate, the latter requiring a compensator, but for the present purpose it can be taken to be a thin, perfectly plane, medium which transmits exactly 50% of the incident radiation and reflects the other 50%. In the focal plane of the interferometer two beams arrive (one from each mirror) and these interfere to produce the interference pattern the central area of which is sampled by the detector.

4.2. DERIVATION OF THE PERFECT INTERFERENCE FUNCTION

4.2.1. Elementary wave theory

The signal arriving at the detector is made up from those in each component beam added algebraically: the intensity is calculated using the theory developed in Chapter 3. Because a monochromatic point source is being considered, the illumination in the focal plane is uniform everywhere within the area covered by the beam. Later these restrictions will be relaxed and realistic sources considered, but one general point will persist: that each Fourier component of the signal may be considered independently of every other. The overall intensity from a polychromatic source is therefore found by calculating the resultant complex spectral amplitude, finding its modulus, and then integrating this over all frequencies present.

4.2.1.1. Monochromatic source

The instantaneous electric fields in the two beams produced by the beam divider may be written

$$\hat{E}_t = \hat{t}\hat{E}_0 \exp(-2\pi i\sigma_0 ct) \quad [\text{V m}^{-1}] \tag{4.7a}$$

and

$$\hat{E}_r = \hat{r}\hat{E}_0 \exp(-2\pi i\sigma_0 ct), \tag{4.7b}$$

where $\hat{E}_0$ is the Hermitian incident complex amplitude, $\hat{t}$ a factor which represents the fraction of this which is transmitted, and $\hat{r}$ a factor which represents the fraction that is reflected. In an idealized interferometer, $\hat{t}$ and

$\hat{r}$ are independent of σ and may be written

$$\hat{t} = |\hat{t}| \exp i\theta_t \quad \text{and} \quad \hat{r} = |\hat{r}| \exp i\theta_r; \qquad (4.8)$$

in the perfect case,

$$|\hat{t}| = |\hat{r}| = 1/\sqrt{2} \quad \text{and} \quad \theta_t - \theta_r = \pi/2.$$

This makes the intensity reflection and transmission factors equal to one another and of magnitude 0.5.

If the signals (4.7) take times t_1 and t_2, respectively, to travel to their respective mirrors and return to the beam divider, the total instantaneous signal travelling towards the focal plane is given by

$$\hat{E}(t_1 - t_2) = \hat{t}\hat{r}\hat{E}_0 \exp\left[-2\pi i\sigma_0 c(t - t_1)\right] + \hat{r}\hat{t}\hat{E}_0 \exp\left[-2\pi i\sigma_0 c(t - t_2)\right]$$
$$= \hat{t}\hat{r}\hat{E}_0[1 + \exp\left[-2\pi i\sigma_0 c(t_1 - t_2)\right] \exp\left[-2\pi i\sigma_0 c(t - t_1)\right]]. \quad (4.9a)$$

The corresponding intensity is (via (3.33))

$$\mathcal{I}(\tau) = \tfrac{1}{2}\varepsilon_0 c \,\overline{|\hat{E}(t_1 - t_2)|^2} \qquad (4.10a)$$

$$= \tfrac{1}{2}\mathcal{I}_0[1 + \cos 2\pi\sigma_0 c\tau] \quad [\text{W m}^{-2}]. \qquad (4.10b)$$

In practice, of course, it is power rather than intensity that is detected, but with the assumption of uniform illumination across the beam, the power is proportional to $\mathcal{I}(\tau)$ and the power passing through the vanishingly small image of the source produced in the focal plane will be given simply by

$$I(\tau) = \tfrac{1}{2}I_0[1 + \cos 2\pi\sigma_0 c\tau] \qquad (4.11)$$

If one can neglect the refractive index effects of the medium—as would for example be appropriate were the interferometer evacuated—the path difference in the instrument is simply given by $x = c\tau$ and (4.11) can be written

$$I(x) = \tfrac{1}{2}I_0 + \tfrac{1}{2}I_0 \cos 2\pi\sigma_0 x. \qquad (4.12)$$

Thus, in the ideal case, the fluctuating (or x-dependent) part of the detected power, that is the interferogram, consists of a cosinusoidal variation that extends to infinite positive or negative path differences. This is superposed on a constant background, $\tfrac{1}{2}I_0$.

The complementary signal that leaves the beam divider in the direction of the source is given analogously by

$$\hat{E}(t_1 - t_2) = \hat{E}_0\{\hat{t}^2 + \hat{r}^2 \exp\left[-2\pi i\sigma_0 c\tau\right]\} \exp\left[-2\pi i\sigma_0 c(t - t_1)\right], \qquad (4.13)$$

which, bearing in mind the $\pi/2$ phase shift between θ_t and θ_r, leads to

$$\mathcal{I}(\tau) = \tfrac{1}{2}\mathcal{I}_0[1 - \cos 2\pi\sigma_0 c\tau] \qquad (4.14)$$

and to

$$I(x) = \tfrac{1}{2}I_0 - \tfrac{1}{2}I_0 \cos 2\pi\sigma_0 x. \qquad (4.15)$$

Thus, in the ideal loss-free interferometer, energy is conserved and the sum of the powers going to the detector and back to the source equals that propagated initially. The peaks and troughs in the interference functions (4.12) and (4.15) are usually called fringes and, by loose analogy with interferometry in the visible region, are usually called *bright* and *dark* fringes, respectively. The fringes produced in the source plane are not usually detected, though with a suitably modified interferometer this becomes possible. From (4.12) and (4.15) it will be seen that a bright fringe in the detector plane corresponds to a dark fringe in the source plane and vice versa. All the bright fringes, for this simple ideal case, are of equal power; in particular, that for $x = 0$ (zero path difference) is no brighter than the others and all the dark fringes are devoid of power. Neither of these conditions applies in realistic polychromatic interferometry, but the result that the *difference* of the detector plane signal and the source plane signal, i.e. (4.12)–(4.15), is twice the interferogram is general and does offer a means of eliminating some of the difficulties associated with the presence of the 'constant' term $(\frac{1}{2}I_0)$ in the usual expression.

4.2.1.2. *Polychromatic source*

If the source spectrum contains several monochromatic lines, then, by the principle of independent superposition, the detected power will be given by the sum of the contributions of each line separately. Therefore if the power from the ith line is I_i one will have

$$I(x) = \tfrac{1}{2} \sum_i I_i + \tfrac{1}{2} \sum_i I_i \cos 2\pi\sigma_i x. \tag{4.16}$$

The argument is now easily generalized to the real world, where even 'monochromatic' lines have in fact a continuous spectral distribution. In place of I_i (that is, the power radiated *at* σ_i), one introduces $L(\sigma)\,d\sigma$ (that is, the power radiated between σ and $\sigma + d\sigma$) and the total detected power becomes

$$I(x) = \tfrac{1}{2} \int_0^\infty L(\sigma)\,d\sigma + \tfrac{1}{2} \int_0^\infty L(\sigma) \cos 2\pi\sigma x\,d\sigma. \tag{4.17}$$

The power returning to the source is, analogously,

$$I(x) = \tfrac{1}{2} \int_0^\infty L(\sigma)\,d\sigma - \tfrac{1}{2} \int_0^\infty L(\sigma) \cos 2\pi\sigma x\,d\sigma. \tag{4.18}$$

The important point to note now is that the interference fringes given by the second terms in the right-hand sides of these equations are no longer of equal contrast. When $x = 0$, there is a uniquely bright fringe—the so-called 'zero path difference fringe' or grand maximum in the detector plane—and a completely dark fringe in the source plane.

These results, for the interference functions, can also be derived by considering the temporal fluctuations arriving in the two planes from each arm of the interferometer. This method is worth exposition in some depth because similar arguments will be appropriate in later discussions. One begins, just as one did in the case of the polychromatic wave-field, by introducing an analytic signal:

$$\hat{E}(t) = \int_0^{+\infty} \hat{a}(\sigma) \exp(-2\pi i \sigma c t) \, d\sigma. \tag{4.19}$$

The signal reaching the detector from the r (reflection) t (transmission) arm is

$$\hat{E}_{rt}(t - t_2) = \int_0^\infty \hat{t}(\sigma)\hat{r}(\sigma)\hat{a}(\sigma) \exp[-2\pi i \sigma c (t - t_2)] \, d\sigma, \tag{4.20}$$

and that from the t r arm is

$$\hat{E}_{tr}(t - t_1) = \int_0^\infty \hat{r}(\sigma)\hat{t}(\sigma)\hat{a}(\sigma) \exp[-2\pi i \sigma c (t - t_1)] \, d\sigma. \tag{4.21}$$

The total signal is therefore the sum of (4.20) and (4.21), i.e.

$$\hat{E}_{tot}(t_1, t_2) = \int_0^\infty \hat{r}(\sigma)\hat{t}(\sigma)\hat{a}(\sigma) \exp[-2\pi i \sigma c t][\exp(2\pi i \sigma c t_1) + \exp(2\pi i \sigma c t_2)] \, d\sigma \tag{4.22a}.$$

$$= \int_0^\infty \hat{d}(\sigma) \exp[-2\pi i \sigma c t] \, d\sigma. \tag{4.22b}$$

The corresponding intensity in the detector plane is

$$\mathscr{I}(\tau) = \tfrac{1}{2}\varepsilon_0 c \lim_{T \to \infty} \frac{1}{2T} \int_0^\infty \overline{|\hat{d}(\sigma, t_1, t_2)|^2} \, d\sigma. \tag{4.23}$$

The detected power is therefore

$$I_D(\tau) = \tfrac{1}{2}A\varepsilon_0 c \lim_{T \to \infty} \frac{1}{2T} \int_0^\infty 2 |\hat{r}(\sigma)|^2 |\hat{t}(\sigma)|^2 \overline{|\hat{a}(\sigma)|^2} (1 + \cos 2\pi \sigma c \tau) \, d\sigma. \tag{4.24}$$

Under the usual assumption that $|\hat{r}(\sigma)|^2 = |\hat{t}(\sigma)|^2 = \tfrac{1}{2}$, this becomes

$$I_D(\tau) = \tfrac{1}{2}A\varepsilon_0 c \lim_{T \to \infty} \frac{1}{2T} \int_0^\infty \overline{\tfrac{1}{2}|\hat{a}(\sigma)|^2} \, d\sigma + \tfrac{1}{2}A\varepsilon_0 c \lim_{T \to \infty} \frac{1}{2T} \int_0^\infty \overline{\tfrac{1}{2}|\hat{a}(\sigma)|^2} \cos 2\pi \sigma c \tau \, d\sigma \tag{4.25}$$

$$= \tfrac{1}{2} \int_0^\infty L(\sigma) \, d\sigma + \tfrac{1}{2} \int_0^\infty L(\sigma) \cos 2\pi \sigma c \tau \, d\sigma, \tag{4.26}$$

where

$$L(\sigma) = \tfrac{1}{2}\varepsilon_0 c A \lim_{T \to \infty} \frac{\overline{|\hat{a}(\sigma)|^2}}{2T}.$$

Since $c\tau = x$, this is identical with (4.17). In an analogous way, the power in the source plane is found to be

$$I_S(\tau) = \int_0^\infty [|\hat{r}(\sigma)|^4 + |\hat{t}(\sigma)|^4] L(\sigma) \, d\sigma - 2 \int_0^\infty |\hat{r}(\sigma)|^2 |\hat{t}(\sigma)|^2 L(\sigma) \cos 2\pi\sigma c\tau \, d\sigma,$$

(4.27)

which is identical to (4.18) in the ideal case.

4.2.2. Coherence theory

The intensity at some point Q irradiated by two small sources P_1, P_2 is (from 3.68)

$$\mathscr{I}(Q) = |\hat{f}_1|^2 \, \mathscr{I}_1 + |\hat{f}_2|^2 \, \mathscr{I}_2 + \hat{f}_1 \hat{f}_2^* \hat{\Gamma}_{12}(\tau) + \hat{f}_1^* \hat{f}_2 \hat{\Gamma}_{12}^*(\tau),$$

(4.28)

where $\hat{\Gamma}_{12}(\tau)$ is the mutual coherence of the fields they produce at Q, $\hat{f}_1$ and $\hat{f}_2$ are the complex amplitude attenuation factors between P_1 and Q and between P_2 and Q, and $\tau = t_2 - t_1$ is the difference in transit times. For the moment, $\hat{f}_1$ and $\hat{f}_2$ will be assumed to be independent of ν or σ. In the Michelson interferometer, an observer viewing the instrument from the detector sees two images of the source, produced by M_1 and M_2. The images appear superimposed at infinity but separated by the distance $\frac{1}{2}(x_2 - x_1)$. These images can be thought of as the secondary sources which give rise to an equation like (4.28). However, since they are directly derived from the primary source, the mutual coherence must be replaced by self-coherence. Moreover, each image is the result of one reflection and one transmission at

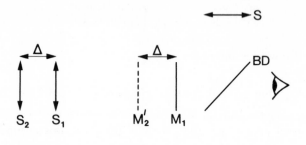

Figure 4.2. Schematic diagram of the optical arrangement of a Michelson interferometer from the point of view of an observer at the detector. M_2' is the image of M_2 as seen via the beam divider BD and the two images of the source S (S_1 and S_2) are separated by the distance Δ, which is one half of the optical path difference between the two arms. For convenience the uncollimated case has been shown: with collimated radiation the two images are at infinity but are still separated by Δ

the beam divider and $\hat{f}$ is the same for each 'source'. Thus (4.28) becomes

$$\mathscr{I}(Q) = 2\,|\hat{f}|^2\,\mathscr{I}_1 + 2\,|\hat{f}|^2\,\Gamma_{11}(t_2 - t_1) \qquad [\text{W m}^{-2}] \qquad (4.29a)$$

or

$$\mathscr{I}(x) = 2\,|\hat{f}|^2\,\mathscr{I}_1 + 2\,|\hat{f}|^2\,\Gamma_{11}(x/c) \qquad [\text{W m}^{-2}], \qquad (4.29b)$$

where

$$x = x_2 - x_1 = c(t_2 - t_1) = c\tau \qquad [\text{m}]$$

and

$$|\hat{f}_1|^2 = |\hat{f}_2|^2 = |\hat{f}|^2 = |\hat{r}\hat{t}|^2. \qquad (4.30)$$

In the ideal case, $|\hat{f}|^2 = \tfrac{1}{4}$ and

$$\mathscr{I}_p(x) = \tfrac{1}{2}\mathscr{I}_1 + \tfrac{1}{2}\Gamma_{11}(x/c) \qquad [\text{W m}^{-2}]. \qquad (4.31)$$

The detected power

$$I_p(x) = A\mathscr{I}_p(x) \qquad [\text{W}] \qquad (4.32)$$

is therefore (as expected) made up of an invariant part equal to half the power incident on the interferometer and an x-dependent part, the interferogram

$$F_p(x) = \tfrac{1}{2}A\Gamma_{11}(x/c) \qquad [\text{W}], \qquad (4.33)$$

which is now revealed as being proportional to the autocorrelation function (the self-coherence) of the electromagnetic waves reaching the detector via the two arms of the interferometer.

According to (3.46), the self-coherence is given by the Fourier integral

$$\Gamma_{11}\!\left(\frac{x}{c}\right) = \operatorname{Re}\int_0^\infty G_{11}(\sigma c)\exp(-2\pi i\sigma x)\,\mathrm{d}(\sigma c) \qquad [\text{W}], \qquad (4.34)$$

so, from (4.33),

$$F_p(x) = \tfrac{1}{2}\operatorname{Re}\int_0^\infty AG_{11}(\sigma c)\exp(-2\pi i\sigma x)\,\mathrm{d}(\sigma c) \qquad (4.35a)$$

$$= \tfrac{1}{2}\int_0^\infty AcG_{11}(\sigma c)\cos 2\pi\sigma x\,\mathrm{d}\sigma \qquad (4.35b)$$

$$= \tfrac{1}{2}\int_0^\infty L(\sigma)\cos 2\pi\sigma x\,\mathrm{d}\sigma \qquad [\text{W}], \qquad (4.35c)$$

where we have put

$$L(\sigma) \equiv AcG_{11}(\sigma c) \qquad [\text{W m}]. \qquad (4.36)$$

The detected power is therefore given by

$$I_p(x) = \tfrac{1}{2}\int_0^\infty L(\sigma)\,\mathrm{d}\sigma + \tfrac{1}{2}\int_0^\infty L(\sigma)\cos 2\pi\sigma x\,\mathrm{d}\sigma \qquad [\text{W}]. \qquad (4.37)$$

Equation (4.37) is identical to the result (4.26) obtained earlier from the elementary theory.

4.2.3. Impulse theory

We suppose an interferometer to be subject to an impulse of radiation, limited in the time domain. The form of the signal received at the detector represents the impulse response $R(t)$ of the system (see sections 2.9.4 and 2.9.7.1). From this we can calculate the transfer function and, hence, the form of the output spectrum.

Let the electric field input be a Dirac delta pulse $\delta(t)$ of unit area. If we measure the zero of time from the instant the pulse is divided at the beam divider, the interferometer produces at the output two pulses having time delays x_1/c and x_2/c occasioned by the transit through, respectively, the dynamic and static arms. The path difference in the interferometer is $x = x_1 - x_2$. The time dependence of the detector signal will therefore be given by

$$\hat{R}(t) = \hat{f}\delta(t - x_1/c) + \hat{f}\delta(t - x_2/c) \qquad [\text{s}^{-1}], \qquad (4.38)$$

where $\hat{f}$ is our attenuation factor describing reflection at and transmission by the beam divider. The electric field pulse at the detector for an arbitrary input $E(t)$ is (see equation (2.122))

$$E_{\text{D}}(t) = \int_{-\infty}^{\infty} E(t')R(t - t')\,dt' = E(t) * R(t) \qquad [\text{V m}^{-1}], \qquad (4.39)$$

which is equivalent to the spectral amplitude

$$\hat{a}_{\text{D}}(\nu) = \hat{a}(\nu)\hat{T}(\nu) \qquad [\text{V m}^{-1}\,\text{Hz}^{-1}] \qquad (4.40)$$

in the frequency domain (see equation (2.121)), where

$$\hat{a}(\nu) = \int_{-\infty}^{\infty} E(t)\exp(-2\pi i\nu t)\,dt \qquad [\text{V m}^{-1}\,\text{Hz}^{-1}] \qquad (4.41)$$

is the Fourier transform of the input and

$$\hat{T}(\nu) = \int_{-\infty}^{\infty} R(t)\exp(-2\pi i\nu t)\,dt, \qquad (4.42)$$

the transfer function, is the Fourier transform of the impulse response. The detected spectral power is, from (3.31),

$$\frac{1}{2}\lim_{T\to\infty}\int_0^{\infty} \varepsilon_0 cA\,\frac{|\hat{a}_{\text{D}}(\nu)|^2}{2T}\,d\nu \qquad [\text{W Hz}^{-1}] \qquad (4.43)$$

and the detected power is

$$I(c\tau) = \frac{1}{2}\varepsilon_0 cA \int_0^{\infty} |\hat{a}(\nu)\hat{T}(\nu)|^2\,d\nu \qquad [\text{W}] \qquad (4.44)$$

after use of (4.40).

From (4.38), we see that the Fourier transform (4.42) of $R(t)$ is

$$\hat{T}(\nu) = \hat{f} \exp(-2\pi i \nu t_1) + \hat{f} \exp(-2\pi i \nu t_2) \qquad (4.45a)$$
$$= \hat{f} \exp(-2\pi i \nu t_1)[1 + \exp(-2\pi i \nu \tau)]. \qquad (4.45b)$$

Consequently,

$$|\hat{T}(\nu)|^2 = 2 |\hat{f}|^2 (1 + \cos 2\pi \nu \tau) \qquad (4.46)$$

and the detected power is therefore

$$I(c\tau) = \tfrac{1}{2}\varepsilon_0 cA \lim_{T \to \infty} \frac{2}{2T} \int_0^\infty \frac{|\hat{a}(\nu)|^2}{2T} |\hat{f}|^2 (1 + \cos 2\pi \nu \tau)\, d\nu \qquad [\text{W}] \quad (4.47)$$

or, in the ideal case when $|\hat{f}|^2 = \tfrac{1}{4}$,

$$I_p(c\tau) = \tfrac{1}{4}\varepsilon_0 cA \lim_{T \to \infty} \frac{1}{2T} \int_0^\infty \frac{|\hat{a}(\nu)|^2}{2T} (1 + \cos 2\pi \nu \tau)\, d\nu \qquad [\text{W}]. \quad (4.48)$$

By substituting

$$L(\sigma) = \tfrac{1}{2}\varepsilon_0 c^2 A \lim_{T \to \infty} \frac{1}{2T} \frac{|\hat{a}(c\sigma)|^2}{2T} \qquad [\text{W m}], \qquad (4.49)$$

as before, we again recover the relation

$$I_p(c\tau) = \tfrac{1}{2}\int_0^\infty L(\sigma)\, d\sigma + \tfrac{1}{2}\int_0^\infty L(\sigma) \cos 2\pi \sigma c\tau\, d\sigma \qquad [\text{W}] \qquad (4.50)$$

for the detected power in terms of the spectral power $L(\sigma)$ incident on the ideal interferometer. With the final substitution $x = c\tau$, this becomes

$$I_p(x) = \tfrac{1}{2}\int_0^\infty L(\sigma)\, d\sigma + \tfrac{1}{2}\int_0^\infty L(\sigma) \cos 2\pi \sigma x\, d\sigma \qquad [\text{W m}], \qquad (4.51)$$

exactly as in (4.17).

4.3. THE PRACTICAL INTERFERENCE FUNCTION

The three differing derivations we have used all lead to the expression

$$I_p(x) = I_p + F_p(x) \qquad [\text{W}], \qquad (4.52)$$

with

$$I_p = \tfrac{1}{2}\int_0^\infty L(\sigma)\, d\sigma \qquad [\text{W}] \qquad (4.53)$$

and

$$F_p(x) = \tfrac{1}{2}\int_0^\infty L(\sigma) \cos 2\pi \sigma x\, d\sigma \qquad [\text{W}] \qquad (4.54)$$

for the detected power, as a function of x, at the exit aperture of an ideal, loss-free interferometer irradiated from a collimated point source. This is the basic relation of Fourier transform spectrometry, for it is possible to recover the form of $L(\sigma)$ from measurements on $F_p(x)$. In practice, of course, the interferometer is not ideal and the source is finite. It is convenient to introduce the spectral dependence of the instrumental throughput at an early stage; the effect of a finite source may be regarded as a correction, not essential to the basic theory, and will be introduced later.

If we put $\tau_0(\sigma)$ as the overall transmission factor of the instrument, the interference function at the detector is

$$I_D(x) = I_D + F(x) \qquad [\text{W}], \qquad (4.55)$$

where

$$I_D = \int_0^\infty B(\sigma)\,d\sigma \qquad [\text{W}], \qquad (4.56)$$

$$F(x) = \int_0^\infty B(\sigma)\cos 2\pi\sigma x\,d\sigma \qquad [\text{W}], \qquad (4.57)$$

$$B(\sigma) = \tfrac{1}{2}L(\sigma)\tau_0(\sigma) \qquad [\text{W m}] \qquad (4.58)$$

is the *detected* spectral power. In this more realistic case, $|\hat{r}|^2$ and $|\hat{t}|^2$ are less than their ideal value of $\tfrac{1}{4}$. The departure is governed by the quantity $\tau_0(\sigma)$, where

$$\tau_0(\sigma) = 4|\hat{r}(\sigma)|^2\,|\hat{t}(\sigma)|^2. \qquad (4.59)$$

The explicit form of $\tau_0(\sigma)$ depends on the type of interferometer that is employed; it will be calculated later for some of the more important arrangements.

The interference function returned to the source is

$$I_s(s) = I_s - F(x) \qquad [\text{W}], \qquad (4.60)$$

where, from (4.27), the constant term is

$$I_s = \int_0^\infty [|\hat{r}(\sigma)|^4 + |\hat{t}(\sigma)|^4]L(\sigma)\,d\sigma \qquad [\text{W}]. \qquad (4.61)$$

It is to be noted that I_s differs from I_D in all cases except when the interferometer is perfect. We refer to the interference functions as being *unbalanced* when this is the case.

In most cases which follow we shall need to refer only to the interference function (4.55) and, consequently, we shall omit the subscript D except when there is specific need for its presence in the interest of clarity.

CHAPTER 5

The determination of spectral line-shapes from visibility measurements

5.1. PARTIALLY CHARACTERIZED SPECTROMETRY

As outlined in section 1.3.1, Michelson was the first to apply measurements on a two-beam interference pattern to the determination of a spectrum. The spectrum was always highly localized—a single line or group of lines—and the interference fringes had slowly varying visibility. By ignoring the detailed structure of the interference pattern and concentrating on the variation of the visibility as a function of the path difference between the beams, the *profile* of the spectrum could be found. Following a long period during which visibility measurements have been neglected, there has been a revival of the technique in recent years, the most notable practitioners of the art being Terrien and his colleagues. Because the details of the interference fringes are not recorded, but only their gross, long-term variation, and because it is only a normalized spectral distribution that is obtained, spectral determination from visibility measurements may be called partially characterized spectrometry (see section 1.2.2).

5.2. RELATION BETWEEN SPECTRAL PROFILE AND VISIBILITY

If the source is confined to a narrow spectral band (quasi-monochromatic) centred at σ_0, the usual situation to which visibility measurements would be applied, then we may conveniently change the wavenumber variable to $s = \sigma - \sigma_0$, which has its origin at σ_0, and write

$$B(\sigma) = b(s) \quad [\text{W m}].\tag{5.1}$$

With this substitution, the first term of equation (1.13) will be unaffected and we may write

$$I(x) = \bar{I} + C(x) \cos 2\pi\sigma_0 x - S(x) \sin 2\pi\sigma_0 x \quad [\text{W}],\tag{5.2}$$

where

$$C(x) = \int_{-\sigma_0}^{\infty} b(s) \cos 2\pi s x \, \mathrm{d}s \equiv \int_{-\infty}^{+\infty} b(s) \cos 2\pi s x \, \mathrm{d}s \quad [\text{W}]\tag{5.3a}$$

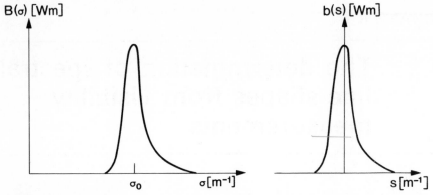

Figure 5.1. Schematic representation of a localized spectrum suitable for study by partially characterized Fourier spectrometry. The effective range σ_{min} to σ_{max} of the spectrum is small compared with the mean wavenumber σ_0

and

$$S(x) = \int_{-\sigma_0}^{\infty} b(s) \sin 2\pi s x \, ds \equiv \int_{-\infty}^{+\infty} b(s) \sin 2\pi s x \, ds \qquad [\text{W}], \qquad (5.3b)$$

and we recover the results of (1.18) and (1.19). Equation (5.2) is quite general, but two special cases are worth picking out. First, if the source is perfectly monochromatic and of mean detected power b_0,

$$b(s) = b_0 \, \delta(s) \qquad (5.4)$$

and it follows from (2.91) that $S(x)$ is zero. We then have

$$I(x) = b_0(1 + \cos 2\pi\sigma_0 x), \qquad (5.5)$$

as expected (Figure 1.8). When the source has a finite, yet small, spectral width, but nevertheless remains symmetrical about $\sigma_0(s = 0)$, so that $b(s) = b(-s)$, then $S(x)$ is again zero and

$$I(x) = \bar{I} + C(x) \cos 2\pi\sigma_0 x \qquad [\text{W}], \qquad (5.6)$$

demonstrating the general result that for a finite line width the fringe amplitude is x-dependent. This can be illustrated by taking a specific form for the line shape, such as the Gaussian profile

$$b(s) = b a \exp(-\pi a^2 s^2) \qquad [\text{W m}]. \qquad (5.7)$$

The amplitude function

$$C(x) = \int_{-\infty}^{\infty} b a \exp(-\pi a^2 s^2) \cos 2\pi s x \, ds \qquad [\text{W}] \qquad (5.8)$$

is simply the Fourier transform (2.127)

$$b \exp(-\pi x^2/a^2) \qquad [\text{W}] \qquad (5.9)$$

of $b(s)$, thus giving

$$I(x) = 6[1 + \exp(-\pi x^2/a^2) \cos 2\pi\sigma_0 x] \qquad [\text{W}] \qquad (5.10)$$

for the interference function. As the parameter a decreases, the spectral line width increases and the fringe amplitude $6\exp(-\pi x^2/a^2)$ falls off more rapidly. However, this fall-off is always slow compared with the period σ_0^{-1} of the interference fringes themselves. We may measure the fall-off by the magnitude of the envelope of the fringes. This is given by the power (or intensity) difference $I_{\max} - I_{\min}$ (see Figure 1.6). Michelson normalized this difference by dividing by the sum $I_{\max} + I_{\min}$ to form the ratio

$$\mathcal{V}(x) = \frac{I_{\max}(x) - I_{\min}(x)}{I_{\max}(x) + I_{\min}(x)}, \qquad (5.11)$$

which he called the visibility. The spectral line shape (5.7), the corresponding intensity function (5.10), and the visibility derived from it are shown schematically in Figure 5.2. This particular case illustrates a general fact that the visibility is not easily found by inspection, because the values of the path difference at which the maxima and minima occur are not given by simple expressions. We need therefore to apply differential calculus to find $I_{\max}$ and $I_{\min}$. In consequence, we differentiate to find the turning points and then choose a value of x midway between a maximum and the following minimum to specify (in a somewhat arbitrary manner) where the visibility ought to be evaluated.

In the general case, the slope of the interference function is

$$\frac{dI(x)}{dx} = \frac{d}{dx}[C(x)\cos 2\pi\sigma_0 x] - \frac{d}{dx}[S(x)\sin 2\pi\sigma_0 x] \qquad [\text{W m}^{-1}], \quad (5.12)$$

which we can write as

$$\frac{dI(x)}{dx} = -2\pi\sigma_0 C(x)\sin 2\pi\sigma_0 x - 2\pi\sigma_0 S(x)\cos 2\pi\sigma_0 x \qquad [\text{W m}^{-1}] \quad (5.13)$$

since we can regard $C(x)$ and $S(x)$ as constant over the interval of interest. The positions of the bright and dark fringes are therefore given by

$$\frac{C(x)}{S(x)}\tan 2\pi\sigma_0 x = -1, \qquad (5.14)$$

which may also be written as

$$\cos 2\pi\sigma_0 x = \frac{\pm C(x)}{\{[C(x)]^2 + [S(x)]^2\}^{1/2}} \qquad (5.15a)$$

and

$$\sin 2\pi\sigma_0 x = \frac{\mp S(x)}{\{[C(x)]^2 + [S(x)]^2\}^{1/2}}, \qquad (5.15b)$$

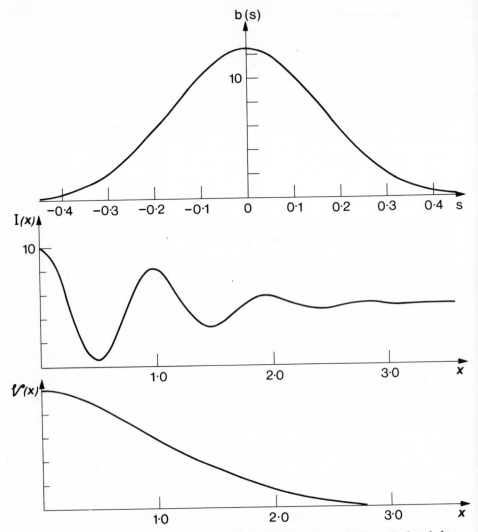

Figure 5.2. Spectral profile, interferogram function, and visibility calculated for a Gaussian line having $a = \sqrt{2\pi}$, $b = 5.0$, and wavenumber and path difference units chosen so that $\sigma_0 = 1.0$

from which it follows that

$$I_{\max} = \bar{I} + \{[C(x)]^2 + [S(x)]^2\}^{1/2} \quad [\text{W}] \quad (5.16a)$$

and

$$I_{\min} = \bar{I} - \{[C(x)]^2 + [S(x)]^2\}^{1/2} \quad [\text{W}]. \quad (5.16b)$$

The visibility, therefore, becomes

$$\mathcal{V}(x) = \{[C(x)]^2 + [S(x)]^2\}^{1/2} (\bar{I})^{-1} \quad (5.17)$$

in the general case and to the limits set by the approximations used in deriving (5.13). This expression, first derived by Michelson, shows that the visibility has a non-linear dependence upon $C(x)$ and $S(x)$.

We now go on to introduce a quantity $\theta(x)$, defined by

$$C(x) = \bar{I}\mathcal{V}(x)\cos\theta(x) \quad [W], \quad (5.18a)$$

$$S(x) = \bar{I}\mathcal{V}(x)\sin\theta(x) \quad [W] \quad (5.18b)$$

and, by applying Fourier's inversion theorem (section 2.2) to equations (5.3), followed by the use of (5.18), one arrives at

$$b_e(s) = \bar{I}\int_{-\infty}^{+\infty}\mathcal{V}(x)\cos\theta(x)\cos 2\pi sx\,dx \quad [W\,m] \quad (5.19a)$$

and

$$b_o(s) = \bar{I}\int_{-\infty}^{+\infty}\mathcal{V}(x)\sin\theta(x)\sin 2\pi sx\,dx \quad [W\,m], \quad (5.19b)$$

which may be combined to give

$$b(s) = b_e(s) + b_o(s) = 6\int_{-\infty}^{+\infty}\mathcal{V}(x)\cos[2\pi sx - \theta(x)]\,dx \quad [W\,m], \quad (5.20)$$

where 6, the mean detected power of the radiation, has been substituted for $\bar{I}$. Equation (5.20) is the real part (see section 2.2) of the full complex Fourier transform

$$6\int_{-\infty}^{+\infty}\mathcal{V}(x)\exp[i\theta(x)]\exp(-2\pi isx)\,dx = 6\int_{-\infty}^{+\infty}\hat{V}(x)\exp(-2\pi isx)\,dx, \quad (5.21)$$

in which we have introduced the *complex visibility*

$$\hat{V}(x) = \mathcal{V}(x)\exp[i\theta(x)]. \quad (5.22)$$

The expressions (5.17), (5.18), and (5.19) show that $\hat{V}(x)$ is Hermitian: i.e.

$$\hat{V}(-x) = \hat{V}^*(x); \quad (5.23)$$

it follows therefore that the normalized distribution of radiance within the (narrow) spectrum is given by the real Fourier transform of the complex visibility and that the Fourier integral (5.21) is *necessarily* real. Thus

$$\frac{b(s)}{6} \equiv \int_{-\infty}^{\infty}\hat{V}(x)\exp(-2\pi isx)\,dx \quad [m]. \quad (5.24)$$

We can now find two useful expressions for the interference function (5.2). Substitution of (5.18) in (5.2) gives

$$I(x) = \bar{I} + \bar{I}\mathcal{V}(x)\cos[2\pi\sigma_0 x + \theta(x)] \quad [W] \quad (5.25)$$

and use of (5.17) leads to

$$I(x) = \bar{I} + \{[C(x)]^2 + [S(x)]^2\}^{1/2}\cos[2\pi\sigma_0 x + \theta(x)] \quad [W]. \quad (5.26)$$

These expressions show that the amplitudes of the fringes are determined by the modulus of the complex visibility and their phases are determined by the phase of the complex visibility.

5.3. THE PHASE OF THE COMPLEX VISIBILITY

5.3.1. Form of the phase

The inverse relation to (5.24),

$$\hat{\mathcal{V}}(x) = \int_{-\infty}^{\infty} \frac{b(s)}{6} \exp 2\pi i s x \, ds, \tag{5.27}$$

shows that if $b(s)$ is symmetrical, that is, $b(s) = b(-s)$, the imaginary part of the right member, the sine Fourier transform, is zero. Consequently, $\hat{\mathcal{V}}(x)$ is of necessity real under these circumstances, $\theta(x) = 0$, and

$$\mathcal{V}(x) = \int_{-\infty}^{\infty} \frac{b(s)}{6} \cos 2\pi s x \, ds. \tag{5.28}$$

It appears, therefore, that the phase of the visibility is an indicator of the asymmetry of the spectrum under consideration. If the required spectrum is symmetrical, the phase is zero, the visibility is real, and it is necessary to record only $\mathcal{V}(x)$ for $x > 0$, whereupon the spectral profile is given by

$$\frac{b(s)}{6} = 2 \int_{0}^{\infty} \mathcal{V}(x) \cos 2\pi s x \, dx \qquad [\text{m}]. \tag{5.29}$$

If the required spectrum is not symmetrical, it is possible to deduce the existence of asymmetry from the visibility, but, without prior knowledge, it is not possible to establish the form of the spectrum uniquely. Consideration of a simple theoretical spectrum elucidates this point.

Suppose two adjacent perfectly monochromatic radiations have mean detected powers 6_0 and 6_1 and are situated at σ_0 and $\sigma_1 = \sigma_0 + \Delta\sigma$ so that the detected spectral feature is (Figure 5.3)

$$B(\sigma) = 6_0 \, \delta(\sigma - \sigma_0) + 6_1 \, \delta(\sigma - \sigma_1) \qquad [\text{W m}] \tag{5.30}$$

and the total mean detected power is

$$6 = \int_{0}^{\infty} B(\sigma) \, d\sigma = 6_0 + 6_1 \qquad [\text{W}]. \tag{5.31}$$

Through use of the principle of superposition, we derive an interference function

$$I(x) = (6_0 + 6_1) + 6_0 \cos 2\pi\sigma_0 x + 6_1 \cos 2\pi(\sigma_0 + \Delta\sigma)x \tag{5.32a}$$

$$= (6_0 + 6_1) + (6_0 - 6_1) \cos 2\pi\sigma_0 x + 6_1[\cos 2\pi(\sigma_0 + \Delta\sigma)x + \cos 2\pi\sigma_0 x] \tag{5.32b}$$

$$= (6_0 + 6_1) + (6_0 - 6_1) \cos 2\pi\sigma_0 x$$
$$+ 26_1 \cos \pi x \Delta\sigma \cos 2\pi\left(\sigma_0 + \frac{\Delta\sigma}{2}\right)x \qquad [\text{W}], \tag{5.32c}$$

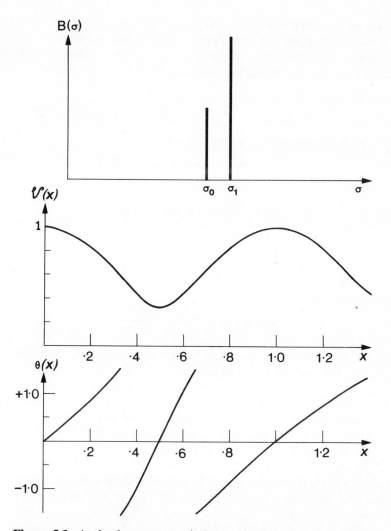

Figure 5.3. A simple unsymmetrical spectrum of small bandwidth shown for the case $\sigma_1 > \sigma_0$: the corresponding visibility and phase are shown below. The breaks in the phase curve at $x = \frac{1}{3}, \frac{2}{3}$, etc., arise from the principal value convention—the curve could be regarded as monotonically rising

which consists of a constant term $(\sigma_0 + \sigma_1)$ equal to the integral of the detected spectral power, a pure cosine term of amplitude $(\sigma_0 - \sigma_1)$ equal to the difference between the powers of the two spectral lines, and a modulated cosine term of amplitude $2\sigma_1 \cos \pi x \Delta\sigma$ and slightly modified period $(\sigma_0 + \Delta\sigma/2)^{-1}$. It appears that as x increases from zero the last two terms in the right member of (5.32c) pass into and out of phase, generating thereby the phenomenon of 'beats' and thus considerably altering the interference function by comparison with the simple cosine term $\cos 2\pi\sigma_0 x$. This alteration is best appreciated by considering the complex visibility.

Equation (5.32) may be rearranged into a form different from those developed above: expanding the last term in (5.32a), we obtain

$$I(x) = (6_0 + 6_1) + (6_0 + 6_1 \cos 2\pi x \,\Delta\sigma) \cos 2\pi\sigma_0 x$$
$$- 6_1 \sin 2\pi x \Delta\sigma \sin 2\pi\sigma_0 x \qquad [\text{W}], \quad (5.33)$$

which shows, in comparison with (5.2), that

$$C(x) = 6_0 + 6_1 \cos 2\pi x \Delta\sigma \qquad [\text{W}] \qquad (5.34a)$$

and

$$S(x) = 6_1 \sin 2\pi x \Delta\sigma \qquad [\text{W}], \qquad (5.34b)$$

which combine to give the visibility

$$\mathcal{V}(x) = \left[1 + \frac{26_0 6_1}{(6_0 + 6_1)^2} (\cos 2\pi x \Delta\sigma - 1) \right]^{1/2} \qquad (5.35a)$$

and the phase function

$$\tan \theta(x) = \frac{6_1 \sin 2\pi x \Delta\sigma}{6_0 + 6_1 \cos 2\pi x \Delta\sigma}. \qquad (5.35b)$$

These are also illustrated, in Figure 5.3, for the case $6_1 > 6_0$. It is interesting to consider the limiting forms of equations (5.35). Thus if we make $6_1 \ll 6_0$, $\tan \theta(x) \to 0$, whereas if, on the other hand, we make $6_1 \gg 6_0$, $\tan \theta(x) \to \tan 2\pi x \Delta\sigma$. This shows that $\theta(x)$ depends strongly and uniquely on the relative powers (intensities) $6_1/6_0$ of the two components of the doublet. However, the absolute value of the visibility $|\hat{\mathcal{V}}(x)|$ depends on $6_0 6_1/(6_0 + 6_1)^2$, which is invariant with respect to an interchange of 6_0 and 6_1. It will be seen that even for relatively simple spectral distributions, such as that chosen here, the modulus and the phase of the visibility can be formally complicated. Moreover, the physical significance of $\theta(x)$ is not immediately apparent.

5.3.2. Meaning of the phase

An appreciation of the meaning of $\theta(x)$ may be gained from an examination of the expression (see equation (5.25))

$$F(x) = I(x) - \bar{I} = \bar{I}\mathcal{V}(x) \cos [2\pi\sigma_0 x + \theta(x)] \qquad [\text{W}] \qquad (5.36)$$

for the interferogram observed in partially characterized spectrometry. This corresponds to the general case when $b(s)$ is unsymmetrical. In the particular case when $b(x)$ *is* symmetrical, (5.36) reduces to

$$F_s(x) = 6_s \mathcal{V}_s(x) \cos 2\pi\sigma_0 x \qquad [\text{W}]. \qquad (5.37)$$

Therefore we see that the fringes observed when the spectrum centred at σ_0

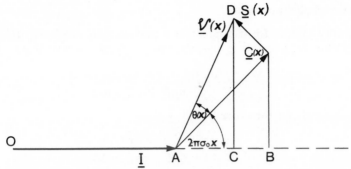

Figure 5.4. Vectorial illustration of the meaning of the phase factor. In labelling the figure, the modulus of **I**, namely $\bar{I}$, has been taken to be unity

is unsymmetrical are out of phase by $\theta(x)$ with respect to the fringes that would be observed for a similar symmetrical spectrum centred at σ_0. Thus to observe $\theta(x)$ it is necessary to make measurements of the *positions* of fringes relative to the positions occupied by the corresponding fringes produced by a symmetrical spectrum. Such measurements are very difficult to make. In fact, their difficulty is in striking contrast to the relative ease with which the visibility may be measured. Michelson did not in fact measure these, but simply derived relative spectral distributions from his visibility measurements.

The meaning of the phase $\theta(x)$ can also be presented using a vector representation in which the detected power $I(x)$ is considered to be the orthogonal projection on an axis of the sum of three vectors. Suppose that $\bar{I}$ is the magnitude of the vector **I**, which defines an arbitrary direction in abstract space, then the projection of a vector $\mathbf{C}(x)$ inclined to **I** at an angle $2\pi\sigma_0 x$ is $\mathbf{C}(x)\cos 2\pi\sigma_0 x$ and the projection of a vector $\mathbf{S}(x)$ inclined at an angle $2\pi\sigma_0 x + \frac{1}{2}\pi$ is $-\mathbf{S}(x)\sin 2\pi\sigma_0 x$. As $\mathbf{C}(x)$ and $\mathbf{S}(x)$ are mutually perpendicular, $\{[C(x)]^2 + [S(x)]^2\}^{1/2}$ represents the magnitude and $\tan^{-1}[S(x)/C(x)]$ the phase of the vector given by the sum $\mathbf{C}(x) + \mathbf{S}(x)$. From (5.17) and (5.18), it follows that the vector $\mathcal{V}(x)$ is defined by

$$\bar{I}\mathcal{V}(x) = \mathbf{C}(x) + \mathbf{S}(x) \qquad [\text{W}]. \qquad (5.38)$$

The projection of the vector sum $\mathbf{I} + \mathbf{C}(x) + \mathbf{S}(x)$ in the direction of **I** is therefore a vector with magnitude $\bar{I} + \bar{I}\mathcal{V}(x)\cos[2\pi\sigma_0 x + \theta(x)]$, which, from (5.36), is simply $\mathbf{I}(x)$, that is, a vector representing the detected power. The x-dependent part $\bar{I}\mathcal{V}(x)\cos[2\pi\sigma_0 x + \theta(x)]$ is the interferogram, $F(x)$, and this is the magnitude of the projection of the vector $\bar{I}\mathcal{V}(x)$, whose length represents the fringe amplitude. For the symmetrical and unsymmetrical spectral distributions already mentioned, this vector is given by

$$6_s\mathcal{V}_s(x) = \mathbf{C}_s(x) \qquad (5.39)$$

and by (5.38), respectively; $\theta(x)$ is the phase angle between these vectors.

5.3.3. Michelson's procedure

The visibility (5.35a) for a doublet of separation $\Delta\sigma$ can be written in the form

$$[\mathcal{V}(x)]^2 = \frac{1 + (\sigma_1/\sigma_0)^2 + 2(\sigma_1/\sigma_0)\cos 2\pi x \Delta\sigma}{1 + (\sigma_1/\sigma_0)^2 + 2(\sigma_1/\sigma_0)} \tag{5.40a}$$

given by Michelson; the corresponding phase is

$$\tan\theta(x) = \frac{(\sigma_1/\sigma_0)\sin 2\pi x \Delta\sigma}{1 + (\sigma_1/\sigma_0)\cos 2\pi x \Delta\sigma}. \tag{5.40b}$$

The procedure adopted by Michelson was to calculate visibility curves for a number of functions, including doublets. He then compared the shapes of the experimental and theoretical curves and selected one of the latter that was most like the former; this gave a general picture of the spectrum. Adjustment of σ_1/σ_0 and $\Delta\sigma$ resulted in a theoretical curve of $[\mathcal{V}(x)]^2$ which matched the experimental data. The theoretical curves were calculated using a mechanical harmonic analyser. We note that $[\mathcal{V}(x)]^2$ is unaltered by change of sign of $\Delta\sigma$, again emphasizing the lack of dependence of the visibility on whichever of the components is at the higher wavenumber. Thus, having found σ_1/σ_0 and $|\Delta\sigma|$, there still remains the problem $+\Delta\sigma$ or $-\Delta\sigma$? Michelson 'guessed' the sign of $\Delta\sigma$ from measurements on the positions of the fringes relative to the positions of the cosinusoids corresponding to σ_0 alone at $s = 0$. When $\Delta\sigma > 0$ (and $\sigma_1 > \sigma_0$) the phase $\theta(x)$ is given by (5.40b), illustrated in Figure 5.3, so that as x is increased from zero $\theta(x)$ increases. As bright fringes occur when $2\pi\sigma_0 x + \theta(x) = 0$, $2\pi, 4\pi, \ldots$, etc., it follows that when $\theta(x)$ behaves as described the bright fringes occur at slightly lower values of x than would be the case if $\theta(x)$ were zero. Thus, when the fringes are displaced, near $x = 0$, towards shorter path differences, the stronger part of the spectrum lies at larger wavenumbers. The converse is true when $\Delta\sigma < 0$ (and $\sigma_1 > \sigma_0$).

5.4. RELATION BETWEEN THE COMPLEX VISIBILITY AND THE COMPLEX DEGREE OF COHERENCE

From (4.29), the power detected at the exit stop of an idealized two-beam interferometer is

$$I(x) = 2\,|\hat{f}|^2\, A\Gamma_{11}(0)[1 + \mathrm{Re}\,\hat{\gamma}_{11}(x/c)] \quad [\mathrm{W}], \tag{5.41}$$

where

$$2|\hat{f}|^2 A\Gamma_{11}(0) = \sigma \quad [\mathrm{W}] \tag{5.42}$$

is the power detected in the partial beam from either arm alone and $\hat{\gamma}_{11}(x/c)$ is the complex degree of coherence (i.e. $\hat{\gamma}_{11}(x/c) = \hat{\Gamma}_{11}(x/c)/\Gamma_{11}(0)$) of the partial beams. It follows from substitution of (5.42) in (5.41) and comparison with (5.25) that

$$\text{Re } \hat{\gamma}_{11}(x/c) = \mathcal{V}(x) \cos \left[2\pi\sigma_0 x + \theta(x) \right] \tag{5.43}$$

and

$$|\hat{\gamma}_{11}(x/c)| = \mathcal{V}(x). \tag{5.44}$$

This shows the intimate link between the modulus of the complex degree of coherence and the visibility, thus the greater the coherence the better the visibility.

5.5. THE VISIBILITY FUNCTION FOR SOME SIMPLE SPECTRAL PROFILES

5.5.1. Simple ideal cases

The two simplest, but extremely unrealistic, profiles are the Dirac impulse $6_0 \delta(s)$ representing ideal monochromatic radiation and the box-car $B \sqcap (s/2\Delta\sigma)$ representing a narrow 'white' band. The first has already been discussed in Chapter 1 and in section 5.2. The visibility for the narrow white band

$$b(s) = B \sqcap (s/2\Delta\sigma) \qquad [\text{W m}] \tag{5.45}$$

is, from (5.27) and (2.76),

$$\mathcal{V}(x) = \int_{-\infty}^{\infty} \frac{B}{6} \sqcap \left(\frac{s}{2\Delta\sigma} \right) \exp 2\pi i s x \, ds \tag{5.46a}$$

$$= |\text{sinc} (2x\Delta\sigma)|, \tag{5.46b}$$

since $6 = 2B\Delta\delta$. We take the modulus of the integral in (5.46) because $\mathcal{V}(x)$ is, by definition, positive (or equal to zero).† The visibility (5.46) is everywhere less than 1, apart from where $x = 0$, and falls as $|x|$ increases. The first zero is at $x = 1/(2\Delta\sigma)$ and so the visibility falls off less slowly the narrower the spectral band. As x increases, further zeros occur and $\mathcal{V}(x)$ becomes smaller between the zeros tending towards zero for infinitely large x.

5.5.2. Realistic spectral profiles

We can now turn to more realistic profiles. There are two main classes of problem in emission spectroscopy to which visibility measurements may be applied: the determination of the profile of a single narrow feature and the revelation of fine structure within a narrow spectral region. Apart from a few important examples, Michelson concentrated mainly on the latter, while Terrien and his colleagues have developed techniques for the former. There

---

† According to Michelson's definition (5.11) the visibility cannot be negative. In fact, negative values can occur, even for symmetrical spectra, and these correspond to ranges of x over which there is a shift of π in the phase of the complex visibility. This is discussed in section 5.5.2.3.

116

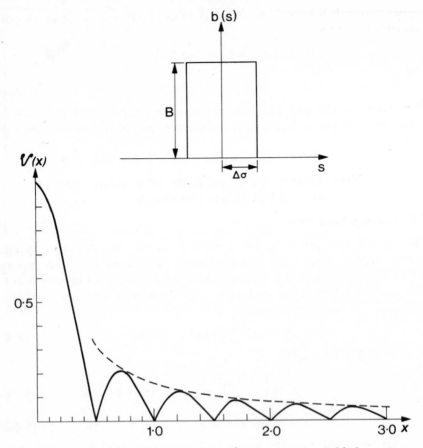

Figure 5.5. Visibility $\mathcal{V}(x) = |\text{sinc}\,(2x\Delta\sigma)|$ for a narrow 'white' spectrum
$b(s) = B\,\Pi(s/2\Delta\sigma)$

are two simple line profiles of particular importance to the emission spectroscopist: the Doppler shape in which the finite width is due to the thermal agitation and the Lorentzian resonance shape in which the broadening is due to such phenomena as interatomic collisions or quantum resonance effects.

Explicitly, these may be written as

$$b_D(s) = \left[\frac{4\bar{b}_D^2 \ln 2}{\pi(2\gamma_D)^2}\right]^{1/2} \exp\left[\frac{-4\ln 2 s^2}{(2\gamma_D)^2}\right] \quad [\text{W m}] \tag{5.47}$$

and

$$b_L(s) = \frac{2\bar{b}_L}{\pi(2\gamma_L)}\frac{(2\gamma_L)^2}{4s^2 + (2\gamma_L)^2} \quad [\text{W m}), \tag{5.48}$$

where the numerical factors are introduced so that $\int_{-\infty}^{\infty} b(s)\,ds = \bar{b}$. Since each of these profiles is symmetrical, we may put $\theta(x) = 0$ and evaluate the visibility from the real expression (5.28).

5.5.2.1. The Doppler profile

The Doppler profile is, in fact, of Gaussian form, as we see by putting

$$\frac{4\ln 2}{(2\gamma_D)^2} = \pi a^2 \quad [\text{m}^2] \tag{5.49}$$

in (5.47) and comparing with (2.124). Therefore, for the visibility, we have the Fourier transform (see 2.127)

$$\mathcal{V}_D(x) = a \int_{-\infty}^{\infty} \exp\left(-\pi a^2 s^2\right) \exp 2\pi i s x \, ds \tag{5.50a}$$

$$= \exp\left(-\frac{\pi x^2}{a^2}\right) \tag{5.50b}$$

$$= \exp\left[\frac{-\pi^2 (2\gamma_D)^2 x^2}{4\ln 2}\right], \tag{5.50c}$$

which is equal to 1 at $x = 0$ and less than one elsewhere, falling to zero as x tends to infinity. Since $2\gamma_D$ is the width at half-height of the profile, increase of this width causes $\mathcal{V}_D(x)$ to fall off more rapidly. The form of the visibility is shown in Figure 5.6.

5.5.2.2. The Lorentz profile

The Lorentz profile is the classical resonance curve. The Fourier transform is

$$\mathcal{V}_L(x) = \frac{2(2\gamma_L)^2}{(2\gamma_L)\pi} \int_{-\infty}^{\infty} \frac{\exp 2\pi i s x}{4s^2 + (2\gamma_L)^2} \, ds \tag{5.51a}$$

$$= \exp\left[-|\pi(2\gamma_L)x|\right], \tag{5.51b}$$

which is also shown in Figure 5.6 for the case where both the Doppler and the resonance curve have identical half-widths. These profiles are very similar near the peak at σ_0, the differences occurring in the wings, where the signal is relatively small; thus reliable distinction between the profiles is difficult in practice. As the resonance profile is the broader of the two in the wings, it gives rise to a visibility curve that falls off more rapidly than in the Doppler case. This relatively rapid fall-off leads to a measurable difference in visibility relatively close to $x = 0$, which enables the distinction between the spectral profiles to be perceived much more easily than would be the case if they were being observed by direct spectroscopy.

5.5.2.3. Self-reversed profile

An important problem both in laboratory and astronomical sources is that of self-absorption, by cold gas, of the radiation emitted from a hotter region of

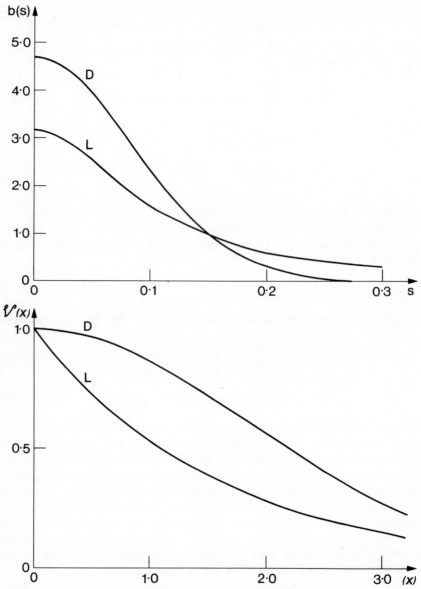

Figure 5.6. Spectral and visibility curves for the Doppler (D) and Lorentz (L) profiles

the same gas. Self-absorption has a marked effect on the form of the visibility as the following considerations show.

Consider a simple model of a region of hot gas having an emission power spectrum $b(s)$ surrounded by a cooler region with an optical depth $[\alpha(s)]^{-1}$.† The overall power spectrum radiated via an interferometer to a detector is then

$$b'(s) = b(s) \exp [-\alpha(s)l] \quad \text{[W m]}, \tag{5.52a}$$

where l is the thickness of the cooler region. This simplifies to

$$b'(s) = b(s)[1 - \alpha(s)l] \quad \text{[W m]} \tag{5.52b}$$

when $\alpha(s)l$ is small. The visibility corresponding to this spectrum is

$$\mathscr{V}'(x) = \int_{-\infty}^{\infty} \frac{b(s)}{6'} \exp 2\pi i s x \, ds - \int_{-\infty}^{\infty} \frac{b(s)\alpha(s)l}{6'} \exp 2\pi i s x \, ds,$$

where

$$6' = \int_{-\infty}^{\infty} b'(s) \, ds \quad \text{[W]} \tag{5.53}$$

is the total radiant power that is detected.

At optical frequencies the case of a resonance profile altered by self-absorption is of little practical interest since Doppler broadening is always present and generally dominant. Consequently, we shall only quote the results for this after we have developed the visibility for the more important Doppler-broadened case. Suppose spectral power collected from the hot gas is

$$b_D(s) = a6_D \exp (-\pi a^2 s^2) \quad \text{[W m]}, \tag{5.54}$$

as in (5.47), and suppose that the absorption coefficient of the cooler layer

$$\alpha(s) = A' \exp (-\pi a'^2 s^2) \quad \text{[m}^{-1}\text{]} \tag{5.55}$$

is also Gaussian, with

$$\pi a'^2 = \frac{4 \ln 2}{(2\gamma'_D)^2} \quad \text{[m}^2\text{]} \tag{5.56}$$

and with the same centre ($s = 0$) as the emission profile (5.54). The total power collected from the gas in the optically thin case $[\exp (\alpha l) \sim (1 - \alpha l)]$ is

$$6'_D = \int_{-\infty}^{\infty} b'(s) \, ds$$

$$= a6_D \int_{-\infty}^{\infty} \exp (-\pi a^2 s^2) \, ds - a6_D A'l \int_{-\infty}^{\infty} \exp [-\pi (a^2 + a'^2)s^2] \, ds$$

$$= 6_D \left[1 - A'l \sqrt{\frac{a^2}{a^2 + a'^2}} \right] = 6_D[1 - \beta] \quad \text{[W]} \tag{5.57}$$

† Astronomers frequently use the symbol τ for optical depth, but this practice is undesirable since τ is used to mean many other things and is also generally understood to indicate a temporal rather than a spatial parameter or variable.

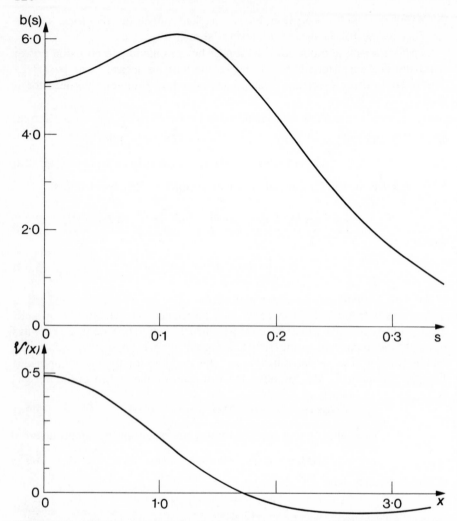

Figure 5.7. Spectral and visibility curves for the self-reversed profile: the upper curve is calculated exactly for a $6_D = 1.0$, $a = \sqrt{2}\pi$, $a' = 2\sqrt{2}\pi$, and $A' = a3\sqrt{5}$. The lower curve is calculated using the optically thin approximation (equation (5.52b))

and the visibility function is, from (5.27) and (2.127)

$$\hat{V}'_D(x) = (1-\beta)\left\{1 - \beta \exp\left[\frac{\pi a'^2 x^2}{(a^2 + a'^2)a^2}\right]\right\} \exp\left(-\frac{\pi x^2}{a^2}\right). \qquad (5.58)$$

Bearing in mind that β, by definition and assumption, must be less than unity, we find that the visibility may become negative where

$$1 - \beta \exp\frac{\pi a'^2 x^2}{(a^2 + a'^2)a^2} < 0, \qquad (5.59a)$$

which occurs where

$$x^2 = x_0^2 \equiv \frac{a^2(a^2 + a'^2)}{\pi a'^2} \ln \frac{1}{\beta} . \tag{5.59b}$$

$\mathcal{V}_D'(x)$ is positive for $|x| < x_0$ and negative for $|x| > x_0$. Of course, strictly speaking, the visibility may not assume negative values; their existence in practice means that there is a change from 0 to π at $x = x_0$ in the phase of the complex visibility; dark fringes become bright and bright fringes become dark. The existence of a single value x_0 of x for which the visibility is zero is characteristic of a self-reversed line spectrum, but it must be remarked that the converse is not necessarily true; that is, it is not a necessary and a sufficient condition for the existence of a self-reversed line when a single zero is recorded in the visibility profile. In fact, no general rule can be formulated, but the vanishing of the visibility is a useful indication of self-absorption in Gaussian sources.

5.5.2.4. The mixed profile

In practice, it is rare for a spectral profile to be described simply by a single expression such as (5.47) or (5.48) above. A common combination usually called the Voigt profile is a mixture of Lorentzian and Doppler, represented by the convolution $b_L(s) * b_D(s)$. Because of the property of the convolution (section 2.3) the visibility of the fringes produced by the mixed spectral profile is the product of the component visibilities

$$\mathcal{V}_L(x)\mathcal{V}_D(x) \overset{\triangle}{=} \mathcal{V}_{LD}(x). \tag{5.60}$$

Thus

$$\mathcal{V}_{LD}(x) = \exp\left[-\frac{\pi^2(2\gamma_D)^2 x^2}{4 \ln 2}\right] \exp\left[-|\pi(2\gamma_L)x|\right], \tag{5.61}$$

so that

$$\ln \mathcal{V}_{LD}(x) = -\frac{\pi^2(2\gamma_D)^2}{4 \ln 2} x^2 - \pi(2\gamma_L)|x| \tag{5.62a}$$

or

$$-\frac{1}{|x|} \ln \mathcal{V}_{LD}(x) = \frac{\pi^2(2\gamma_D)^2}{4 \ln 2} |x| + (2\gamma_L)\pi, \tag{5.62b}$$

showing that if $(-1/|x|) \ln \mathcal{V}_{LD}(x)$ is plotted versus $|x|$ a straight line is obtained.

The slope of the line yields the Doppler half-width $2\gamma_D$ and the intercept gives the Lorentz half-width $2\gamma_L$. Although the line has a simple shape, the mathematical description $b_L(s) * b_D(s)$ is not expressible simply; the visibility, on the other hand, has the advantages of both graphical and mathematical simplicity which enables the form of the spectrum and the component widths to be deduced.

5.6. THE COMPUTED SPECTRAL DISTRIBUTION

It is clear that in the three cases cited above a suitable plot of appropriate combinations of $\ln \mathcal{V}(x)$ and $|x|$ gives a straight line from which half-widths can be evaluated. It must be emphasized that this simple procedure is of value *only* if the spectral form has been anticipated; when this is the case there is no need to evaluate the Fourier transform of the visibility curve. Terrien has shown the merits of this approach for the study of simple narrow lines. In more complicated cases, where the line shape is not of a well-known form or where there is structure giving rise to asymmetry for the feature as a whole, a Fourier transformation is essential.

The ideal and exact relation between the complex visibility $\hat{\mathcal{V}}(x)$ and the spectrum is

$$r(s) \triangleq \frac{b(s)}{6} = \int_{-\infty}^{\infty} \hat{\mathcal{V}}(x) \exp(-2\pi isx) \, dx \qquad [\text{m}]. \qquad (5.63)$$

In practice however this function cannot be observed to infinite values of x, neither is it recorded as a continuous function of x within the range of measurement. It is therefore necessary to consider the effects of truncating the integral (5.63) and of replacing it, as one must, by a summation.

5.6.1. Finite range for the visibility function

We evaluate a computed spectrum which, taking account of the finiteness of x, is given by

$$r^c(s) = \int_{-D}^{D} \hat{\mathcal{V}}(x) \exp(-2\pi isx) \, dx \qquad [\text{m}] \qquad (5.64a)$$

$$\equiv \int_{-\infty}^{\infty} \hat{\mathcal{V}}(x) \sqcap \left(\frac{x}{2D}\right) \exp(-2\pi isx) \, dx \qquad [\text{m}], \qquad (5.64b)$$

where the range of observation $-D \le x \le D$ is denoted by the box-car $\sqcap(x/2D)$ (see section 2.9.1). We use the superscript c to denote *calculated* or *computed*. The convolution theorem enables this relation to be written

$$r^c(s) = 2Dr(s) * \text{sinc}(2sD) \qquad [\text{m}] \qquad (5.65)$$

since $r(s) \supset \mathcal{V}(x)$ and $2D \, \text{sinc}(2sD) \supset \sqcap(x/2D)$. The significance of the computed spectrum being the convolution of the true spectrum with a sinc function may be appreciated by allowing $r(s)$ to be an ideal monochromatic

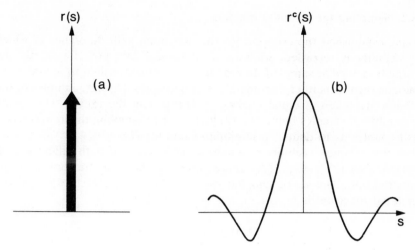

Figure 5.8. Ideal delta function (a) and the result (b) of computing it from a truncated visibility function

feature, that is, $r(s) = 6_0 \, \delta(s)$. When this is the case, the computed spectrum is

$$r^c(s) = 2D6_0 \int_{-\infty}^{\infty} \text{sinc} \, (2s'D) \, \delta(s - s') \, ds \qquad (5.66a)$$

$$= 2D6_0 \, \text{sinc} \, (2sD) \qquad [\text{m}], \qquad (5.66b)$$

which represents a function of finite width with side lobes that are alternately positive and negative in sign; thus the truncation of the visibility function leads to a computed spectrum having finite resolution. Taking the position of the first zeros in the sinc function as an (arbitrary) criterion for spectral width leads to a value D^{-1} for this since $\text{sinc} \, (2sD) = 0$ when $s = \pm 1/(2D)$; the resolving power $\mathscr{R} = \sigma_0/R$ is $\mathscr{R} = \sigma_0 D$.

The computed spectral profile closely resembles the true profile if the natural width 2γ is considerably greater than D^{-1}. As the independent variable is D rather than 2γ, we can write this statement as a condition to be satisfied by D, namely $D \gg (2\gamma)^{-1}$ for small distortion. In practice, this requirement is fairly easily satisfied even for the sharpest lines. For example, the $2p_{10} - 5d_5$ line of krypton 86 at $\lambda_0 = 605.8$ nm ($\sigma_0 = 1.6 \times 10^6 \, \text{m}^{-1}$) has an exceptionally small half-width of less than 2 m^{-1}, but nevertheless, since path differences of the order 1 m can be realized without too much difficulty in the laboratory, one can determine the width of the line and achieve a resolving power of about 10^6. Measurements at $D \approx 1$ m are not usual with astronomical sources because these have greater spectral width. They also have much lower power than do laboratory sources and the resulting weaker fringes tend to get rapidly lost in the noise as the visibility falls. Thus maximum path differences of 10 to 20 mm are often adequate.

5.6.2. Sampling the visibility function

Having established the criterion for the maximum path difference at which the visibility is recorded, we now need to ascertain how frequently the information need be sampled. In practice, it is not convenient to record as an analogue record; neither, fortunately, is it necessary. Discrete samples of the visibility are taken at equal intervals β throughout the range $-D \leq x \leq D$, the number of samples being $N = (2D/\beta) + 1$. The sampling theorem (section 2.10) enables us to deduce a minimum value for N.

Let the complex visibility be recorded at a series of path differences $r\beta$ ($r = -N, -N+1, \ldots, 0, \ldots, N$) represented by the shah function $\text{Ш}(x/\beta)$ (section 2.9.9). The set of sampled values is $\hat{\mathcal{V}}(x) \sqcap (x/2D) \text{Ш}(x/\beta)$ and the computed spectrum is

$$r^c(s) = \int_{-\infty}^{\infty} \hat{\mathcal{V}}(x) \sqcap (x/2D) \text{Ш}\left(\frac{x}{\beta}\right) \exp(-2\pi isx) \, dx \qquad [\text{m}], \qquad (5.67a)$$

which is expressible as the convolution (section 2.3)

$$r^c(s) = 2D\beta r(s) * \text{sinc}(2sD) * \text{Ш}(s\beta) \qquad [\text{m}] \qquad (5.67b)$$

after noting that $r(s) \supset \hat{\mathcal{V}}(x)$; $2D \text{ sinc}(2sD) \supset \sqcap(x/2D)$ (see section 2.9.1) and $\beta \text{Ш}(s\beta) \supset \text{Ш}(x/\beta)$ (see section 2.9.9).

As $\text{Ш}(s\beta)$ is merely a replicating function, (5.67b) shows that the spectrum $r^c(s)$ is repeated at intervals $2m/\beta$ along the s-axis. We can use a box-car function to restrict (5.67b) to the range of interest near $s = 0$:

$$r^c(s) = 2D\beta[r(s) * \text{sinc}(2sD) * \text{Ш}(s\beta)]\sqcap(s\beta/2) \qquad (5.68a)$$

$$\equiv 2Dr(s) * \text{sinc}(2sD), \qquad |s| < 2/\beta \qquad [\text{m}]. \qquad (5.68b)$$

Recalling the statement (section 2.10) of the sampling theorem, we see that there is no overlapping of computed spectra if $(2\beta)^{-1}$ is equal to the highest recorded frequency, say s_M (where $s_M \ll \sigma_0$). Hence, the condition to be satisfied by β is that $\beta < 1/(2s_M)$.

Returning to the example given in section 5.6.1 above, we see that since the half-width is 2 m^{-1} one can in practice take s_M to be about 4 m^{-1}, thus $\beta < 0.125$ m. This corresponds to $N = 17$ samples. Even fewer samples are required for narrower spectral regions. Thus we see that an acceptably undistorted computed spectral profile is obtained from a visibility curve sampled at a few relatively coarse intervals up to a fairly large path-difference. Very small oscillations may appear in the wings of the computed line, but these are generally of little practical inconvenience (see Figure 5.8).

5.7. PROCEDURES FOR FINDING SPECTRAL DISTRIBUTIONS FROM THE VISIBILITY

The procedure adopted in practice in partially characterized Fourier spectrometry depends very much on whether the form of the spectrum can be anticipated and whether a simple examination of the visibility supports the anticipation.

1. If the spectrum consists of a single Gaussian or Lorentzian feature, the visibility falls monotonically. Only a few values need be recorded and a plot of $(-1/|x|) \ln \mathcal{V}(x)$ against $|x|$ drawn to yield the line parameters. If the lines are self-reversed this is apparent from the non-linearity of the graph and the falling of the visibility to zero.
2. If the spectrum is more complicated but, nevertheless, expected to be symmetrical, N values of $\mathcal{V}(x)$ are recorded at discrete intervals β at $x = r\beta$ $(r = 0, 1, \ldots, N)$ and subject to Fourier transformation to give the relative spectral distribution $r(s)$. Should $r(s)$ not turn out to be symmetrical, the unique spectrum can be found from an examination of the phase shift near $x = 0$ if the spectrum is of doublet form.
3. If the spectrum is more complicated, it is essential to measure $\mathcal{V}(x)$ and $\theta(x)$ to evaluate $r(s)$ from a full Fourier transform.

The observations needed at any one sampling point are measurements of the amplitude $I(x)$ and the phase $\theta(x)$ of the quasi-cosinusoidal interferogram (5.36) produced by the quasi-monochromatic spectrum $b(s)$. The amplitude can be obtained from four or five measurements of detected power taken over a range of about a wavelength centred on x; absolute phase is deduced by measuring interferometrically the path difference at which the observations are made. This interferometric measurement is done by deducing the path difference from the near cosinusoidal fringes arising from the spectrally narrowest available source. The visibility of these fringes is nearly constant and their phase is zero. Gas lasers are near ideal reference sources since they emit virtually monochromatic radiations. It goes without saying that, with such narrowness, partially characterized spectrometry is not of any use to measure the spectrum of the laser itself because the attainment of the enormous path differences necessary to produce significant change from unity for the visibility is not practical. Laser line widths are best deduced from heterodyne or correlation measurement made with a square law detector. These systems generate difference frequencies which when measured give directly the line profile. However, it is necessary that the highest frequency to be measured pass through the subsequent electronic system and the method is limited therefore to lines with widths less than a few gigahertz ($\sim 5 \text{ m}^{-1}$). However, the linewidths of CW lasers are often only a few kilohertz, so there is no limitation in practice.

CHAPTER 6

The determination of the power spectrum by fully characterized Fourier spectrometry

Part 1

The Single-sided Interferogram

6.1. FULLY CHARACTERIZED FOURIER SPECTROMETRY

In Chapter 4 we established, by a variety of arguments, the expression

$$F(x) = \int_0^\infty B(\sigma) \cos 2\pi\sigma x \, d\sigma \qquad [\text{W}] \qquad (6.1)$$

for the interferogram $F(x)$ produced at the output of an ideal two-beam interferometer which is irradiated by the power spectrum $B(\sigma)$. The observation of the fine detail of such an interferogram, rather than its coarse long term variation, is the basis of *fully characterized Fourier spectrometry*. From its definition, (6.1), $F(x)$ is necessarily perfectly symmetrical about $x = 0$, but some experimental considerations enter which may modify this conclusion. Principal amongst these is that it is almost universal practice to modulate the radiation beam at a low audio frequency in order to secure the advantages accruing from a.c. amplification and phase-synchronous rectification. Depending on the type of modulation employed, one may get either a perfectly symmetric interferogram (with amplitude modulation) or else a perfectly antisymmetric interferogram (with phase modulation). Second, the practical interferometer will always have residual imperfections which will lead to the interferogram being slightly asymmetric. However, this will usually be a small effect and one can develop the theory best by starting with the perfect case and introducing corrections later. Given this, one has the situation that because of the symmetry about $x = 0$, either side of the interferogram (that is, either $x \geq 0$ or $x \leq 0$) contains all the information and one need only record the interferogram over a semi-range of either all positive or all negative values. We call an interferogram that is recorded only over a semi-range a *single-sided interferogram*, although other authors use alternatively the phrase *one-sided interferogram*. Fourier spectrometry commonly consists of recording such single-sided interferograms and then inverting the digitized versions, in a computer, to retrieve the spectra.

6.2. THE PHYSICAL SPECTRUM AND THE MATHEMATICAL SPECTRUM

The physical spectrum exists, of course, only for σ values greater than zero, but it is convenient to introduce a mathematical spectrum which can be resolved into odd and even components (see section 2.1) which are defined for all values of σ. One has then

$$B(\sigma) = B_e(\sigma) + B_o(\sigma) \qquad [\text{W m}], \tag{6.2}$$

where, because of the perfect one-sidedness of $B(\sigma)$, $B_e(\sigma)$, and $B_o(\sigma)$ are simply related (see Figure 2.1). The integral (6.1) can therefore be written

$$F_e(x) = \int_{-\infty}^{+\infty} [B_e(\sigma) + B_o(\sigma)] \cos 2\pi\sigma x \, d\sigma = 2 \int_0^\infty B_e(\sigma) \cos 2\pi\sigma x \, d\sigma \qquad [\text{W}];$$
$$\tag{6.3}$$

application of Fourier's inversion theorem to (6.3) gives

$$B_e(\sigma) = 2 \int_0^\infty F_e(x) \cos 2\pi\sigma x \, dx \qquad [\text{W m}]. \tag{6.4}$$

Hence the cosine Fourier transform of the observed even interferogram $F_e(x)$, over positive values of x, gives the even part of the mathematical spectrum, which is seen to be related to the physical spectrum by

$$B_e(\sigma) = \tfrac{1}{2}[B(\sigma) + B(-\sigma)]$$
$$= \tfrac{1}{2}B(\sigma), \qquad \sigma > 0 \qquad [\text{W m}]. \tag{6.5}$$

We can therefore write

$$B(\sigma) = 4 \int_0^\infty F_e(x) \cos 2\pi\sigma x \, dx, \qquad \sigma > 0 \qquad [\text{W m}] \tag{6.6}$$

as the exact relation between the desired spectrum $B(\sigma)$ and the measured quantity $F_e(x)$. However, it must be borne in mind that $B(\sigma)$, defined by (6.6), has physical meaning only for positive values of σ. Entirely equivalent arguments apply for the perfectly antisymmetric case, the only difference being that sine transforms replace the cosine transforms. One would therefore have

$$B(\sigma) = 4 \int_0^\infty F_o(x) \sin 2\pi\sigma x \, dx, \qquad \sigma > 0 \qquad [\text{W m}]. \tag{6.7}$$

When a homogeneous specimen is placed in the beam, the interferometer will still be optically perfect, but its transmission will become frequency dependent. One would then have that the power spectrum observed would be related to that incident on the interferometer by

$$B_s(\sigma) = \tau_s(\sigma)B(\sigma), \tag{6.8}$$

where the subscript s denotes specimen. It follows therefore that, if one observes a *background* interferogram $F(x)$ and a *specimen* interferogram $F_s(x)$ under identical conditions,† then the ratio of their transforms gives the transmission spectrum of the specimen. Similar arguments apply for power reflection measurements in which $\tau(\sigma)$ is replaced by the power reflectivity $R(\sigma)$.

6.3. THE SPECTRAL WINDOW

In fully characterized spectrometry one has to face the problem, just as one does in partly characterized spectrometry, that one cannot observe the interferogram (or visibility curve) over the entire semi-infinite range $0 \leqslant x \leqslant \infty$. The practical result, once again, is that the calculated mathematical spectrum differs from the true physical spectrum. The nature of this difference is best brought out by considering the simplest possible incident spectrum, namely, that of monochromatic radiation. The corresponding interferogram is

$$F_e(x) = 6_0 \cos 2\pi\sigma_0 x = 26_e(\sigma_0) \cos 2\pi\sigma_0 x. \tag{6.9}$$

The Fourier transform of this interferogram, over the observed range of x, $0 \leqslant x \leqslant D$, is given by

$$\begin{aligned}
B_e^c(\sigma) &= 4 \int_0^D [6_e(\sigma_0) \cos 2\pi\sigma_0 x] \cos 2\pi\sigma x \, \mathrm{d}x \\
&= 2D6_e(\sigma_0)[\mathrm{sinc}\, 2(\sigma_0 + \sigma)D + \mathrm{sinc}\, 2(\sigma_0 - \sigma)D].
\end{aligned} \tag{6.10}$$

We see that the calculated spectrum is even and consists of two features, one centred at $\sigma = +\sigma_0$ and the other at $\sigma = -\sigma_0$. Neither of these has the delta form of the incident spectrum, but rather each shows the broad central feature plus accompanying side lobes of the sinc function (compare Figure 5.8). In physical terms, the result of a calculation over a finite range of interferogram is a reduction in the resolution and the introduction of troublesome spurious features. There is nothing that can be done about the loss of resolution, for this has its origin in the loss of information caused by the sharp cut-off at $x = D$. However, techniques for mitigating the effects of the spurious features are available and will be discussed in the next section.

---

† Of course in practice one may wish to change amplifier gain between the two runs in order to minimize experimental error. This is permissible provided the electronic system is properly calibrated.

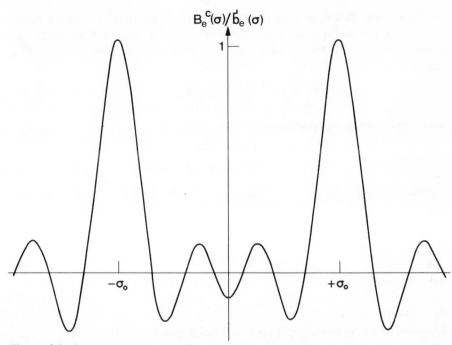

$B_e^c(\sigma)/\breve{b}_e(\sigma)$

1

$-\sigma_0$

$+\sigma_0$

Figure 6.1. Spectral window resulting from the observation of a monochromatic line over a finite range of mirror travel

Normalizing equation (6.10) to the case of unit incident power, one has

$$B_e^c(\sigma)/\breve{b}_e(\sigma) = 2D[\operatorname{sinc} 2(\sigma_0 + \sigma)D + \operatorname{sinc} 2(\sigma_0 - \sigma)D], \qquad (6.11)$$

which can be written

$$B_e^c(\sigma)/\breve{b}_e(\sigma) = A_0((\sigma_0 + \sigma)D) + A_0((\sigma_0 - \sigma)D), \qquad (6.12)$$

where $A_0(\sigma D)$ is a function, called the *spectral window* or *apparatus function*. It is analogous to the *slit function* of a dispersive spectrometer and is similarly defined—namely that it is the spectrum produced by the instrument when observing monochromatic radiation of unit power. The properties of the sinc function $A_0(\sigma D)$ have been discussed at some length previously; all that will be reiterated here is that the amplitude of the side lobes falls off relatively slowly with increase (or decrease) of argument away from the peak at $\sigma D = 0$, that the half-width of the centre lobe is about $1.2/(2D)$ at half height and $1.4/(2D)$ at half amplitude, and that as D increases the peak height increases and the side lobes move in closer to the central peak. Nevertheless, the amplitude ratio of the first (or any subsequent) lobe to the central lobe is independent of D and the side lobes therefore pose a serious threat for they could, in practical spectroscopy, be confused with real spectral features.

The meaning of the apparatus function can be approached in another way and one which illustrates rather better its effect in practical, i.e. broad-band, spectroscopy. One may write the basic truncated Fourier transform as the infinite integral of the product of the interferogram and a truncating function, i.e.

$$B_e^c(\sigma) = 2 \int_0^\infty F_e(x) \, \sqcap\!\left(\frac{x}{D}\right) \cos 2\pi\sigma x \, dx \qquad [\text{W m}], \qquad (6.13)$$

where the truncating function

$$\sqcap\!\left(\frac{x}{D}\right) = \begin{cases} 1, & 0 \leqslant x \leqslant D, \\ 0, & x < 0, x > D, \end{cases} \qquad (6.14)$$

represents the cut-off at $x = D$. The cosine Fourier transform of this function is

$$2 \int_0^\infty \sqcap\!\left(\frac{x}{D}\right) \cos 2\pi\sigma x \, dx = 2D \, \text{sinc} \, 2\sigma D = A_0(\sigma D). \qquad (6.15)$$

That is, it is the apparatus function for the hypothetical case of irradiation by zero-frequency radiation. By noting that $\sqcap(x/D)$ may be written† as

$$\sqcap(x/D) = \sqcap(x/2D)H(x), \qquad (6.16)$$

equation (6.15) may be expressed as a full Fourier transform:

$$2 \int_0^\infty \sqcap\!\left(\frac{x}{D}\right) \cos 2\pi\sigma x \, dx = \int_{-\infty}^\infty \sqcap\!\left(\frac{x}{2D}\right) \exp\left(-2\pi i\sigma x\right) dx \qquad [\text{m}].$$

Equation (6.13) can therefore be written

$$B_e^c(\sigma) = \int_{-\infty}^{+\infty} F_e(x) \sqcap\!\left(\frac{x}{2D}\right) \exp\left(-2\pi i\sigma x\right) dx \qquad [\text{W m}], \qquad (6.18)$$

which, after application of the convolution theorem (section 2.3), becomes

$$B_e^c(\sigma) = B_e(\sigma) * A_0(\sigma D). \qquad (6.19)$$

The calculated spectrum is, therefore, the convolution of the true spectrum with the apparatus function. By writing out (6.19) explicitly,

$$B_e^c(\sigma) = \int_{-\infty}^{+\infty} B_e(\sigma') A_0((\sigma - \sigma')D) \, d\sigma' \qquad [\text{W m}], \qquad (6.20)$$

with σ' representing the 'dummy' variable for wavenumber, we recover the relation

$$B_e^c(\sigma) = \tfrac{1}{2}\delta_0[A_0((\sigma_0 + \sigma)D) + A_0((\sigma_0 - \sigma)D)] \qquad [\text{W m}] \qquad (6.21)$$

when $B_e(\sigma)$ represents an even 'monochromatic' spectrum centred at $\pm\sigma_0$.

† The truncating function may also be written

$$\sqcap(x/D) = \sqcap(x/2D) * \delta(x - \tfrac{1}{2}D).$$

6.4. WEIGHTING AND APODIZATION

6.4.1. Weighting

One may reduce the abruptness of the truncation at $x = D$ by multiplying the cut-off function by a slowly falling *weighting function* $W(x/D)$. When this is done, the calculated spectrum, formerly given by (6.13), becomes

$$B_e^c(\sigma) = 2 \int_0^\infty F_e(x) W\left(\frac{x}{D}\right) \sqcap\left(\frac{x}{D}\right) \cos 2\pi\sigma x \, dx \quad [\text{W m}]. \quad (6.22)$$

If the interferogram function $F_e(x)$ corresponds to monochromatic irradiation, then the evaluation of (6.22) leads to

$$B_e^c(\sigma)/\mathbf{6}_e(\sigma) = A((\sigma + \sigma_0)D) + A((\sigma - \sigma_0)D), \quad (6.23)$$

where

$$A(\sigma D) = 2 \int_0^\infty W\left(\frac{x}{D}\right) \sqcap\left(\frac{x}{D}\right) \cos 2\pi\sigma x \, dx \quad [\text{m}] \quad (6.24)$$

represents the new apparatus function. Since both $W(x/D)$ and $\sqcap(x/2D)$ are even functions, (6.24) may be written as the full Fourier transform:

$$A(\sigma D) = \int_{-\infty}^{+\infty} W\left(\frac{x}{D}\right) \sqcap\left(\frac{x}{2D}\right) \exp(-2\pi i\sigma x) \, dx \quad [\text{m}]. \quad (6.25)$$

$A(\sigma D)$ is usually called the *apodized spectral window*, whilst the Fourier transform of the weighting function

$$\Omega(\sigma D) \triangleq \int_{-\infty}^{+\infty} W\left(\frac{x}{D}\right) \exp(-2\pi i\sigma x) \, dx \quad [\text{m}] \quad (6.26)$$

is called the *apodizing function*. The convolution theorem permits us to write (6.25) as

$$A(\sigma D) = \Omega(\sigma D) * A_0(\sigma D) \quad (6.27a)$$

$$= 2D\Omega(\sigma D) * \text{sinc}(2\sigma D). \quad (6.27b)$$

When the weighting function is itself zero beyond $x = D$, in other words it already contains the box-car function, the apodized spectral window $A(\sigma D)$ and the apodizing function $\Omega(\sigma D)$ become identical.

The relation between the calculated spectrum and the ideal spectrum, in the apodized case, becomes

$$B_e^c(\sigma) = 2 \int_0^\infty F_e(x) W\left(\frac{x}{D}\right) \sqcap\left(\frac{x}{D}\right) \cos 2\pi\sigma x \, dx \quad (6.28a)$$

$$= B_e(\sigma) * A(\sigma D). \quad (6.28b)$$

This is a more general version of (6.19), which itself is now seen to be a special case of (6.28b) corresponding to unit weighting applied over the range $0 \leqslant x \leqslant D$.

A great number of weighting functions, of diverse efficacies, have been proposed from time to time. We will consider in detail only the more important and useful of these, but it will emerge that the more useful a function, in the sense of efficiency of suppression of the side lobes (Greek $\alpha\pi o\delta$ = 'removal of the feet'), the greater is the width of the central lobe and the greater therefore is the loss of resolution.

6.4.2. The normalized spectral window

In comparing one weighting function with another, it is convenient to normalize the corresponding spectral windows thus:

$$\mathscr{A}(\sigma D) \triangleq \frac{A(\sigma D)}{A_{max}} = \frac{A(\sigma D)}{A(0)}. \tag{6.29}$$

From the definite integral theorem, one can write $A(0)$, which is the maximum value of A, in the form

$$A(0) = \int_{-\infty}^{+\infty} W\left(\frac{x}{D}\right) \sqcap\left(\frac{x}{2D}\right) dx. \tag{6.30}$$

A little rearrangement permits this to be written

$$A(0) = 2Dq, \tag{6.31a}$$

where

$$q = \frac{1}{2D} \int_{-D}^{+D} W\left(\frac{x}{D}\right) dx \tag{6.31b}$$

is the mean value of the truncated weighting function. It follows immediately that q is less than unity for all weighting functions except the simple box-car, which is, of course, only a weighting function in a formal sense. In terms of q, equation (6.29) may be rewritten

$$A(\sigma D) = 2Dq\mathscr{A}(\sigma D) \tag{6.32}$$

and applying this relation to the unapodized (i.e. $q = 1$) spectral window, as a simple check, we find (from 6.15)

$$\mathscr{A}(\sigma D) = \text{sinc}\,(2\sigma D) \tag{6.33}$$

as expected.

6.4.3. Weighting and apodizing functions

The simplest non-trivial weighting function is the triangular function (see section 2.9.3) $\bigwedge(x/D)$, which is defined as

$$\bigwedge\left(\frac{x}{D}\right) = \begin{cases} 1 - \dfrac{|x|}{D}, & |x| \leqslant D, \\ 0, & |x| > D, \end{cases} \tag{6.34}$$

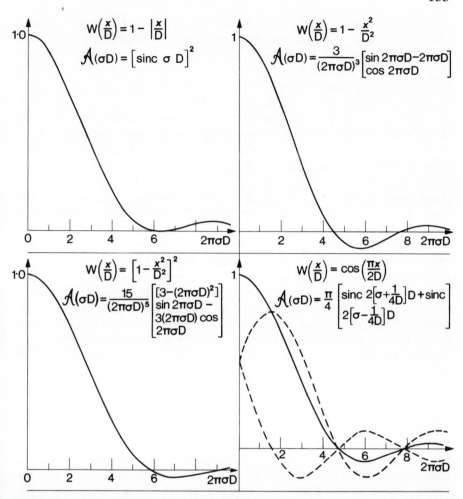

Figure 6.2. Some practical weighting functions and their resulting normalized apodized spectral windows. The dashed curves in the lower right inset are the component sinc functions which add to give the solid curve

This gives a linear taper from 1 at $x = 0$ to 0 at $|x| = D$. The apodized spectral window is (from 2.86)

$$A(\sigma D) = D[\text{sinc } \sigma D]^2 \qquad (6.35a)$$

and the normalized form (see Figure 6.2) is

$$\mathscr{A}(\sigma D) = [\text{sinc } \sigma D]^2. \qquad (6.35b)$$

The amplitude of the oscillations in the side lobes of the sinc^2 function falls off as the square of the wavenumber separation compared with the merely σ^{-1} dependence of the sinc function itself. This more rapid attenuation, i.e.

apodization, is the result of the taper. All effective weighting functions eliminate the slow σ^{-1} dependence and replace this with a σ^{-2} or even more rapid attenuation. The $\text{sinc}^2(\sigma D)$ function has been illustrated already (Figure 2.9) and is shown again in Figure 6.2, so all that will be noted here is that the first side lobes are only 4.7% of the central maximum, that the peak height (unnormalized) is only half that of the unapodized spectral window, and that the whole function has a more 'spread-out' nature than does the $\text{sinc}(2\sigma D)$ function. The side lobe spacing is nearly twice as great and the width at half-height of the dominant central lobe is nearly 1.47 times greater. Thus we have illustrated the general theorem that apodization (which is equivalent to suppressing information) produces spectra which show poorer resolution and this is the price that has to be paid for the suppression of the troublesome side lobes. It is interesting to observe, in passing, that side lobes are not a phenomenon restricted to two-beam interferometric spectrometers. All spectrometers which use the delay principle (i.e. do not measure time-frequency directly) operate by invoking interference between two or more beams, and the resulting pattern is necessarily restricted by practical considerations. In the case of a dispersive monochromator, the side lobes arise from diffraction at the various apertures in the instrument (for example, the grating and the slits) and it is worth noting that the slit function of a monochromator is also of the sinc^2 form.

There are clearly very many weighting functions which may be used to produce apodized spectra. The function

$$W\left(\frac{x}{D}\right) = \begin{cases} 1 - \left(\dfrac{x}{D}\right)^2, & |x| \leq D, \\ \\ 0, & |x| > D, \end{cases} \tag{6.36}$$

has the merit of zero slope at $x = 0$, so that it does not introduce a discontinuity of its own and, furthermore, does not seriously attenuate the larger ordinates which tend to occur near $x = 0$. This is an important point in practical (i.e. noise-limited) interferometry. The apodized spectral window is

$$A(\sigma D) = \frac{4D}{(2\pi\sigma D)^2} [\text{sinc}(2\sigma D) - \cos 2\pi\sigma D]. \tag{6.37}$$

The peak height $(4D/3)$ is larger than that produced by the linear taper and the width of the central lobe is less: correspondingly, the amplitudes of the side lobes are somewhat greater; the first, depending on how it is defined has an amplitude of about 9% of the main lobe.

The next step in developing functions of this kind is to remove the discontinuity at $x = \pm D$. This leads to

$$W\left(\frac{x}{D}\right) = \begin{cases} \left[1 - \left(\dfrac{x}{D}\right)^2\right]^2, & |x| \leq D, \\ \\ 0, & |x| > D. \end{cases} \tag{6.38}$$

The corresponding apodized spectral window is

$$A(\sigma D) = \frac{15D}{(2\pi\sigma D)^5} \left\lVert [3 - (2\pi\sigma D)^2] \sin 2\pi\sigma D - 3(2\pi\sigma D) \cos 2\pi\sigma D \right\rVert. \quad (6.39)$$

This function (see Figure 6.2) has the widest central lobe of all the functions so far considered, but on the other hand its side lobes are hardly perceptible. It is particularly useful in routine spectroscopy where the available resolution is usually ample.

A rather different approach to apodization involves the use of trigonometric weighting functions. The two most popular are

$$W(x/D) = \cos(\pi x/2D) \quad (6.40a)$$

and

$$W(x/D) = \cos^2(\pi x/2D). \quad (6.41a)$$

The latter, like (6.38), has the merits of zero slope at $x = 0$ and $x = D$. The corresponding apodized spectral windows are

$$A(\sigma D) = D \text{ sinc } 2\left[\sigma + \frac{1}{4D}\right]D + D \text{ sinc } 2\left[\sigma - \frac{1}{4D}\right]D \quad (6.40b)$$

and

$$A(\sigma D) = D \text{ sinc } 2\sigma D + \tfrac{1}{2}D \text{ sinc } 2[\tfrac{1}{2} + \sigma D] + \tfrac{1}{2}D \text{ sinc } 2[\tfrac{1}{2} - \sigma D]. \quad (6.41b)$$

The first of these is shown in the fourth inset of Figure 6.2 and the second is shown, together with its components, in Figure 6.3. It will be seen from these figures that the mechanism of apodization is quite different from that which is introduced by algebraic weighting functions. The slowly varying sinc functions are not removed, instead one deliberately adds further sinc functions and the observed attenuation of side lobes then arises from interference between their out-of-phase oscillations. This is clearly brought out in Figure 6.3. This figure also illustrates the principle of *post-apodization*. Under normal circumstances, the spectrum will be calculated at intervals of σ which equal the resolution limit, namely $1/2D$. The ordinates are, therefore, calculated at intervals of $\frac{1}{2}$ in the variable σD. The three components which make up (6.41b) are fundamentally the same function but referred to different origins, namely $-\frac{1}{2}$, 0, and $\frac{1}{2}$, and having different amplitudes, namely $\frac{1}{2}$, 1, $\frac{1}{2}$. If one therefore calculates an *unapodized* spectrum, one can arrive at an *apodized* spectrum merely by adding to each ordinate one half of those on each side and dividing by two. This is called *post*-apodization since it is carried out *after* the transformation, but it is completely equivalent to *pre*-apodization by the $\cos^2(\pi x/2D)$ function.

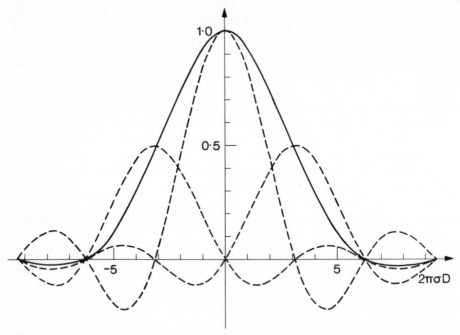

Figure 6.3. Mechanism of apodization by the $\cos^2(\pi x/2D)$ weighting function. The three dotted sinc functions add to give a much less oscillatory resultant shown in the solid curve

6.4.4. Bessel function notation for spectral windows

The Bessel functions are a class of higher transcendental functions which occur frequently in mathematical physics. They arise for example in the solution of the differential equation

$$x^2 f''(x) + x f'(x) + (x^2 - n^2)f(x) = 0. \tag{6.42}$$

They also occur as the Fourier transforms of some algebraic functions, for example

$$J_0(x) \supset (1 - 4\pi^2\sigma^2)^{-1/2},$$

and as coefficients in the expansion of some periodic functions-of-a-function as a Fourier series, for example

$$\cos[a \cos \omega t] = J_0(a) - 2J_2(a) \cos 2\omega t + 2J_4(a) \cos 4\omega t, \quad \text{etc.} \tag{6.43}$$

The parameter n which appears in the notation for the Bessel function $J_n(x)$ is called the order of the function and may have any value, positive or negative, but for the great majority of physical applications it is either zero or else a small positive integer. The Bessel functions can be thought of as generalizations of the sine and cosine functions and can likewise be expanded in a power series.

$$J_n(x) = \sum_{s=0}^{s=\infty} \frac{(-1)^s}{s!(n+s)!} x^{n+2s}. \tag{6.44}$$

Like the sine and cosine functions, the Bessel functions oscillate but they are not periodic (see Figure 6.4). $J_0(x)$ has something in common with the sinc function and for large x approaches the asymptotic form

$$J_0(x) = \sqrt{\frac{2}{\pi x}} \cos [x - \pi/4]. \tag{6.45}$$

There are several relations amongst the Bessel functions, just as there are amongst the trigonometrical functions, but possibly the most useful of these is the recurrence relation

$$J_{n-1}(x) + J_{n+1}(x) = \frac{2n}{x} J_n(x), \tag{6.46}$$

which is used, amongst other things, in the practical computation of these functions.

Equation (6.44) was originally derived on the understanding that the

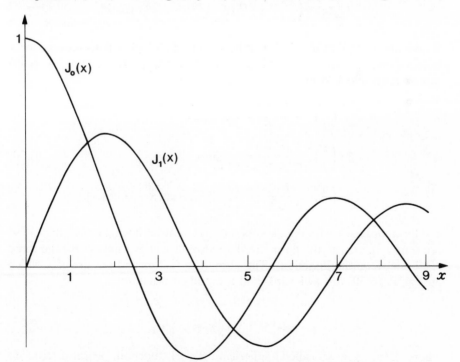

Figure 6.4. The two lowest integral order Bessel functions

denominators were the ordinary factorial expressions or, in other words, that n was an integer. However, by invoking the extended factorial function $v!$, which is defined for any value of v by the relation

$$v! = \Gamma(1+v), \tag{6.47}$$

where Γ is the gamma function, the equation can be reinterpreted to *define* Bessel functions of any order. Of particular interest in the present context is the class of Bessel functions of half integral order, $J_{(2n+1)/2}(x)$. They are readily derived in series form by making use of the relation

$$(m+\tfrac{1}{2})! = (m+\tfrac{1}{2})(m-1+\tfrac{1}{2})!, \qquad \tfrac{1}{2}! = \tfrac{1}{2}\pi^{-1/2}. \tag{6.48}$$

It will be found, however (after some considerable manipulation!!), that the series can be rearranged into simple combinations of the familiar trigonometric series, that is, these particular Bessel functions degenerate into simple analytical forms. In particular,

$$J_{1/2}(x) = \left(\frac{2}{\pi}\right)^{1/2} x^{-1/2}[\sin x],$$

$$J_{3/2}(x) = \left(\frac{2}{\pi}\right)^{1/2} x^{-3/2}[\sin x - x \cos x], \tag{6.49}$$

$$J_{5/2}(x) = \left(\frac{2}{\pi}\right)^{1/2} x^{-5/2}[(3-x^2)\sin x - 3x \cos x].$$

By the application of (6.46), it will be seen that all higher half-integral order Bessel functions will also be simple analytical forms. The equations (6.49) can be rearranged to give

$$\left(\frac{\pi}{2}\right)^{1/2} \cdot \frac{J_{1/2}(x)}{x^{1/2}} = \mathrm{sinc}\,(x/\pi),$$

$$3 \cdot \left(\frac{\pi}{2}\right)^{1/2} \cdot \frac{J_{3/2}(x)}{x^{3/2}} = \frac{3}{x^3}[\sin x - x \cos x], \tag{6.50}$$

$$15 \cdot \left(\frac{\pi}{2}\right)^{1/2} \cdot \frac{J_{5/2}(x)}{x^{5/2}} = \frac{15}{x^5}[(3-x^2)\sin x - 3x \cos x],$$

and by comparison with the expressions given earlier it will be seen that these are simply related to the normalized apodized spectral windows for the three algebraic weighting functions. The numerical factors and powers of x which appear in (6.50) are awkward, and a modified notation,

$$j_n(x) = \sqrt{\frac{\pi}{2x}}\, J_{n+1/2}(x), \tag{6.51}$$

where $j_n(x)$ is a so-called 'spherical' Bessel function, is introduced to simplify matters. The weighting functions and their corresponding normalized apodized spectral windows are then

$$W\left(\frac{x}{D}\right) = 1 - \frac{|x|}{D}, \qquad \mathcal{A}(\sigma D) = [j_0(2\pi\sigma D)]^2;$$

$$W\left(\frac{x}{D}\right) = 1 - \left(\frac{x}{D}\right)^2, \qquad \mathcal{A}(\sigma D) = \left(\frac{3}{2\pi\sigma D}\right) j_1(2\pi\sigma D); \qquad (6.52)$$

$$W\left(\frac{x}{D}\right) = \left[1 - \left(\frac{x}{D}\right)^2\right]^2, \qquad \mathcal{A}(\sigma D) = \frac{15}{(2\pi\sigma D)^2} j_2(2\pi\sigma D).$$

This is a convenient notation and is often used, but it fails to convey what is actually happening in the apodization process to all save experts in Bessel function theory. It has also another disadvantage: that there is no universally agreed notation for the Bessel functions and considerable confusion can arise.

6.5. RESOLUTION AND RESOLVING POWER

The subject of resolution has occurred several times in earlier sections, but here it will be considered in more detail and the threads drawn together. There are two aspects: first, those which are common to all spectroscopic systems and, second, those which are peculiar to two-beam interferometric spectroscopy.

The common problem which all spectroscopists face is to define what they mean by resolution. This ultimately has to be an arbitrary definition since the question of what separation, of two close lines, is sufficient for them to be considered resolved, cannot have an objective answer. The factors which enter into the judgement are (1) the distance apart of the lines, (2) the observed line widths, (3) the form of the instrumental spectral window, and (4) the relative intensities of the lines. The simplest situation to consider is a doublet well separated from other lines. For this case one might suggest that one would regard the components as resolved when there was a distinct minimum within the combined profile. Clearly, however, this definition would only be appropriate when the components were of comparable intensity. The situation where the components are of greatly differing intensity presents severe problems and the reported observation of what are known as 'shoulders', in spectroscopic parlance, is properly regarded with caution. However, given that the components are of roughly equal intensity, a frequently used touchstone is the Rayleigh criterion. This was proposed by Lord Rayleigh at the time when only dispersive spectrometers were in use and these, as remarked earlier, usually have an instrumental response function, or spectral window, which is of the sinc^2 form. Lord Rayleigh proposed that two equal intensity lines be regarded as resolved when the grand maximum of the sinc^2 pattern due to one line lay on the first zero of that due to the other. To consider this in more detail, we will approach the problem analytically. One may write the spectrum in the form

$$I(u) = \left[\frac{\sin(u-\Delta)}{u-\Delta}\right]^2 + \left[\frac{\sin(u+\Delta)}{u+\Delta}\right]^2, \qquad (6.53)$$

where u and Δ are reduced wavenumber variables given by

$$u = 2\pi D\left[\sigma - \left(\frac{\sigma_1+\sigma_2}{2}\right)\right] \quad \text{and} \quad \Delta = 2\pi D\left[\frac{\sigma_1-\sigma_2}{2}\right],$$

respectively; σ_1 and σ_2 are the line wavenumbers and D is an instrumental dimension which determines the form of the detector plane diffraction pattern. Because of the form chosen for $I(u)$, this is symmetrical about $u = 0$ and its first derivative always vanishes at $u = 0$. What we are interested in is the value of Δ which corresponds to a switch from a maximum to a minimum at $u = 0$, that is, the value of u where the second derivative of $I(u)$ vanishes. This is given by the lowest root of the transcendental equation

$$3 - [3 - 2\Delta^2]\cos 2\Delta - 4\Delta \sin 2\Delta = 0, \qquad (6.54)$$

namely $\Delta = 1.303\,081\,8$. The Rayleigh criterion corresponds to $\Delta = \pi/2 = 1.570\,796\,3$. The effect of increasing values of Δ on the form of the combined spectral profile is shown in Figure 6.5. It will be seen that there is a shallow minimum when $\Delta = 1.4$ and that this has deepened to 19% of the maximum value when $\Delta = \pi/2$, a result similar to that obtained in applying the Rayleigh criterion to the Fabry–Perot interferometer.

From the point of view of resolution, interferometric spectroscopy is more complex than dispersive spectroscopy[36] because of the wide variety of spectral windows that may be used. This, as pointed out in the previous section, comes about because each weighting function produces a different apodized spectral window. Experimental spectroscopists are always faced with the problem of striking a balance between the need for the best resolution and the need for the reduction of spurious side lobes to the minimum possible size. In Fourier spectroscopy this balancing takes the form of choosing the best compromise weighting function, and we begin by considering the simplest case, that is, unapodized spectroscopy. The equivalent of (6.53) for the interferometric case is

$$I(u) = \frac{\sin(u-\Delta)}{u-\Delta} + \frac{\sin(u+\Delta)}{u+\Delta}. \qquad (6.55)$$

The quantities u and Δ are defined as before, but D is now taken to be the maximum optical path difference introduced. The critical condition of switchover from a maximum to a minimum at $u = 0$ is given once again by a transcendental equation, namely

$$\tan \Delta = 2\Delta[2 - \Delta^2]^{-1}. \qquad (6.56)$$

The lowest root of this equation is $\Delta = 2.081\,576\,4$. The interferometric case

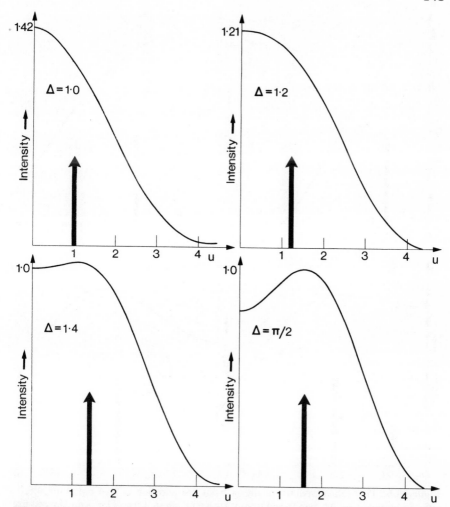

Figure 6.5. Illustration of the resolution criterion. The insets show the results of combining two sinc² functions of equal amplitude but gradually increasing separation. The diagrams are symmetrical about $u = 0$, where u is a reduced wavenumber variable. The critical separation occurs for $\Delta = 1.30308$ and the Rayleigh criterion (inset 4) occurs for a line separation of $u_1 - u_2 = \pi$

with a spectral window of the sinc form is shown in Figure 6.6. As Δ increases, the profile broadens, and a minimum appears as soon as Δ exceeds the critical value. The most important point to notice is that the Rayleigh criterion no longer works. At $\Delta = \pi/2$ the profile still shows a maximum. In fact, the critical wavenumber separation $\delta\sigma$ is given by

$$2\pi\,\delta\sigma D = 2\Delta_{\text{crit}} = 2 \cdot 2.081\,576\,4 \qquad (6.57a)$$

and therefore

$$\delta\sigma \simeq 2.08/\pi \cdot D \simeq 0.66/D. \qquad (6.57b)$$

142

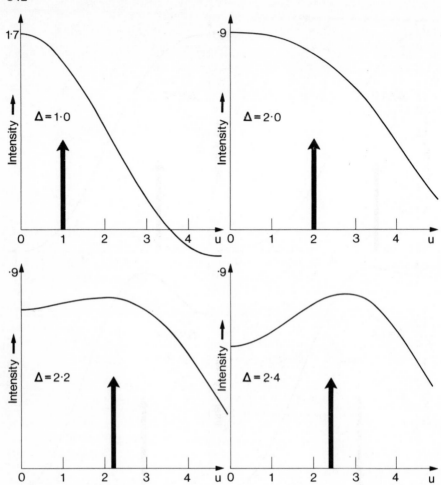

Figure 6.6. Theoretical spectra resulting from the addition of two equal amplitude sinc functions $(u-\Delta)^{-1}\sin(u-\Delta)$ and $(u+\Delta)^{-1}\sin(u+\Delta)$. The four insets are symmetrical about $u = 0$ and correspond to $\Delta = 1.0$, 2.0, 2.2, and 2.4. The reduced variables u and Δ are related to the original line wavenumbers σ_1 and σ_2 by $u = 2\pi D[\sigma - (\sigma_1 + \sigma_2)/2]$ and $\Delta = 2\pi D[(\sigma_1 - \sigma_2)/2]$, where D is the maximum optical path difference introduced

This is the best resolution possible and for real instruments one must expect $\delta\sigma$ to be somewhat larger; as a rule of thumb, one can say, therefore, that

$$\delta\sigma = 0.7/D. \tag{6.58}$$

As a practical illustration, with a total mirror travel of 5 cm, D would have a maximum value of 10 cm and $\delta\sigma$ would be $0.07\ \mathrm{cm}^{-1}$. Some confusion on this point exists in the literature and several authors quote rather more optimistic versions of equation (6.58).

In the case of triangular apodization, the spectral window is of the sinc^2 form and the analysis given above applies. However, it must be remembered that the argument of the sinc^2 function is one half that of the sinc function in unapodized computation and one has, therefore,

$$2\pi\,\delta\sigma D = 2\cdot 2\cdot\Delta_{crit} = 5.212\,327\,2$$

and therefore

$$\delta\sigma \simeq 5.21/2\pi D \simeq 0.83/D.$$

Again allowing for practical effects it would be wise to quote

$$\delta\sigma = 0.9/D. \tag{6.59}$$

From this it will be seen that the loss of resolution in going to the apodized case is not as great as is often quoted.

For other spectral windows, the numerical analysis follows similar, if more tedious, lines. The amount of algebra can, however, be somewhat shortened by a theorem which follows from the general symmetry of the situation. This is that Δ_{crit} is given by that value of the wavenumber separation from the peak where the slope of the profile has its maximum value.

It has been assumed throughout this section that the resolution is limited by instrumental effects. When the widths of the lines themselves become comparable to the instrumental resolution one has a complicated convolution problem to be considered. If the spectral lines are broader than the spectral window, they will, in essence, apodize *it* and further apodization will be redundant. However, for the situation that is the main theme of this section, i.e. where the spectral window is the limiting feature, one must next go on to consider the case of an isolated narrow line and define what one means by the instrumental line width. Several definitions spring to mind but the one with the greatest general support is to define the line width as the full-width at half maximum (FWHM). For the two cases we have considered so far this gives

$$\text{unapodized,} \quad \text{FWHM} = \frac{2.145\,116}{\pi D} \simeq \frac{0.68}{D}; \tag{6.60a}$$

$$\text{triangular apodization,} \quad \text{FWHM} = \frac{2\times 1.391\,55}{\pi D} \simeq \frac{0.89}{D}. \tag{6.60b}$$

It will be seen, therefore, that the line widths so defined agree closely in value with the resolutions defined earlier and this is the basis for the common practice of spectroscopists which is to regard the two concepts as interchangeable. Mathematically, this interchangeability arises from the fact that for most spectral windows, line shapes, etc., the value of frequency displacement where the line shape slope is a maximum corresponds closely to that where the function has reached half amplitude.

The nature of the interferometric process makes possible another approach to the question of the meaning of resolution: this is to consider the problem *not* in the spectral domain but in the interferogram, i.e. spatial, domain. The two monochromatic lines of separation $\delta\sigma$ will give rise to two endless cosine waves which, because of their slightly different periods, will become progressively more and more out of step as the path difference increases. When the phase difference reaches π, i.e. when

$$2\pi\,\delta\sigma x = \pi \quad \text{and} \quad \delta\sigma = 0.5/x \tag{6.61a}$$

the envelope of the interferogram will vanish, that is to say the visibility will fall to zero. Now, as will be recalled from Chapter 5, it is necessary to go beyond the position where the visibility has vanished and into a region where it is once more finite before the doublet nature of the line can be deduced. Therefore $\delta\sigma$ must be greater than $0.5/D$ before the line can be resolved. This analysis agrees with that given above and contradicts the commonly quoted relation

$$\delta\sigma = 0.5/D \tag{6.61b}$$

for the resolution in unapodized spectroscopy.

6.6. THE PRACTICAL CALCULATION OF THE ABSORPTION SPECTRUM

The absorption spectrum is calculated by applying Lambert's law (see section 1.1.1) to the transmission spectrum which is itself obtained from the ratio of a specimen spectrum to a background spectrum. The calculated background spectrum is given by (6.28b), namely

$$B_e^c(\sigma) = B_e(\sigma) * A(\sigma D), \qquad \sigma > 0. \tag{6.62a}$$

The corresponding specimen spectrum is given by

$$B_{e(s)}^c(\sigma) = B_{e(s)}(\sigma) * A(\sigma D), \qquad \sigma > 0. \tag{6.62b}$$

The transmission factor is therefore given by

$$\tau_s^c(\sigma) = \frac{B_{e(s)}(\sigma) * A(\sigma D)}{B_e(\sigma) * A(\sigma D)}. \tag{6.63}$$

The calculated transmission spectrum will agree with the true transmission spectrum provided that the effects of the convolutions in (6.63) do not amount to significant spectral line distortion. This is tantamount to saying that the width of the spectral window, and therefore the resolution limit, should be considerably less than the spectral line widths. This conclusion is not special to Fourier transform spectroscopy; equivalents of (6.63) occur in all branches of absorption spectroscopy and the same conclusion emerges, that the determination of true absorption spectra requires the use of a resolution high enough that the spectral profile may be followed without distortion.

6.7. SAMPLING AND ITS EFFECT ON THE COMPUTED SPECTRUM; THE PHENOMENON OF ALIASING

The interferogram function actually used in the evaluation of an integral such as (6.18) is not usually continuous, for, just as discussed in Chapter 5, it is necessary to sample it at discrete intervals if it is to be transformed by digital computation. (This is the commonly used approach. There is an alternative described in Chapter 8.) These intervals should be of exactly equal size and the first sample should coincide with zero path difference. The frequency of sampling and, consequently, the number of samples taken in the interval $0 \leqslant x \leqslant D$ require some critical discussion.

We assume that the interferogram $F_e(x)$ is sampled at discrete intervals of path difference β up to a maximum number $N+1$, such that

$$x = r\beta \quad \text{and} \quad D = N\beta \quad \text{[m]}, \tag{6.64}$$

where r is an integer $(0 \leqslant r \leqslant N)$. The sampling may be represented by a finite number of equidistant Dirac delta functions selected from the infinite array denoted by the shah symbol (see section 2.9.9):

$$\text{Ш}\left(\frac{x}{\beta}\right) = \beta \sum_{r=-\infty}^{\infty} \delta(x - r\beta). \tag{6.65}$$

The sampling function is therefore $\text{Ш}(x/\beta)$, confined to the range $0 \leqslant x \leqslant D$ by the application of the truncating function $\Pi(x/D)$, that is

$$\text{Ш}(x/\beta)\,\Pi(x/D). \tag{6.66}$$

Consequently, the expression for the computed spectrum becomes

$$B_e^c(\sigma) = 2\int_0^\infty F_e(x)\,W\left(\frac{x}{D}\right)\text{Ш}\left(\frac{x}{\beta}\right)\Pi\left(\frac{x}{D}\right)\cos 2\pi\sigma x\,dx \quad \text{[W m]}. \tag{6.67}$$

This equation may be treated in a number of different ways. First, we modify the limits in (6.67) to take account of the truncation and substitute (6.65) to give

$$B_e^c(\sigma) = 2\int_0^{N\beta} F_e(x)\,W\left(\frac{x}{D}\right)\left[\beta\sum_{r=-\infty}^{\infty}\delta(x - r\beta)\right]\cos 2\pi\sigma x\,dx$$

$$= 2\beta\sum_{r=-\infty}^{\infty} F_e(r\beta)\,W\left(\frac{r}{N}\right)\cos 2\pi\sigma r\beta\int_0^{N\beta}\delta(x - r\beta)\,dx \tag{6.68a}$$

$$= \beta F_e(0) + 2\beta\sum_{r=1}^{N-1} F_e(r\beta)\,W\left(\frac{r}{N}\right)\cos 2\pi\sigma r\beta$$

$$+ \beta F_e(N\beta)\,W(1)\cos 2\pi\sigma N\beta \quad \text{[W m]}. \tag{6.68b}$$

The form (6.68b) follows from (6.68a) because

$$\int_0^{N\beta}\delta(x - r\beta)\,dx \simeq \int_{-N\beta}^{N\beta}\delta(x - r\beta)\,dx \simeq \int_{-\infty}^{\infty}\delta(x - r\beta)\,dx = 1 \tag{6.69a}$$

146

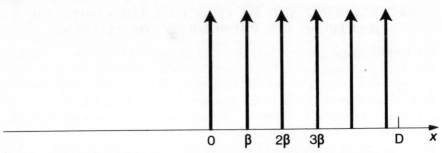

Figure 6.7. The one-sided sampling function $\sqcup\!\sqcup(x/\beta)\,\sqcap(x/D)$

for all values of r except 0 and N for which the integral is

$$\int_0^{N\beta} \delta(x)\,\mathrm{d}x \simeq \int_0^{\infty} \delta(x)\,\mathrm{d}x \simeq \tfrac{1}{2}\int_{-\infty}^{\infty} \delta(x)\,\mathrm{d}x = \tfrac{1}{2} \qquad (6.69b)$$

and

$$\int_0^{N\beta} \delta(x-N\beta)\,\mathrm{d}x \simeq \int_{-\infty}^{0} \delta(x)\,\mathrm{d}x = \tfrac{1}{2}, \qquad (6.69c)$$

respectively. Equation (6.68) may be expressed more briefly as

$$B_e^c(\sigma) = 2\beta \sum_{r=0}^{N} F_e(r\beta)[1 - \tfrac{1}{2}\delta(r;0) - \tfrac{1}{2}\delta(r;N)]W\!\left(\frac{r}{N}\right)\cos 2\pi\sigma r\beta \qquad [\text{W m}],$$

$$(6.70)$$

where $\delta(r;0)$ and $\delta(r;N)$ are Kronecker deltas of the type

$$\delta(r;m) = \begin{cases} 1, & r=m, \\ 0, & r\neq m. \end{cases}$$

Equation (6.70) can be interpreted in a simple way, if the special case $\sigma = 0$ is considered. $B_e^c(0)$ is then the area under the interferogram envelope and one might think of approximating this, given the set of sampled ordinates, by taking the average of the upper and lower bounds,

$$\text{upper bound} = 2\beta \sum_{r=0}^{N-1} F_e(r\beta), \qquad (6.71a)$$

$$\text{lower bound} = 2\beta \sum_{r=1}^{r=N} F_e(r\beta), \qquad (6.71b)$$

namely

$$\text{average} = F_e(0) + 2 \sum_{r=1}^{r=N-1} F_e(r\beta) + F_e(N\beta). \qquad (6.71c)$$

When, as is usually the case, the spectrum is apodized, the weighting function used to taper the interferogram falls to zero at $x = D$; thus $W(1) = 0$, the $r = N$ term in (6.68) and (6.70) is zero, and we can write

$$B_e^c(\sigma) = \beta F_e(0) + 2\beta \sum_{r=1}^{N} F_e(r\beta) W\left(\frac{r}{N}\right) \cos 2\pi\sigma r\beta$$

$$= 2\beta \sum_{r=0}^{N-1} F_e(r\beta)$$

$$\times [1 - \tfrac{1}{2}\delta(r; 0)] W\left(\frac{r}{N}\right) \cos 2\pi\sigma r\beta \qquad [\text{W m}] \qquad (6.72)$$

for the computed spectrum. This summation is used as the basis of numerical computation of $B_e^c(\sigma)$ from the observed sequence of data $\{F_s(r\beta)\}$—see Chapter 10.

We can now return to (6.67) and derive an alternative representation. Since the Fourier transform of a periodic Dirac function of period β is another periodic Dirac function (see equations (2.133) and (2.135) and Figure 2.14),

$$\beta \, \mathrm{III}(\sigma\beta) = \sum_{r=-\infty}^{\infty} \delta\left(\sigma - \frac{r}{\beta}\right), \qquad (6.73)$$

of period $1/\beta$, we can apply the convolution theorem to show that the computed spectrum (6.67) is

$$B_e^c(\sigma) = \beta B_e(\sigma) * A(\sigma N\beta) * \mathrm{III}(\sigma\beta) = B_e(\sigma) * \Lambda(\sigma N\beta) \qquad [\text{W m}], \quad (6.74)$$

which consists of a series of identical spectral contours repeated at intervals $1/\beta$ in the wavenumber domain. This is because the apodizing function $A(\sigma D)$ appropriate to the continuously recorded interferogram is replaced by the replicated function

$$\Lambda(\sigma N\beta) = \beta A(\sigma N\beta) * \mathrm{III}(\sigma\beta) \qquad (6.75a)$$

$$= \sum_{m=-\infty}^{\infty} A(\sigma_m N\beta), \qquad \sigma_m = \sigma + \frac{m}{\beta} \qquad [\text{m}], \qquad (6.75b)$$

which we can regard as the new apparatus function. We call $\Lambda(\sigma N\beta)$ the *replicated (even) apparatus function*.

If the incident spectrum is strictly monochromatic, $B(\sigma) = 6_0\delta(\sigma - \sigma_0)$, then the computed (unapodized) spectrum

$$B_e^c\sigma = N\beta 6_0[\mathrm{sinc}\, 2(\sigma + \sigma_0)N\beta + \mathrm{sinc}\, 2(\sigma - \sigma_0)N\beta] * \mathrm{III}(\sigma\beta)$$

$$= N\beta 6_0 \sum_{m=-\infty}^{\infty} \left\{ \mathrm{sinc}\, 2\left[\left(\sigma - \frac{m}{\beta} + \sigma_0\right)N\beta\right] + \mathrm{sinc}\left[2\left(\sigma - \frac{m}{\beta} - \sigma_0\right)N\beta\right] \right\}$$

$$(6.76)$$

consists of a series of identical features which appear at

$$\sigma = \frac{m}{\beta} \mp \sigma_0 \qquad [\text{m}^{-1}]. \qquad (6.77)$$

148

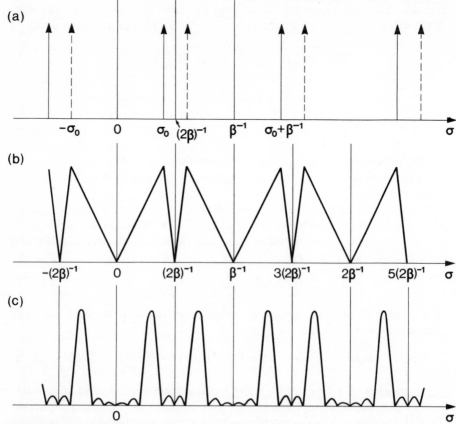

Figure 6.8. The phenomenon of aliasing. Each spectral element (a) at $+\sigma_0$ is endlessly replicated (solid arrows) at wavenumbers $(\sigma_0 \pm m/\beta)$. Likewise, each image element at $-\sigma_0$ is replicated (broken arrows) at wavenumbers $(-\sigma_0 \pm m/\beta)$. The spectral domain therefore has mirror symmetry at $\sigma = \pm m/2\beta$ and is divided up into two sets of aliases (b) which are mirror images of one another. Reliable spectroscopy depends on alias overlap (c) being reduced to a minimum

These repeated features represent, apart from a factor $\frac{1}{2}\delta_0$, the replicated apparatus function.

$$\Lambda_0(\sigma N\beta) = 2N\beta \sum_{m=-\infty}^{\infty} \left\{ \mathrm{sinc}\left[2\left(\sigma - \frac{m}{\beta} + \sigma_0\right)N\beta\right] \right.$$
$$\left. + \mathrm{sinc}\left[2\left(\sigma - \frac{m}{\beta} - \sigma_0\right)N\beta\right] \right\} \qquad [\mathrm{m}] \qquad (6.78)$$

is therefore Figure 6.1 endlessly repeated. Only one of these apparatus functions is of interest and this is the one whose wavenumber range coincides with the range occupied by the physical spectrum $\beta(\sigma)$. The repetition of the spectrum along the wavenumber axis is known as *aliasing*. It is essential that none of the *aliases* overlap if unique spectra are to be computed.

This concept can be used to extend the treatment given in section 2.10 of the problem of reconstructing a continuous function from a set of its sampled ordinates. Ignoring Fourier transforms for the moment, this is equivalent to interpolation for points which lie within the span of the ordinates and to extrapolation for those which do not. One can now see that if the Fourier transform is *not* band-limited then there will inevitably be alias overlap and the reconstruction will be subject to error. However, the more finely the original is sampled the more widely separated will be the aliases and the less the overlap and the accuracy of the reconstruction will increase as expected. It is interesting to note that, from this argument, all the spectral window functions can be uniquely interpolated.

6.8. RELATION BETWEEN THE SPECTRAL INTERVAL AND THE SAMPLING INTERVAL

When the spectrum is finite in extent, one has an analogous problem to that encountered in section 5.6.2, and one must be careful that the sampling interval be chosen sufficiently small to prevent the computed aliases from overlapping and becoming entangled. We will suppose that the physical spectrum extends from a lower limit of σ_1 to an upper limit of σ_2. Two cases can now be distinguished: (1) when σ_1 is zero and (2) when it is not. In the first case, β must equal $(2\sigma_2)^{-1}$, that is, $\frac{1}{2}\lambda_m$, where λ_m is the wavelength corresponding to σ_2 and is by definition the shortest wavelength present in the radiation being studied. This case is the normal one encountered in far infrared spectroscopy. When σ_1 is greater than zero, one can of course continue to sample at $\beta = (2\sigma_2)^{-1}$, but this is a wasteful procedure since more data than necessary are used to compute the spectrum. If $\sigma_2 - \sigma_1$ is an exact integral submultiple of σ_2, then β may be chosen to be $[2(\sigma_2 - \sigma_1)]^{-1}$. This will lead to a strongly aliased spectrum (see Figure 6.9), but because of the integral condition the aliases slot together perfectly without overlap. The integral condition is not much of a handicap in practice since one only knows σ_2 and σ_1 approximately anyway. One therefore chooses 'safe' values of σ_2 and σ_1, i.e. values where $I(\sigma)$ will certainly be zero and such that $\sigma_2 - \sigma_1$ is an integral submultiple of σ_2. As an example, if one were fairly certain that one was studying a band of radiation between 800 and 950 cm^{-1}, one could take $\sigma_2 = 1000$, $\sigma_1 = 750$ and $(\sigma_2 - \sigma_1) = \frac{1}{4}\sigma_2$. The interferogram would then only need to be sampled at intervals $\beta = 1/500$ cm. In this case, as in all others, one need not instruct the computer to produce what one may call the 'prime' alias, i.e. that lying between 750 and 1000 cm^{-1}; one could just as well compute the alias lying between 0 and 250 cm^{-1}, or any other for that matter, since with due regard to mirror symmetry, they are, identical. This is the basis of the method used some years ago to observe middle infrared spectra using an interferometer designed for the far infrared.

Attractive though this reduction in data collection and processing is, some words of caution need be sounded. First, the interferogram will certainly

(a)

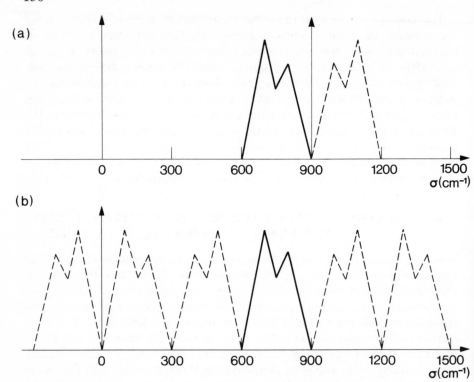

(b)

Figure 6.9. A band-limited spectrum (a) can be reconstructed by sampling the inter-
ferogram at $\beta = (2\sigma_2)^{-1}$ or much more economically (b) by sampling at $\beta = [2(\sigma_2 - \sigma_1)]^{-1}$
provided that σ_2 is an exact integral multiple of $\sigma_2 - \sigma_1$

have 'noise' superimposed on it and under this condition there may be an
advantage in recording and computing excess data. Second, one has to be
really sure that there is no radiant power, at the detector, for wavenumbers
less than σ_1. Third, the quality of mirror drive and of path difference
measurement have to be adequate for the *highest* wavenumber present, i.e.
σ_2. It is these potential sources of error which have often persuaded
experimentalists not to use far infrared instruments in the middle infrared.
However, if one is using a high quality interferometer in the region for
which it was designed and if the bandwidth is known to be limited and if
noise is not a problem, the use of this trick can be very welcome, especially
if the power of the computer is limited.

6.9. EFFECT OF ASYMMETRY IN THE RECORDED INTERFEROGRAM

So far it has been assumed that the interferogram $F(x)$ is an even (that is,
symmetrical) function. With this assumption one need only carry out a

cosine Fourier transform for positive values of x alone and then double the result to get the effective full transform; in other words, the interferogram need only be recorded on the positive side of zero path difference. This procedure leads to a reliable computed spectrum only if both the interferogram itself *and* the recorded version $F(x) \sqcap (x/D)$ are exactly even. In practice, the infinite *analogue interferogram* may not be symmetrical about $x = 0$ due to imperfect adjustment or a lack of symmetry between the partial beams, while the truncated interferogram $F(x) \sqcap (x/D)$ may not be properly one-sided due to a displacement from true zero path difference of the truncating function, even when $F(x)$ is itself perfectly even.

Suppose that the analogue interferogram is truly symmetrical but that the truncating function is displaced by a small amount δ so that the first value of $F(x)$ is obtained at $x = \delta$ rather than $x = 0$ and that subsequent values are obtained at $x + \delta$ up to $D + \delta$ (Figure 6.10). It is clear that according to whether δ is positive or negative the function used for the cosine Fourier transform is part of an even function of one of the two forms of Figure 6.10, neither of which is the same as the true interferogram. Before seeing how to overcome the problem furnished by this discrepancy, it is instructive to inspect the effects on the spectral window and the calculated spectrum of a displacement error δ in the range of the observed interferogram.

6.9.1. The unsymmetrical apparatus function

If $\delta > 0$, we do not calculate from $x = 0$ but from $x = \delta$, using $x' = x - \sigma$ as variable, and instead of observing $F(x) \sqcap (x/D)$ we observe

$$F(x) \sqcap [(x - \delta)/D] = F(x' + \delta) \sqcap (x'/D) \quad [\text{W}] \quad (6.79\text{a})$$

or, in terms of convolutions,

$$F(x)[\sqcap (x/D) * \delta(x - \delta)] = [F(x') * \delta(x' + \delta)] \sqcap (x'/D) \quad [\text{W}]. \quad (6.79\text{b})$$

Correspondingly, we do not evaluate

$$B_e^c(\sigma) = 2 \int_0^\infty F(x) \sqcap \left(\frac{x}{D}\right) W\left(\frac{x}{D}\right) \cos 2\pi\sigma x \, dx \quad [\text{W m}] \quad (6.80)$$

but

$$B'^c(\sigma) = 2 \int_0^\infty F(x' + \delta) \sqcap \left(\frac{x'}{D}\right) W\left(\frac{x'}{D}\right) \cos 2\pi\sigma x' \, dx' \quad [\text{W m}]. \quad (6.81)$$

To assess the consequences of this substitution of incorrect data into the Fourier integral, we first of all suppose that the incident spectrum is truly monochromatic, then

$$F(x' + \delta) = 6_0 \cos 2\pi\sigma(x' + \delta) \quad [\text{W}] \quad (6.82)$$

152

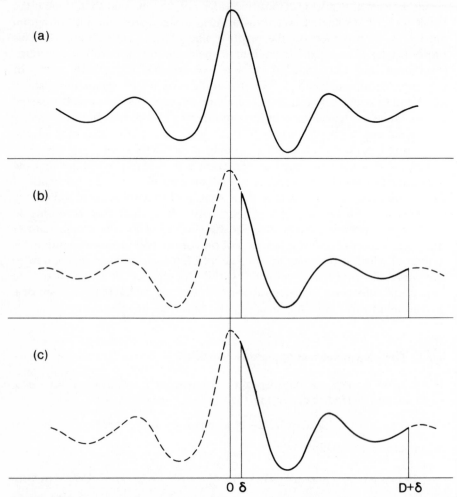

Figure 6.10. A one-sided interferogram may be unsymmetrical because (a) the original interferogram is unsymmetrical, (b) the sampling comb is displaced, or (c) both phenomena are occurring

is the recorded interferogram and, in the absence of weighting,

$$B'^c(\sigma) = 2 \int_0^\infty б_0 \sqcap\!\left(\frac{x'}{D}\right) \cos 2\pi\sigma_0(x' + \delta)\, \cos 2\pi\sigma x'\, dx' \qquad (6.83a)$$

$$= 2б_0 \cos 2\pi\sigma_0\delta \int_0^\infty \sqcap\!\left(\frac{x'}{D}\right) \cos 2\pi\sigma_0 x'\, \cos 2\pi\sigma x'\, dx'$$

$$- 2б_0 \sin 2\pi\sigma_0\delta \int_0^\infty \sqcap\!\left(\frac{x'}{D}\right) \sin 2\pi\sigma_0 x'\, \cos 2\pi\sigma x'\, dx' \qquad [\text{W m}]$$

$$(6.83b)$$

is the calculated spectrum. On evaluating the integrals in (6.83), we find that they are, respectively, even and odd:

$$B'^c(\sigma) = D\mathfrak{b}_0 \cos 2\pi\sigma_0\delta\{\text{sinc}\,[2(\sigma+\sigma_0)D]+\text{sinc}\,[2(\sigma-\sigma_0)D]\}$$

$$- D\mathfrak{b}_0 \sin 2\pi\sigma_0\delta\left[\frac{1-\cos 2\pi(\sigma+\sigma_0)D}{2\pi(\sigma+\sigma_0)D}-\frac{1-\cos 2\pi(\sigma-\sigma_0)D}{2\pi(\sigma-\sigma_0)D}\right] \quad [\text{W m}].$$

(6.84a)

This expression becomes

$$B'^c(\sigma) = D\mathfrak{b}_0\left[\cos 2\pi\sigma_0\delta \,\text{sinc}\,2(\sigma-\sigma_0)D\right.$$

$$\left. +\sin 2\pi\sigma_0\delta\frac{1-\cos 2\pi(\sigma-\sigma_0)D}{2\pi(\sigma-\sigma_0)D}\right] \quad [\text{W m}] \quad (6.84b)$$

for $\sigma \gg 0$ (Figure 6.11). We call the odd function

$$U_o(\sigma D) = 2D\frac{1-\cos 2\pi\sigma D}{2\pi\sigma D} \quad [\text{m}] \tag{6.85}$$

the *unsymmetrical apparatus function* and note that it is the sine Fourier transform

$$U_o(\sigma D) = 2\int_0^\infty \sqcap\left(\frac{x'}{D}\right) \sin 2\pi\sigma x'\,dx' \quad [\text{m}] \tag{6.86}$$

of the rectangular truncation function. When the weighting function is included, it is evident that the unsymmetrical apparatus function is

$$U(\sigma D) = 2\int_0^\infty \sqcap\left(\frac{x'}{D}\right)W\left(\frac{x'}{D}\right) \sin 2\pi\sigma x'\,dx' \quad [\text{m}], \tag{6.87}$$

which may be written as the full-range integral

$$U(\sigma D) = -i\int_{-\infty}^\infty \text{sgn}\,(x')\sqcap\left(\frac{x'}{2D}\right)W\left(\frac{x'}{D}\right) \exp\,(-2\pi i\sigma x')\,dx' \tag{6.88}$$

since the cosine Fourier transform is zero. The unsymmetrical apparatus function is, therefore,

$$U(\sigma D) = -i\mathscr{F}\{\text{sgn}\,(x')\sqcap(x'/2D)W(x'/D)\}$$

$$= 1/\pi\sigma * A(\sigma D) \quad [\text{m}] \tag{6.89}$$

since (see 2.105)

$$\mathscr{F}\{\text{sgn}\,(x')\} = -i/\pi\sigma \quad [\text{m}] \tag{6.90}$$

and

$$\mathscr{F}\{\sqcap(x'/2D)W(x'/D)\} = A(\sigma D) \quad [\text{m}].$$

The convolution of (6.89) is equivalent to the Hilbert transform (2.108) of $-A(\sigma D)$ so

$$U(\sigma D) = -\mathscr{H}\{A(\sigma D)\} \quad [\text{m}] \tag{6.91}$$

154

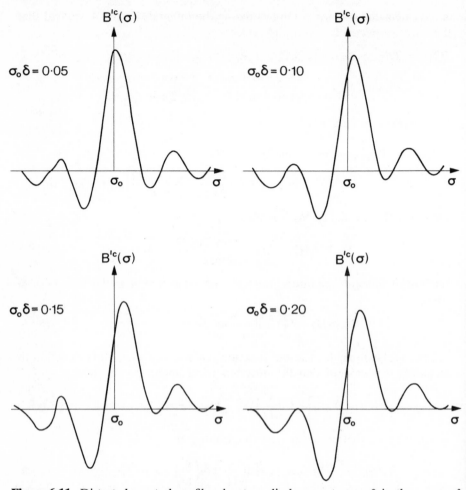

Figure 6.11. Distorted spectral profiles due to a displacement error δ in the range of the observed interferogram. The incident radiation is monochromatic, so for $\sigma_0\delta = 0$ the profile would be the even spectral window sinc $2(\sigma' - \sigma_0)D$, whereas for $\sigma_0\delta = \frac{1}{4}$ it would be the odd spectral window illustrated in Figure 6.12

and, from the properties of Hilbert transforms, we conclude that the even and odd spectral windows are Hilbert transforms of one another. This result also follows from noting that they are respectively the cosine and sine transforms of a one-sided function. We can also relate $U(\sigma D)$ to $U_o(\sigma D)$ for, from (6.27) and (6.89)

$$U(\sigma D) = (1/\pi\sigma) * A_0(\sigma D) * \Omega(\sigma D)$$

$$= U_o(\sigma D) * \Omega(\sigma D) \quad [\text{m}] \quad (6.92)$$

where

$$U_o(\sigma D) = (1/\pi\sigma) * A_0(\sigma D) \quad [\text{m}]. \quad (6.93)$$

(6.93) is readily deduced from (6.89) with the weighting function excluded.

We can now return to the case of the monochromatic spectrum and include weighting so that (6.83a) becomes

$$B'^{c}(\sigma) = 2 \int_{0}^{\infty} 6_{0} W\left(\frac{x'}{D}\right) \sqcap \left(\frac{x'}{D}\right) \cos 2\pi\sigma_0(x'+\delta) \cos 2\pi\sigma x' \, dx'$$

$$= \tfrac{1}{2}6_0 \cos 2\pi\sigma_0\delta[A((\sigma+\sigma_0)D) + A((\sigma-\sigma_0)D)]$$

$$- \tfrac{1}{2}6_0 \sin 2\pi\sigma_0\delta[U((\sigma+\sigma_0)D) - U((\sigma-\sigma_0)D)] \qquad [\text{m}],$$
$$(6.94)$$

which may be written

$$B'^{c}(\sigma)/\tfrac{1}{2}6_0 = \cos 2\pi\sigma_0\delta A((\sigma-\sigma_0)D) + \sin 2\pi\sigma_0\delta U((\sigma-\sigma_0)D) \qquad [\text{m}]$$
$$(6.95)$$

for $\sigma \gg 0$ after normalizing to the value $\tfrac{1}{2}6_0$. It is clear that the form of (6.84) and (6.94) depends strongly on the wavenumber analysed in the spectrum as well as on the magnitude of δ. The quantity $\sigma_0\delta$ is a measure of the importance of the error made in the choice of the origin of the range of the recorded interferogram. When $\sigma_0\delta = 0$, $2B'^{c}(\sigma)/6_0 = A((\sigma-\sigma_0)D)$, which is an even function, and when $\sigma_0\delta = \tfrac{1}{4}$, $2B'^{c}(\sigma)/6_0 = U((\sigma-\sigma_0)D)$, which is an odd function. Between these two extreme cases all intervening ones are possible. The asymmetry of $2B'^{c}(\sigma)/6_0$ increases as $\sigma_0\delta$ increases from 0 and we see that strong negative ordinates can occur in the observed spectral window (Figure 6.11 and 6.13). The relative magnitude of the negative minimum increases as $\sigma_0\delta$ increases and negative ordinates can occur for very small values of $\sigma_0\delta$—a result first obtained by J. Connes.[16]

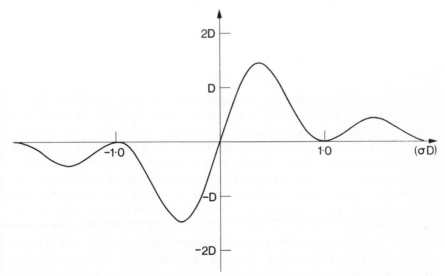

Figure 6.12. The unsymmetrical apparatus function $U_o(\sigma D) = 2D(1-\cos 2\pi\sigma D)/2\pi\sigma D$. From its definition, this function is the Hilbert transform of the even spectral window $A(\sigma D)$

156

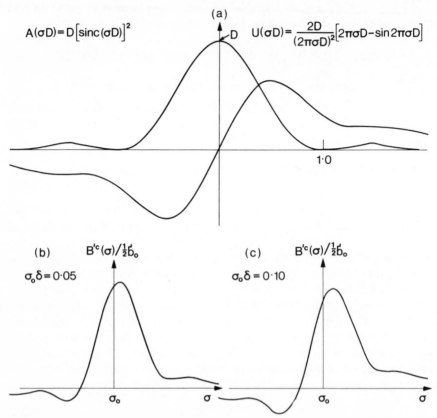

$$A(\sigma D) = D\left[\text{sinc}(\sigma D)\right]^2 \qquad (a) \qquad U(\sigma D) = \frac{2D}{(2\pi\sigma D)^2}\left[2\pi\sigma D - \sin 2\pi\sigma D\right]$$

(b) $B'^c(\sigma)/\tfrac{1}{2}b_0$ $\qquad$ (c) $B'^c(\sigma)/\tfrac{1}{2}b_0$

$\sigma_0\delta = 0.05$ $\qquad\qquad\qquad\qquad$ $\sigma_0\delta = 0.10$

Figure 6.13. The even and odd apodized spectral windows corresponding to the linear taper weighting function. Insets (b) and (c) show how negative ordinates can occur in the spectral profile even for very small values of $\sigma_0\delta$

The development of the fully unsymmetrical apparatus function from the fully symmetrical function as $\sigma_0\delta$ increases can be described in terms of a *coefficient of asymmetry* $C(\sigma_0\delta)$. This is defined by J. Connes as the ratio of the height h of the negative minimum to that H of the central maximum. From this definition, $C(\sigma_0\delta)$ is only zero in exceptional cases and, since it will mostly be used to quantify small departures from symmetry, this is a disadvantage. Ideally one would like $C = 0$ to signify total symmetry and $C = 1$ to signify total antisymmetry, but it is rather difficult to specify a unique objective way of calculating $C(\sigma_0\delta)$ such that these criteria are satisfied. Mme Connes' definition is well behaved at $\sigma_0\delta = 0.25$, but one can transfer this good behaviour to the more significant $\sigma_0\delta \approx 0$ region by the simple modification

$$C(\sigma_0\delta) = \left[\frac{h(\sigma_0\delta) - h_0}{H(\sigma_0\delta)}\right], \qquad (6.96a)$$

where h_0 is the value of the perfectly symmetric profile at the value of $\sigma\delta$ where the principal minimum occurs. Another possible definition,

$$C(\sigma_0\delta) = \left[\frac{h_-(\sigma_0\delta) - h_+(\sigma_0\delta)}{H(\sigma_0\delta)}\right], \tag{6.96b}$$

where h_- and h_+ are the depths of the minima each side of the principal maximum, gives a more linear dependence (see Figure 6.14) of C upon $\sigma_0\delta$. Neither (6.96a) nor (6.96b) approaches the ideal value of 1.0 at $\sigma_0\delta = 0.25$, and for each of them some mental contortion is necessary to arrive at a satisfactory definition for $\sigma_0\delta$ values lying outside the 0–0.25 range. Another criticism of this approach is that it assumes that the absolute values of the ordinates and the position of the origin on the abscissa scale are both known. Neither would necessarily be known in real spectroscopy, so one is led to a purely pragmatic definition

$$C(\sigma_0\delta) = \left[\frac{H_- - H_+}{H_\pm}\right], \tag{6.96c}$$

where H_- and H_+ are the amplitudes from main peak to the minima each side and H_+ is whichever of these is the greater. This definition again works well for $\sigma_0\delta \approx 0$ but does not approach unity as $\sigma_0\delta$ approaches 0.25.

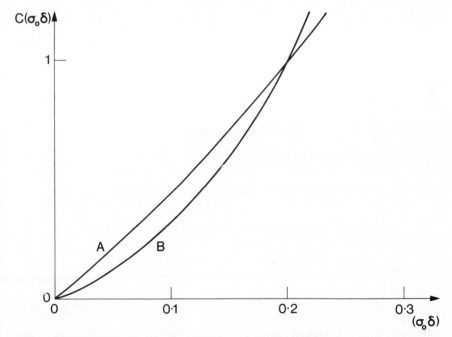

Figure 6.14. Graphs showing the variation of the coefficient of asymmetry $C(\sigma_0\delta)$ with $(\sigma_0\delta)$. Curve A corresponds to the definition (6.96b) whereas curve B corresponds to that given by (6.96c)

Let us now consider the more general case of the spectrum which has an extended range. The quantity calculated from a displaced interferogram $F(x'+\delta)$ is given by (6.81), while that required is (as in 6.80)

$$B^c(\sigma) = 2 \int_0^\infty F(x) W\left(\frac{x}{D}\right) \sqcap\left(\frac{x}{D}\right) \cos 2\pi\sigma x \, dx \qquad [\text{W m}], \qquad (6.97)$$

whilst, of course, the true, or ideal, spectrum is

$$B_e(\sigma) = 2 \int_0^\infty F(x) \cos 2\pi\sigma x \, dx \qquad [\text{W m}]. \qquad (6.98)$$

Application of Fourier's inversion theorem to (6.98) gives

$$F(x) = 2 \int_0^\infty B_e(\sigma') \cos 2\pi\sigma' x \, d\sigma' \qquad [\text{W}], \qquad (6.99)$$

which, on change of variable to $x'+\delta$, becomes

$$F(x'+\delta) = 2 \int_0^\infty B_e(\sigma') \cos 2\pi\sigma'(x'+\delta) \, d\sigma' \qquad [\text{W}]. \qquad (6.100)$$

Multiplication by $W(x'/D) \sqcap(x'/D) \cos 2\pi\sigma x'$ and integration over x' from 0 to ∞ recovers (6.81) and shows $B'^c(\sigma)$ to be expressible as

$$B'^c(\sigma) = 4 \int_0^\infty d\sigma' \int_0^\infty dx' B_e(\sigma') W\left(\frac{x'}{D}\right) \sqcap\left(\frac{x'}{D}\right) \cos 2\pi\sigma'(x'+\delta) \cos 2\pi\sigma x'$$

$$\qquad (6.101a)$$

$$= 2 \int_0^\infty d\sigma' \int_0^\infty dx' B_e(\sigma') W\left(\frac{x'}{D}\right) \sqcap\left(\frac{x'}{D}\right) [\cos 2\pi(\sigma'+\sigma)x'$$

$$+\cos 2\pi(\sigma'-\sigma)x'] \cos 2\pi\sigma'\delta$$

$$-2 \int_0^\infty d\sigma' \int_0^\infty dx' B_e(\sigma') W\left(\frac{x'}{D}\right) \sqcap\left(\frac{x'}{D}\right) [\sin 2\pi(\sigma'+\sigma)x'$$

$$+\sin 2\pi(\sigma'-\sigma)x'] \sin 2\pi x'\delta$$

$$= \int_0^\infty B_e(\sigma') \cos 2\pi\sigma'\delta[A((\sigma'+\sigma)D) + A((\sigma'-\sigma))] \, d\sigma'$$

$$- \int_0^\infty B_e(\sigma') \sin 2\pi\sigma'\delta[U((\sigma'+\sigma)D) + U((\sigma'-\sigma)D)] \, d\sigma'$$

$$= [B_e(\sigma) \cos 2\pi\sigma\delta] * A(\sigma D) + [B_e(\sigma) \sin 2\pi\sigma\delta] * U(\sigma D) \qquad [\text{W m}]$$

$$\qquad (6.101b)$$

in terms of the true spectrum $B_e(\sigma)$. The relationship between the computed spectrum and the true spectrum is rather complicated, being the sum of $B_e(\sigma) \cos 2\pi\sigma\delta$ convolved with the even spectral window $A(\sigma D)$ and $B_e(\sigma) \sin 2\pi\sigma\delta$ convolved with the odd spectral window $U(\sigma D)$. The spectral window varies throughout the range and the effect of this variation is to produce power distortions in the computed spectrum even to the extent of providing negative values. An example of such a spectrum is shown in Figure 6.15. It is noteworthy that the positions of the sharp features are hardly altered and one-sided transforms, with a small δ error, may be used for the study of band positions if absolute power levels are not of interest. Care must be taken, however, to ensure that the product $\sigma_M \delta$ is sufficiently small to prevent strong negative minima from appearing in the spectral window, thus giving an impression of spectral absorption. It should be noted that the algebraic area of the unsymmetrical spectral window is zero, thus ensuring that it makes little contribution to a 'slowly-varying' spectrum as the convolution of $B_e(\sigma) \sin 2\pi\sigma\delta$ with $U(\sigma D)$ is practically zero. The computed spectrum is, therefore, approximately $B_e(\sigma) \cos 2\pi\sigma\delta$ (as $A(\sigma D) \sim 1$), which is equal to the true (even) spectrum $B_e(\sigma)$ multiplied by a frequency-dependent periodic attenuating factor $\cos 2\pi\sigma\delta$. In general, however, effects due to both windows are present, details at the limit of resolution being more or less deprived of symmetry and slowly varying parts of the spectrum being more or less attenuated, in each case according to their greater or lesser distance from the wavenumber origin.

6.10. EFFECT OF DISPLACED TRUNCATION

It has been assumed, so far, that the truncation function is chosen to be finite only over the semi-range $0 \le x \le D$. The effect of an error in recording the interferogram was considered in section 6.9, where, it should be noted, *incorrect data* were introduced into the Fourier integral. The purpose of the present discussion is to consider the effect of evaluating a Fourier integral containing *correct* data over a displaced truncation interval. In other words, to see the effect of deliberately excluding the features near zero path difference. This is of interest in practice because the fluctuations of power near zero path difference are large compared with those at more remote values and a large dynamic range is required to accommodate these satisfactorily. If a satisfactory spectrum can be obtained with these features excluded, then this could make the task of the practical spectroscopist considerably easier.

We do not evaluate

$$B_e^c(\sigma) = 2 \int_0^\infty F(x) \, \sqcap\left(\frac{x}{D}\right) W\left(\frac{x}{D}\right) \cos 2\pi\sigma x \, \mathrm{d}x \qquad [\text{W m}] \qquad (6.102)$$

but

$$B_e^c(\sigma) = 2 \int_0^\infty F(x) \, \sqcap\left(\frac{x-\Delta}{D}\right) W\left(\frac{x}{D}\right) \cos 2\pi\sigma x \, \mathrm{d}x \qquad [\text{W m}], \qquad (6.103)$$

which represents an integration over the displaced semi-range $\Delta \le x \le D + \Delta$. However, the expression (6.103) cannot be used as its stands because $W(x/D)$ is not defined outside the interval $|x| \le D$, thus we shall have to

160

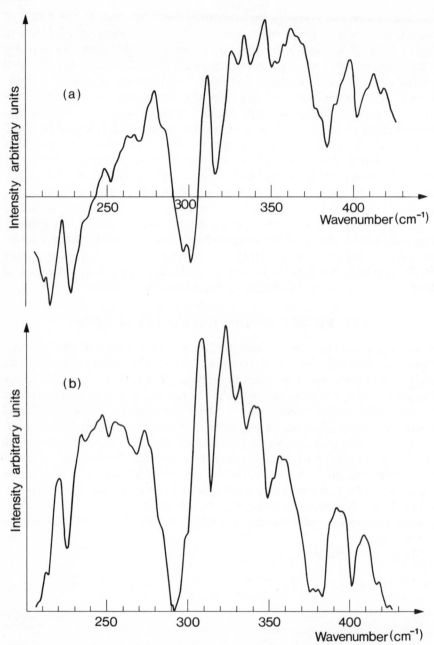

Figure 6.15. One-sided (a) and two-sided (b) transforms of an interferogram sampled with an offset comb. The offset is the maximum possible (one half of a sampling interval) and the range involves the higher frequencies, so the distortions and negative ordinates in (a) are more noticeable than would usually be the case. It is to be observed that the sharp features occur more or less in their correct positions, but broad features may suffer considerable shifts

dispense with weighting for the present discussion. When this is so, (6.103) becomes

$$B'^c(\sigma) = 2 \int_0^\infty F(x'+\Delta) \, \Pi\!\left(\frac{x'}{D}\right) \cos 2\pi\sigma(x'+\Delta) \, dx' \qquad [\text{W m}], \quad (6.104)$$

where

$$x' = x - \Delta \qquad [\text{m}].$$

If the detected radiation is monochromatic,

$$F(x'+\Delta) = \mathfrak{b}_0 \cos 2\pi\sigma_0(x'+\Delta) \qquad [\text{W}] \qquad (6.105)$$

and

$$B'^c(\sigma) = 2\mathfrak{b}_0 \int_0^\infty \Pi\!\left(\frac{x'}{D}\right) \cos 2\pi\sigma_0(x'+\Delta) \cos 2\pi\sigma(x'+\Delta) \, dx' \qquad (6.106a)$$

$$= \mathfrak{b}_0 \cos 2\pi(\sigma+\sigma_0)\Delta \int_0^D \cos 2\pi(\sigma+\sigma_0)x' \, dx' - \mathfrak{b}_0 \sin 2\pi(\sigma+\sigma_0)\Delta$$

$$\times \int_0^D \sin 2\pi(\sigma+\sigma_0)x' \, dx'$$

$$+ \mathfrak{b}_0 \cos 2\pi(\sigma-\sigma_0)\Delta \int_0^D \cos 2\pi(\sigma-\sigma_0)x' \, dx' - \mathfrak{b}_0 \sin 2\pi(\sigma-\sigma_0)\Delta$$

$$\times \int_0^D \sin 2\pi(\sigma-\sigma_0)x' \, dx'$$

$$= \tfrac{1}{2}\mathfrak{b}_0[A_0((\sigma+\sigma_0)D) \cos 2\pi(\sigma+\sigma_0)\Delta$$
$$+ U_o((\sigma+\sigma_0)D) \sin 2\pi(\sigma+\sigma_0)\Delta]$$
$$+ \tfrac{1}{2}\mathfrak{b}_0[A_0((\sigma-\sigma_0)D) \cos 2\pi(\sigma-\sigma_0)\Delta$$
$$+ U_o((\sigma-\sigma_0)D) \sin 2\pi(\sigma-\sigma_0)\Delta]. \qquad (6.106b)$$

We see that for $\sigma \gg 0$ the spectral window is changed from $A_0((\sigma-\sigma_0)D)$ to

$$\frac{2B'^c(\sigma)}{\mathfrak{b}_0} = A_0((\sigma-\sigma_0)D) \cos 2\pi(\sigma-\sigma_0)\Delta$$

$$+ U_o((\sigma-\sigma_0)D) \sin 2\pi(\sigma-\sigma_0)\Delta \qquad [\text{m}]. \quad (6.107)$$

The form of this depends strongly on the magnitude of $(\sigma-\sigma_0)\Delta$, for when $(\sigma-\sigma_0)\Delta = 0$ the window is even and when $(\sigma-\sigma_0)\Delta = \tfrac{1}{4}$ it is odd. Between these two extreme cases all intervening ones are possible. It should be noted that the primary determining factor here is the difference $(\sigma-\sigma_0)\Delta$ compared with the simple product $\sigma_0\delta$ in the case of an error, as considered in section 6.9.

If we generalize the study to the case of a spectrum of extended range, then, since

$$F(x'+\Delta) = 2 \int_0^\infty B_e(\sigma') \cos 2\pi\sigma'(x'+\Delta) \, d\sigma' \qquad [\text{W}], \qquad (6.108)$$

(following the argument of (6.101)) we can write the calculated spectrum as

$$B'^c(\sigma) = 4 \int_0^\infty d\sigma' \int_0^\infty dx' B_e(\sigma') \sqcap\left(\frac{x'}{D}\right) \cos 2\pi\sigma'(x'+\Delta) \cos 2\pi\sigma(x'+\Delta)$$

(6.109a)

$$= \int_0^\infty B_e(\sigma')[\cos 2\pi(\sigma'+\sigma)\Delta A_0((\sigma+\sigma')D)$$

$$+ \cos 2\pi(\sigma-\sigma')\Delta A_0((\sigma-\sigma')D)]\, d\sigma'$$

$$+ \int_0^\infty B_e(\sigma')[\sin 2\pi(\sigma'+\sigma)\Delta U_o((\sigma+\sigma')D)$$

$$+ \sin 2\pi(\sigma-\sigma')\Delta U_o((\sigma-\sigma')D)]\, d\sigma'$$

$$= B_e(\sigma) * [A_0(\sigma D) \cos 2\pi\sigma\Delta] + B_e(\sigma) * [U_o(\sigma D) \sin 2\pi\sigma\Delta] \qquad [\text{W m}]$$

(6.109b)

in terms of the true spectrum $B_e(\sigma)$. Thus, the effect of calculating the spectrum from a displaced portion of the interferogram is to produce power distortions due to distortions in the spectral window

$$A_0(\sigma D) \cos 2\pi\sigma\Delta + U_o(\sigma D) \sin 2\pi\sigma\Delta \qquad [\text{m}]. \qquad (6.110)$$

6.10.1. Effect of exclusion of the region near zero path difference

While it would not appear to be satisfactory to displace the truncation function to exclude the region of zero path difference, there is another possibility, namely to keep the truncation function fixed on the interval $0 \le x \le D$ but to exclude from the range of integration the smaller interval $0 \le x \le \Delta$ so that one evaluates

$$B'^c(\sigma) = 2 \int_\Delta^D F_e(x) W\left(\frac{x}{D}\right) \cos 2\pi\sigma x\, dx \qquad (6.111a)$$

$$= 2 \int_0^D F_e(x) W\left(\frac{x}{D}\right) \cos 2\pi\sigma x\, dx - 2 \int_0^\Delta F_e(x) W\left(\frac{x}{D}\right) \cos 2\pi\sigma x\, dx$$

$$= B_e(\sigma) * \Omega(\sigma D) * [A_0(\sigma D) - A_0(\sigma\Delta)] \qquad [\text{W m}], \qquad (6.111b)$$

in which the required spectrum is convolved with a window

$$\Omega(\sigma D) * [A_0(\sigma D) - A_0(\sigma \Delta)] \qquad [\text{m}]. \qquad (6.112)$$

This simple evaluation does not allow one to assess very easily the effect of the exclusion. If we rewrite (6.111b) as

$$B'^c(\sigma) = 2 \,\text{Re} \int_{-\infty}^\infty F_e(x) W\left(\frac{x}{D}\right)$$

$$\times \left[H(x) \sqcap\left(\frac{x}{2(D-\Delta)}\right) * \delta(x-\Delta)\right] \exp(-2\pi i\sigma x)\, dx, \qquad (6.113a)$$

we find

$$B^{rc}(\sigma) = B_e(\sigma) * \Omega(\sigma D) * [A_0(\sigma(D-\Delta)) \cos 2\pi\sigma\Delta$$

$$+ A_0(\sigma(D-\Delta)) \sin 2\pi\sigma\Delta], \quad (6.113b)$$

which shows that when Δ is small the factor of the Hilbert transform term is small and a good approximation to the required spectrum is obtained with slightly less resolution because of the $D-\Delta$ appearing in $A_0(\sigma(D-\Delta))$. However, if Δ is to be small, one will be including much of the large amplitude zero path region and not much will have been gained. The conclusion must be that other, i.e. electrical or mechanical, solutions to the dynamic range problem must be sought.

6.11. CORRECTION FOR PHASE ERROR IN ONE-SIDED INTERFEROGRAMS

The presence of inherent asymmetry in the interferogram $F(x)$ and/or displacement in the cut-off function (and, hence, the sampling) are both equivalent to a wavenumber-dependent phase error $\phi(\sigma)$ in the observed interferogram. The unsymmetrical interferogram can therefore be written

$$F_u(x) = \int_0^\infty B(\sigma) \cos[2\pi\sigma x - \phi(\sigma)] \, d\sigma \quad [\text{W}] \quad (6.114)$$

in the general case. When the error in the record is due solely to a shift δ in the origin of computation, $\phi(\sigma)$ is a linear phase error,

$$\phi(\sigma) = -2\pi\sigma\delta \quad [\text{rad}] \quad (6.115)$$

and

$$F_u(x) = \int_0^\infty B(\sigma) \cos 2\pi\sigma(x+\delta) \, d\sigma = F(x+\delta) \quad [\text{W}]. \quad (6.116)$$

A number of authors have proposed ways in which a phase error may be corrected. Of course, the ideal way in principle is physically to remove the cause in the interferometer itself. In practice, this may be extremely tedious and difficult, if not impossible. Failing the possibility of removing the cause, the next possibility is to make some *post hoc* allowance for (or correction of) the effect. J. Connes,[37] Loewenstein,[38] Mertz,[39] Forman, Steel, and Vanasse,[40] Gibbs and Gebbie,[41] amongst others,[42] have all given descriptions of procedures which correct for the presence of phase error. Of these, the methods of Loewenstein and Gibbs are applicable *only* to the case where the interferogram itself is symmetrical but is sampled with an offset comb. We deal first with Gibbs' treatment and then give that of Forman *et al.*, which is more general and includes linear phase error as a special case.

6.11.1. Correction for phase error due to displaced sampling

The analogue interferogram is $F_e(x)$ and is truly symmetrical. When correctly sampled, with the first $(r = 0)$ sample coinciding with $x = 0$, the rth sample is $F_e(x)\delta[(x/\beta) - r] = \beta F_e(r\beta)$. However, when the first sample occurs at $x = \delta$ and all subsequent samples are displaced by δ from their ideal positions, the rth sample is $F_e(x)\delta\{[(x - \delta)/\beta] - r\} = \beta F_e(r\beta + \delta)$. Consequently, the computed power spectrum is given, on introducing this expression into (6.72), by

$$B'^c(\sigma) = 2\beta\left[\tfrac{1}{2}F(\delta) + \sum_{r=1}^{N} F(r\beta + \delta)W\left(\frac{r}{N}\right)\cos 2\pi\sigma r\beta\right.$$

$$\left. + \tfrac{1}{2}F(N\beta + \delta)W(1)\cos 2\pi\sigma N\beta\right] \quad (6.117)$$

instead of by

$$B^c(\sigma) = 2\beta\left[\tfrac{1}{2}F(0) + \sum_{r=1}^{N} F(r\beta)W\left(\frac{r}{N}\right)\cos 2\pi\sigma r\beta\right.$$

$$\left. + \tfrac{1}{2}F(N\beta)W(1)\cos 2\pi\sigma N\beta\right], \quad (6.118)$$

which applies when there is no sampling error.

The incorrect spectrum represented by the summation (6.117) is, in fact, given by the result (6.101b), which can be derived in an alternative way. We write (6.81) as

$$B'^c(\sigma) = 2\,\text{Re}\int_0^\infty [F(x) * \delta(x + \delta)]W\left(\frac{x}{D}\right)$$

$$\sqcap\left(\frac{x}{D}\right)\sqcup\left(\frac{x}{\beta}\right)\exp(-2\pi i\sigma x)\,dx \quad \text{[W m]} \quad (6.119)$$

and evaluate the Fourier integral to give

$$B'^c(\sigma) = \text{Re}\,[B_e(\sigma)\exp 2\pi i\sigma\delta] * \Omega(\sigma D) * [2D\,\text{sinc}\,(\sigma D)\exp(-i\pi\sigma D)]$$

$$* \beta\sqcup(\sigma\beta)$$

$$= [B_e(\sigma)\cos 2\pi\sigma\delta] * \Omega(\sigma D) * [2D\,\text{sinc}\,(\sigma D)\cos \pi\sigma D] * \beta\sqcup(\sigma\beta)$$

$$+ [B_e(\sigma)\sin 2\pi\sigma\delta] * \Omega(\sigma D) * [2D\,\text{sinc}\,(\sigma D)\sin \pi\sigma D] * \beta\sqcup(\sigma\beta)$$

$$= \{[B_e(\sigma)\cos 2\pi\sigma\delta] * \Omega(\sigma D) * A_0(\sigma D)$$

$$+ [B_e(\sigma)\sin 2\pi\sigma D] * \Omega(\sigma D) * U_0(\sigma D)\} * \beta\sqcup(\sigma\beta), \quad (6.120)$$

where we have noted that

$$\int_{-\infty}^\infty \sqcap\left(\frac{x}{D}\right)\exp(-2\pi i\sigma x)\,dx$$

$$= \int_{-\infty}^\infty \left[\sqcap\left(\frac{x}{D}\right) * \delta(x - \tfrac{1}{2}D)\right]\exp(-2\pi i\sigma x)\,dx$$

$$= D\,\text{sinc}\,(\sigma D)\exp(-i\pi\sigma D). \quad (6.121)$$

By applying the result (6.27a) and confining our attention to the alias which coincides with the known spectral range, we obtain

$$B'^c(\sigma) = [B_e(\sigma) \cos 2\pi\sigma\delta] * A(\sigma D) + [B_e(\sigma) \sin 2\pi\sigma\delta] * U(\sigma D) \qquad [\text{W m}],$$
(6.122)

as in (6.101b). This argument will be of use when we evaluate the form of the spectrum resulting from Fourier transformation of an interferogram to which a phase error correction has been applied.

Gibbs and Gebbie[41] proposed that when $F(x)$ is symmetrical but sampled with an offset sampling comb

$$\beta \sqcup\!\sqcup\!\left(\frac{x-\delta}{\beta}\right) = \beta \sum_{r=-\infty}^{\infty} \delta(x-\delta-r\beta),$$
(6.123)

the computational equation (6.72) should be adjusted to make the values of the weighting function and the kernel of the integral accord with the samples. This correction is, therefore, a *post hoc* correction for error which is carried out at the computational stage.

Following Gibbs' argument, we replace $r\beta$ in the weighting function and the kernel of (6.72) by

$$r\beta + \eta,$$

so that instead of (6.72) we evaluate

$$B''^c(\sigma) = 2\beta \ \text{Re} \sum_{r=0}^{N} [1 - \tfrac{1}{2}\delta(r;0) - \tfrac{1}{2}\delta(N;0)] F(r\beta + \delta)$$
$$\times W\!\left(\frac{r\beta + \eta}{N\beta}\right) \exp\left[-2\pi i\sigma(r\beta + \eta)\right]$$
$$\equiv 2 \ \text{Re} \int_0^{\infty} F(x+\delta) W\!\left(\frac{x+\eta}{D}\right) \sqcap\!\left(\frac{x}{D}\right) \sqcup\!\sqcup\!\left(\frac{x}{\beta}\right)$$
$$\times \exp\left[-2\pi i\sigma(x+\eta)\right] dx \qquad [\text{W m}].$$
(6.124)

By changing the variable to

$$x'' = x + \eta,$$
(6.125)

we obtain

$$B''^c(\sigma) = 2 \ \text{Re} \int_{-\infty}^{\infty} [F(x'') * \delta(x'' - \eta + \delta)] W\!\left(\frac{x''}{D}\right)$$
$$\times \left\{\left[\sqcap\!\left(\frac{x''}{D}\right) \sqcup\!\sqcup\!\left(\frac{x''}{\beta}\right)\right] * \delta(x'' - \eta)\right\} \exp\left(-2\pi i\sigma x''\right) dx''$$
$$= 2 \ \text{Re} \left[B_e(\sigma) \exp 2\pi i\sigma(\eta - \delta)\right] * \Omega(\sigma D) * \{[D \ \text{sinc} (\sigma D)$$
$$\times \exp(-i\pi\sigma D)] * \beta \sqcup\!\sqcup\!(\sigma\beta)\} \exp(-2\pi i\sigma\eta).$$
(6.126)

If we now use the convolution theorem and put $\eta = \delta$,

$$B'''^c(\sigma) = B_e(\sigma) * \Omega(\sigma D) * \{[A_0(\sigma D) * \beta \, \sqcup\!\sqcup(\sigma\beta)] \cos 2\pi\sigma\delta\}$$
$$- B_e(\sigma) * \Omega(\sigma D) * \{[U_0(\sigma D) * \beta \, \sqcup\!\sqcup(\sigma\beta)] \sin 2\pi\sigma\delta\} \quad \text{[W m]}.$$

$$(6.127)$$

Rigorous analysis of (6.127) shows that if δ is small compared with D, then the positions of spectral features are exactly correct. This result can be derived in an elementary semi-rigorous fashion by considering, as usual, monochromatic irradiation. The corrected computational equation becomes,

$$B'''^c(\sigma) = 2 \operatorname{Re} \int_0^D \mathfrak{b}_0 \cos 2\pi\sigma_0[x + \eta] \cos 2\pi\sigma[x + \eta] \, dx, \quad (6.128)$$

which, for σ and σ_0 both much larger than zero, becomes

$$B'''^c(\sigma) = \mathfrak{b}_0 \left[[D + \eta] \frac{\sin 2\pi[\sigma_0 - \sigma][D + \eta]}{2\pi[\sigma_0 - \sigma][D + \eta]} - \eta \frac{\sin 2\pi[\sigma_0 - \sigma]\eta}{2\pi[\sigma_0 - \sigma]\eta} \right]. \quad (6.129)$$

Therefore, when $D \gg \eta$, the second term becomes negligible compared to the first and we arrive at the result quoted above. The question of the accuracy of absolute power measurements requires a rather more sophisticated analysis. One can do this either by expansion of (6.124) or else, more elegantly, by a consideration of (6.127). Here it will be seen that, if $2\pi\sigma\delta$ is everywhere small, that is, if $\sigma_{max} \ll (2\pi\delta)^{-1}$, then $\cos 2\pi\sigma\delta$ will be a constant (≈ 1) over the full range and therefore can be regarded as a purely multiplicative term in (6.127). Likewise $\sin 2\pi\sigma\delta$ will be everywhere small and

$$B'''^c(\sigma) \simeq B_e(\sigma) * \Omega(\sigma D) * A_0(\sigma D) * \beta \, \sqcup\!\sqcup(\sigma\beta); \quad (6.130)$$

that is, the calculated spectrum will agree with that which would have been calculated from an ideally sampled interferogram. One should note, however, that if the interferogram is sampled economically, i.e. in a fashion dictated by the sampling theorem, one would have $\sigma_{max} = (2\beta)^{-1}$ and the condition on η would be $\eta \ll \pi^{-1}\beta$. One is, therefore, restricted to very small values on η if serious distortions of the absolute power record are to be avoided. In practice, however, this may not turn out to be too much of a difficulty because the distortions introduced are usually most marked at the extremities of the spectral range. Now, by means of filters, the experimentalist will usually arrange interference in the beam divider, etc., to maximize his sensitivity in the middle of the range and to have zero signal at $\sigma = 0$ and $\sigma = \sigma_{max}$. Under these conditions, and bearing in mind that η need never exceed 0.5β, quite reliable spectroscopy may be carried out.

This can be illustrated by considering an example, which, though somewhat artificial, has the merit of numerical simplicity, namely the reconstruction of the triangle function from the cosine Fourier transformation of the sinc^2 function. Since we are here merely trying to illustrate a point, we will compute from a double sided interferogram, ranging from $-(N/2)\beta$ to $+[(N/2)-1]\beta$; first correctly sampled, and second, offset sampled. One has, therefore

$$\bigwedge(\sigma) \supset \text{sinc}^2 x, \qquad \sigma > 0. \qquad (6.131)$$

The reconstruction can be done quite effectively using only eight ordinates, because of the rapid fall-off of the sinc^2 function and to facilitate the computation it is best carried out using the principles of the fast Fourier transform (FFT) (see Chapter 10). This involves a correct choice of sampling interval coupled with a judicious choice of σ values, at which $\bigwedge(\sigma)$ is to be calculated, so that the various terms in the computational equation become related by the symmetry of the cosine function. In fact, if a_n represents the nth ordinate of the function to be transformed, the computational equation becomes

$$\bigwedge(\sigma) = \beta \sum_{n=0}^{n=N-1} a_n \cos \frac{2mn\pi}{N}. \qquad (6.132)$$

In the present case, where we choose units such that σ_{max} is unity, β becomes 0.5 and N is chosen to be eight. The five ordinates of $_c\bigwedge_m(\sigma)$, which span the range $\sigma = 0$ to $\sigma = 1$ (i.e. σ_{max}), are then given by the equations

$$
\begin{aligned}
_c\bigwedge_0 &= A + B \\
_c\bigwedge_1 &= C + D \\
_c\bigwedge_2 &= E - F \qquad \text{and} \qquad
\begin{aligned}
_c\bigwedge_5 &= {}_c\bigwedge_3 \\
_c\bigwedge_6 &= {}_c\bigwedge_2 \\
_c\bigwedge_7 &= {}_c\bigwedge_1
\end{aligned} \\
_c\bigwedge_3 &= C - D \\
_c\bigwedge_4 &= A - B
\end{aligned}
\qquad (6.133)
$$

where $2A = a_0 + a_2 + a_4 + a_6$, $2B = a_1 + a_3 + a_5 + a_7$, $2C = a_0 - a_4$, $2D = (a_1 - a_3 - a_5 + a_7) \cos(\pi/4)$, $2E = a_0 + a_4$, and $2F = a_2 + a_6$. The results are shown in Table 6.1(a) for the cases $\delta = 0$ and $\delta = 0.2$ (i.e. 0.4β). It will be seen that the reconstruction is quite good for the $\delta = 0$ case but rather worse

Table 6.1(a)

Ordinate	$\delta = 0$	$\delta = 0.2$	Ordinate	$\delta = 0$	$\delta = 0.2$	Theoretical
a_0	1.0	0.8751	$_c\bigwedge_0$	0.9503	0.9478	1.000
a_1	0.4053	0.1353	$_c\bigwedge_1$	0.7548	0.7203	0.750
a_2	0	0.0243	$_c\bigwedge_2$	0.5000	0.4017	0.500
a_3	0.0450	0.0229	$_c\bigwedge_3$	0.2452	0.1476	0.250
a_4	0	0.0072	$_c\bigwedge_4$	0.0497	0.0136	0.000
a_5	0.0450	0.0392	$_c\bigwedge_5$	0.2452	0.1476	0.250
a_6	0	0.0547	$_c\bigwedge_6$	0.5000	0.4017	0.500
a_7	0.4053	0.7368	$_c\bigwedge_7$	0.7548	0.7203	0.750

Table 6.1(b)

Ordinate	$\delta = 0.2$	$\cos \dfrac{\pi m}{N}\left(\dfrac{\delta}{\beta}\right) {}_c\!\bigwedge_m$	$\sin \dfrac{\pi m}{N}\left(\dfrac{\delta}{\beta}\right) {}_s\!\bigwedge_m$	Corrected spectrum
${}_s\!\bigwedge_0$	0	0.9478	0	0.9478
${}_s\!\bigwedge_1$	−0.2336	0.6850	−0.0722	0.7572
${}_s\!\bigwedge_2$	−0.2926	0.3250	−0.1720	0.4970
${}_s\!\bigwedge_3$	−0.2032	0.0868	−0.1644	0.2512
${}_s\!\bigwedge_4$	0.000	0.0042	0.0000	0.0042
${}_s\!\bigwedge_5$	+0.2032	0.000	+0.2032	−0.2032
${}_s\!\bigwedge_6$	+0.2926	−0.1241	+0.2783	−0.4024
${}_s\!\bigwedge_7$	+0.2336	−0.4234	+0.1890	−0.6124

for the $\sigma = 0.2$ case. One now goes on to consider the modified computational equation incorporating the Gibbs' amendment

$$ {}_G\!\bigwedge_m = \beta \sum_{n=0}^{n=N-1} a_n \cos\left(\frac{2mn\pi}{N} + \frac{\pi m \delta}{D}\right). \tag{6.134} $$

D is of course $N\beta$, so in the present case it becomes 4. Expansion of equation (6.134) leads to the previously given cosine Fourier transform multiplied by $\cos(\pi m\delta/D)$ and a sine Fourier transform multiplied by $-\sin(\pi m\delta/D)$. The sine Fourier transform can be carried out by an FFT method using the analogous equations

$$
\begin{aligned}
&{}_s\!\bigwedge_0 = 0 \\
&{}_s\!\bigwedge_1 = A' + B' \\
&{}_s\!\bigwedge_2 = C' \qquad\qquad \text{and} \qquad
\begin{aligned}
&{}_s\!\bigwedge_5 = -{}_s\!\bigwedge_3 \\
&{}_s\!\bigwedge_6 = -{}_s\!\bigwedge_2 \\
&{}_s\!\bigwedge_7 = -{}_s\!\bigwedge_1
\end{aligned} \\
&{}_s\!\bigwedge_3 = A' - B' \\
&{}_s\!\bigwedge_4 = 0
\end{aligned}
\tag{6.135}
$$

where $2A' = (a_1 + a_3 - a_5 - a_7)\sin(\pi/4)$, $2B' = (a_2 - a_6)$, and $2C' = (a_1 - a_3 + a_5 - a_7)$. The results of using this modified approach are shown in Table 6.1(b).

The agreement over the region of the first alias (i.e. $m = 0$ to $m = 4$) is very much improved. The method has therefore much to commend it since only linear operations are involved. However it is not easy to adopt this method to the full FFT without having to compute *two* transforms (in the case of the slow FT one may use (6.134) as it stands). If one has computed both the cosine transform and the sine transform, one might as well use the modulus method discussed in Section 6.12. It will be noted that from Table 6.1, $[{}_s\!\bigwedge_m^2 + {}_c\!\bigwedge_m^2]^{1/2}$ is a close approximation to the ideal $\delta = 0$ value.

Having illustrated the basic principles of the method in this way, we now ask the question whether this method can be applied to the practically important case where one has observed only one side of a long interferogram with perhaps just a few points on the shorter side of ZPD. We are constrained to use the slow Fourier transform, but Gibbs and Gebbie[41] suggest that the only modification required is to change the kernels as in (6.134). They write, therefore,

$$_G I_m = 2\beta \left[\tfrac{1}{2} a_0 \cos \frac{\pi m \delta}{N\beta} + \sum_{n=1}^{n=N-1} a_n \cos \left\{ \frac{mn\pi}{N} + \frac{m\pi\delta}{N\beta} \right\} \right.$$

$$\left. + \tfrac{1}{2} a_N \cos \left\{ m\pi + \frac{m\pi\delta}{N\beta} \right\} \right]. \quad (6.134a)$$

Usually the last term will be zero because of the use of apodization. The application of this equation to correct an offset sampled sinc^2 function interferogram computation is shown in Figure 6.16. The corrected version is better, but is a poor approximation to the true curve A. Walmsley, Clark, and Jennings[41a] have shown that the discrepancy is partly due to the presence of a constant term but more to the effect of overlapping aliases. Correction for this is intractable and this combined with the limitation to the slow FT makes the Gibbs' procedure of only limited use.

In practice the value of δ is found by inspecting the ordinates actually sampled in the region of zero path difference and, either by relating them to the analogue record or else by interpolation, finding the displacement of the largest ordinate from the position where the analogue (or interpolated)

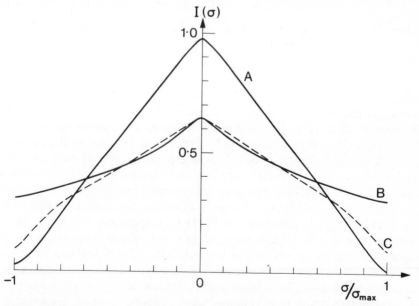

Figure 6.16. The reconstruction of the triangle function by one-sided cosine Fourier transforms of the sinc^2 function. Curve A comes from a correctly sampled version of the sinc^2 function and curve B from one which has an offset in the sampling of 0.2 (i.e. 0.4β). Curve C shows the effect of applying the Gibbs' phase correction procedure

170

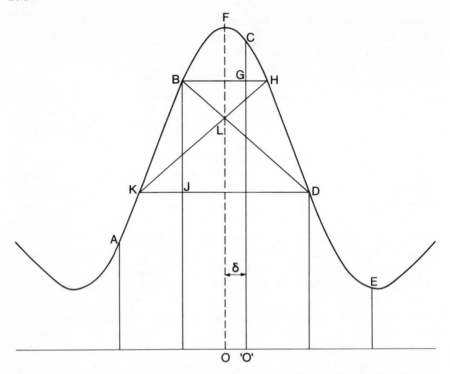

Figure 6.17. Geometrical construction to locate the position of zero path difference from a set of sampled ordinates each of which is displaced an amount δ from its proper position

record would have a maximum. The parameter δ (called η by Gibbs and Gebbie) is then equal to this displacement. The interpolation requires the observation of only a few additional data in the region of zero path difference, so that δ may be found by fitting a polynomial to the profile of the peak of the brightest fringe. The interpolation procedure is generally capable of providing sufficient accuracy and may be executed as shown in Figure 6.17. The full lines represent the analogue interferogram (not generally recorded) and A, B, C, D, and E represent the samples taken in the region of zero path difference at F. The computation will generally be initiated from C, the largest recorded ordinate, which corresponds to the origin '0' displaced by δ from the true origin 0. A horizontal line from B to G intercepts the full curve through ED at H, while a similar line from D to J has an intercept at K. To a good approximation BHDK is a parallelogram whose diagonals intersect at L immediately beneath F; thus 0 is located and may be measured.

Loewenstein has proposed a correction[38] that is similar to Gibbs' (and does, in fact, pre-date it); however, he does not correct the weighting function nor does he adjust the variable of his integration. This introduces additional terms into the instrument function which are negligible only if δ/D is small (say of the order 10^{-3}).[42]

6.11.2. Symmetrizing function for phase error correction

An alternative and more general approach, proposed by Forman, Steel, and Vanasse,[40] is applicable to those cases where $F(x)$ is inherently unsymmetrical due to some lack of symmetry within the interferometer as well as to those where the asymmetry is due to displaced sampling. By noting that the linear phase error (6.115) is odd, that is,

$$\phi(\sigma) = -\phi(-\sigma) \qquad [\text{rad}], \qquad (6.136)$$

we may write (6.114) as

$$F_u(x) = \text{Re} \int_{-\infty}^{\infty} B_e(\sigma) \exp(-i\phi(\sigma)) \exp(2\pi i\sigma x)\,d\sigma \qquad [\text{W}], \quad (6.137)$$

so that application of the full Fourier inversion theorem gives

$$\hat{s}(\sigma) = B_e(\sigma) \exp(-i\phi(\sigma)) = \int_{-\infty}^{\infty} F_u(x) \exp(-2\pi i\sigma x)\,dx \qquad [\text{W m}],$$
$$(6.138)$$
$$= p(\sigma) - iq(\sigma),$$

where

$$p(\sigma) = \int_{-\infty}^{\infty} F_u(x) \cos 2\pi\sigma x\,dx \qquad [\text{W m}] \qquad (6.138a)$$

and

$$q(\sigma) = \int_{-\infty}^{\infty} F_u(x) \sin 2\pi\sigma x\,dx \qquad [\text{W m}]. \qquad (6.138b)$$

However, in addition to being the modulus

$$|\hat{s}(\sigma)| = B_e(\sigma) \qquad [\text{W m}] \qquad (6.139)$$

of the complex spectrum $\hat{s}(\sigma)$, the required spectrum is also represented by the ideal transform

$$B_e(\sigma) = \int_{-\infty}^{\infty} F_e(x) \exp(-2\pi i\sigma x)\,dx \qquad [\text{W m}] \qquad (6.140)$$

of the symmetrical interferogram $F_e(x)$. Inversion of this integral shows that

$$F_e(x) = \int_{-\infty}^{\infty} B_e(\sigma) \exp(2\pi i\sigma x)\,d\sigma$$

$$= \int_{-\infty}^{\infty} [B_e(\sigma) \exp(-i\phi(\sigma))] \exp(+i\phi(\sigma)) \exp(2\pi i\sigma x)\,d\sigma \qquad (6.140a)$$

$$= F_u(x) * \chi(x) \qquad [\text{W}], \qquad (6.140b)$$

where we have used (2.47) and put

$$\chi(x) = \int_{-\infty}^{\infty} \exp +i\phi(\sigma) \exp(2\pi i \sigma x) \, d\sigma \qquad [\text{m}^{-1}]. \qquad (6.141)$$

Since $\exp(+i\phi(\sigma))$ is, from (6.136), Hermitian, $\chi(x)$ is real. Equation (6.140a) shows that the required symmetrical interferogram may be generated from the measured unsymmetrical interferogram by convolution

$$F_e(x) = \int_{-\infty}^{\infty} F_u(x') \chi(x - x') \, dx' \qquad [\text{W}] \qquad (6.142)$$

with a phase correction function which is defined by (6.141).

In a practical situation, we shall have the portion $0 \le x \le D$ of the (unsymmetrical) interferogram $F_u(x)$ recorded and shall need to evaluate $\chi(x)$ in order to effect the convolution (6.142) which yields $F_e(x)$. To do this some additional data are required. When the phase spectrum is only slowly varying, it is not necessary to use high spectral resolution for its determination; in other words, only a small region of the interferogram need be recorded in the vicinity of zero path difference (where the signal-to-noise ratio is high) for the evaluation of $\phi(\sigma)$. We record, therefore, $F_u(x)$ for values of x satisfying $-l \le x \le D$ and take the portion $0 \le x \le D$ as the interferogram to be corrected to give $F_e(x)$. The portion $-l \le x \le +l$ is used for the low-resolution evaluation of the phase. From (6.138), the (complex) spectrum we compute from this short two-sided portion of interferogram is

$$\hat{s}^c(\sigma, l) = \int_{-\infty}^{\infty} F_u(x) W\left(\frac{x}{l}\right) \sqcap\left(\frac{x}{2l}\right) \exp(-2\pi i \sigma x) \, dx \qquad [\text{W m}], \qquad (6.143)$$

where the l in

$$\hat{s}^c(\sigma, l) = B_e^c(\sigma, l) \exp(-i\phi^c(\sigma, l)) \qquad (6.144a)$$

$$= p^c(\sigma, l) - iq^c(\sigma, l) \qquad [\text{W m}] \qquad (6.144b)$$

denotes that the resolution in the computed cosine transform

$$p^c(\sigma, l) = p(\sigma) * A(\sigma l) \qquad [\text{W m}] \qquad (6.145a)$$

and the computed sine transform

$$q^c(\sigma, l) = q(\sigma) * A(\sigma l) \qquad [\text{W m}] \qquad (6.145b)$$

is low and equal to $1/l$. Since $p^c(\sigma, l)$ and $q^c(\sigma, l)$ are non-zero over a finite range $-\sigma_M \le \sigma \le \sigma_M$, the computed low-resolution phase spectrum

$$\phi^c(\sigma, l) = \arctan \frac{q^c(\sigma, l)}{p^c(\sigma, l)} \qquad [\text{rad}] \qquad (6.146)$$

is evaluated over the same finite range and $\exp(-i\phi^c(\sigma, l))$ formed from (6.144) is a truncated function. The computed phase correction function is the Fourier transform

$$\chi^c(x, l, \sigma_M) = \int_{-\infty}^{\infty} \exp\left(+i\phi^c(\sigma, l)\right) \sqcap\left(\frac{\sigma}{2\sigma_M}\right) \exp\left(2\pi i\sigma x\right) d\sigma \quad [\text{m}^{-1}],$$

of the truncated exponential. If we write
(6.147a)

$$\chi^c(x, l) = \int_{-\infty}^{\infty} \exp\left(+i\phi^c(\sigma, l)\right) \exp\left(2\pi i\sigma x\right) d\sigma \quad [\text{m}^{-1}], \quad (6.148)$$

then (6.147a) becomes

$$\chi^c(x, l, \sigma_M) = \chi^c(x, l) * 2\sigma_M \operatorname{sinc}\left(2x\sigma_M\right) \quad [\text{m}^{-1}]. \quad (6.147b)$$

In general, $\phi^c(\sigma, l)$ is slowly varying and the phase correction function $\chi^c(x, l)$ is significantly different from zero only near $x = 0$. The convolution (6.147b) consists, therefore, of a series of fluctuations spreading out from $x = 0$ and slowly falling as $|x|$ increases. The corrected interferogram may now be calculated from

$$F_e^c(x) = F_u(x) * \chi^c(x, l, \sigma_M)$$

$$= \int_{-\infty}^{\infty} F_u(x')\chi^c(x - x', l, \sigma_M) \, dx' \quad [\text{W}]. \quad (6.149)$$

It is desirable to truncate the phase correction function to keep the range of negative values of x' over which $F_u(x')$ need be known small. If, however, $\chi^c(x, l, \sigma_M)$ is truncated at, say, $x = \pm X$ such that the oscillations in the function are abruptly terminated, this, in turn, leads to spurious ripples in the final spectrum derived from $F_e^c(x)$. Consequently, we select a value X of x beyond which the oscillations are only a few per cent of the central value $\chi^c(0, l, \sigma_M)$ and then weight $\chi^c(x, l, \sigma_M)$ over this range to make it fall smoothly to zero at $x = \pm X$. We take, therefore,

$$\chi_X^c(x, l, \sigma) = \chi^c(x, l, \sigma_M) W\left(\frac{x}{X}\right) \sqcap\left(\frac{x}{2X}\right) \quad [\text{m}^{-1}] \quad (6.150)$$

as the computed phase correction to be convolved with the unsymmetrical interferogram according to

$$F_e^c(x) = F_u(x) * \chi_X^c(x, l, \sigma_M) \quad (6.151a)$$

$$= \int_{-\infty}^{\infty} F_u(x')\chi_X^c(x - x', l, \sigma_M) \, dx' \quad [\text{W}]. \quad (6.151b)$$

Since $\chi_X^c(x, l, \sigma_M)$ is very narrow, the integration need be taken over only a small range $-X + x \leq x' \leq x + X$ on either side of the value of x being corrected; thus

$$F_e^c(x) \simeq \int_{-X+x}^{X+x} F_u(x')\chi_X^c(x - x', l, \sigma_M) \, dx' \quad [\text{W}]. \quad (6.152)$$

Since we need to evaluate $F_e^c(x)$ only for $x > 0$ (because we compute a single-sided transformation of the corrected interferogram), the full range over which $F_u(x')$ needs to be known is $-X \leqslant x' \leqslant X + D$.

The observation required is, therefore, a record of the unsymmetrical interferogram $F_u(x)$ over the range $-X \leqslant x \leqslant X + D$. From this, the portion $-l \leqslant x \leqslant l$ (with $l < X \ll D$) is selected and the phase correction function is calculated and convolved according to (6.142) with the entire record of $F_u(x)$. The portion $0 \leqslant x \leqslant D$ of the corrected interferogram is then used for the one-sided cosine Fourier transformation. In practice, as the interferograms are not symmetrical, the transform (6.143) often cannot be taken about a well-defined origin because this origin cannot be recognized. The effect of an error in the origin is the addition of a linear term to the computed phase (see section 6.11). The origin should be adjusted, if necessary, to minimize this linear term so that it is zero at some mean wavenumber $\sigma_{\bar{m}}$ near the centre of the spectral range $(\sigma_{\bar{m}} \sim \frac{1}{2}\sigma_M)$.

6.11.3. Application to the case of a displaced sampling comb

We can now apply this correction to the case, already considered in (section 6.11.1) above, of a symmetrical interferogram sampled at equal intervals β that are off centre by δ. In this case, the phase error

$$\phi(\sigma) = -2\pi\sigma\delta \qquad [\text{rad}]$$

is linear and the phase correction factor at $x = r\beta$ is

$$\chi^c(r\beta, \sigma_M) = \int_{-\infty}^{\infty} \exp(-2\pi i\sigma\delta) \sqcap\left(\frac{\sigma}{2\sigma_M}\right) \exp 2\pi i\sigma r\beta \, d\sigma$$

$$= 2\sigma_M \operatorname{sinc}(2r\beta\sigma_M) * \delta(r\beta - \delta)$$

$$= 2\sigma_M \operatorname{sinc}[2\sigma_M(r\beta - \delta)] \qquad [\text{m}^{-1}]. \tag{6.153}$$

We can now correct each of the sampled values $F(r\beta + \delta)$ by convolution with $\chi^c(r\beta, \sigma_M)$; since both functions are known only at discrete values, the convolution integral (6.152) is, in practice, replaced by a finite summation taken over a range $-p\beta$ to $+p\beta$ centred on $r\beta$:

$$F_e^c(r\beta) = \beta \sum_{s=r-p}^{r+p} F(s\beta + \delta) \, \chi^c(r\beta - s\beta, \sigma_M) \tag{6.154a}$$

$$= 2\sigma_M \beta \sum_{s=r-p}^{r+p} F(s\beta + \delta) \operatorname{sinc}[2\sigma_M(r\beta - s\beta - \delta)] \qquad [\text{W}]. \tag{6.154b}$$

The range of p corresponds to X in (6.152), for $p\beta \equiv X$.

Reference to section 2.10 shows that the expression (6.154b) is nothing other than a statement, in truncated form, of the sampling theorem. Since this is an interpolation formula, there is, as Filler[43] has pointed out, a consequential error. This error decreases as δ decreases and as p increases.

For practical considerations, however, p needs to be fairly small, so a compromise value has to be chosen.

6.11.4. Two-sided Fourier transform for phase correction

An alternative way of making allowance for asymmetry in the interferogram has already been suggested by equation (6.138): from which we see that

$$B_e(\sigma) \exp - i\phi(\sigma) = p(\sigma) - iq(\sigma) = \int_{-\infty}^{\infty} F_u(x) \exp(-2\pi i\sigma x) \, dx \qquad [\text{W m}].$$
(6.155)

The required spectrum is therefore given by

$$B_e(\sigma) = \{[p(\sigma)]^2 + [q(\sigma)]^2\}^{1/2} \qquad [\text{W m}]$$
(6.156)

with no need to know $\phi(\sigma)$. Some aspects and properties of the *two-sided transform* are discussed in the next part of this chapter.

Part 2

The Two-sided Interferogram

6.12. THE TWO-SIDED INTERFEROGRAM

The proposals set out in section 6.11 for the correction of phase error in one-sided interferograms require precise knowledge of δ when $F(x)$ is inherently symmetrical, but incorrectly sampled, and application of an elaborate correction procedure when $F(x)$ is itself unsymmetrical. However, as we have just seen, the evaluation of the full Fourier transform of a two-sided interferogram produces a result which, in principle, is independent of the phase error. It is therefore of considerable interest to explore the range of permissible phase error which the technique can cope with and to investigate the modifications to the conclusions on spectral window resolution, etc., which it introduces. A *two-sided interferogram* is one that is, in principle, available for study over the full range $-\infty < x < \infty$, but in practice of course it will only be available over the finite range $-D \leq x \leq D$. By identifying real and imaginary parts in (6.155), one has

$$p(\sigma) = B_e(\sigma) \cos \phi(\sigma) \quad \text{and} \quad q(\sigma) = B_e(\sigma) \sin \phi(\sigma), \qquad (6.157)$$

from which (6.156) follows immediately; but should one require to know

176

$\phi(\sigma)$ itself, then by division one has

$$\phi(\sigma) = \tan^{-1}(q(\sigma)/p(\sigma)). \qquad (6.158)$$

In the general case, the two-sided interferogram is unsymmetrical. This lack of symmetry can be thought of as arising in three ways which, though apparently quite different, are mathematically equivalent. A physical interpretation would be that the observed interferogram is the superimposition of cosinusoidal waves, each corresponding to a particular wavenumber and each having its own particular phase shift. Alternatively, one may regard the interferogram as the sum of symmetrical and antisymmetrical partial interferograms. A third possibility is to consider the unsymmetrical interferogram as the convolution of a symmetrical interferogram with a distorting function. This latter approach begins with equation (6.137), which can alternatively be written as the convolution

$$F_u(x) = F_e(x) * D(x), \qquad (6.159)$$

where

$$D(x) = \int_{-\infty}^{+\infty} \exp\left[-i\phi(\sigma)\right] \exp\left(2\pi i\sigma x\right) d\sigma \qquad [\text{m}^{-1}] \qquad (6.160)$$

is the *distortion function*. In words, the observed unsymmetrical interferogram is the convolution of the ideal symmetrical interferogram with the distortion function arising from defects in the observing system and collectively described by the phase function $\phi(\sigma)$. It is the existence of this convolution which provides the basis for the phase correction method of Forman et al.[40] It follows from their respective definitions (since $\phi(\sigma)$ is odd) that

$$D(\pm x) = \chi(\mp x) \qquad [\text{m}^{-1}]. \qquad (6.161)$$

Convolution of $F_u(x)$ with $\chi(x)$ then gives

$$\begin{aligned} F_u(x) * \chi(x) &= [F_e(x) * D(x)] * \chi(x) \\ &= F_e(x) * [\chi(-x) * \chi(x)] \\ &= F_e(x) \qquad [\text{W}] \end{aligned} \qquad (6.162)$$

since, from 6.141,

$$\chi(x) * \chi(-x) = \delta(x) \qquad [\text{m}^{-1}]. \qquad (6.163)$$

6.13. THE PHYSICAL SPECTRUM AND THE COMPLEX SPECTRUM

The spectrum which is calculated by the two-sided transform is no longer real but is instead complex:

$$\hat{S}(\sigma) = B_e(\sigma) \exp\left(-i\phi(\sigma)\right). \qquad (6.164)$$

It has a modulus $B_e(\sigma)$ and a phase $-\phi(\sigma)$ or, alternatively, real and imaginary parts $p(\sigma)$ and $-iq(\sigma)$. The physical spectrum exists of course only for $\sigma > 0$ and over this half-range it is equal to $2B_e(\sigma)$. One may therefore write

$$B(\sigma) = 2\,|\hat{S}(\sigma)|\,H(\sigma) \qquad (6.165)$$

and note that from this definition $B(\sigma)$ is always positive. This presents little difficulty with practical physical spectra since their ordinates are necessarily positive, but some care is needed in dealing with model interferograms and spectra where negative ordinates can occur. The occurrence of positive-only ordinates is, however, of considerable significance for experimental (that is, 'noisy') spectroscopy, since the average value of the noise signals will no longer be zero. This point will be taken up again later in the book.

Returning now to equation (6.138), one can write

$$\hat{S}(\sigma) = \int_{-\infty}^{+\infty} F_e(x) \cos 2\pi\sigma x \; dx - i \int_{-\infty}^{+\infty} F_o(x) \sin 2\pi\sigma x \; dx \qquad [\text{W m}] \quad (6.166)$$

by resolving $F_u(x)$ into its even and odd parts. It follows at once therefore that

$$p(\sigma) = p(-\sigma) = \int_{-\infty}^{+\infty} F_e(x) \cos 2\pi\sigma x \; dx \qquad [\text{W m}] \qquad (6.167a)$$

is the cosine Fourier transform of the even part of the interferogram and

$$q(\sigma) = -q(-\sigma) = \int_{-\infty}^{+\infty} F_o(x) \sin 2\pi\sigma x \; dx \qquad [\text{W m}] \qquad (6.167b)$$

is the sine Fourier transform of the odd part. From these two relations it follows that $\hat{S}(\sigma)$ is Hermitian or complex symmetric, that is,

$$\hat{S}(\sigma) = \hat{S}^*(-\sigma) \qquad [\text{W m}]. \qquad (6.168)$$

Therefore from the properties of Hermitian functions it follows that the Fourier transform of $\hat{S}(\sigma)$ is necessarily real and, hence,

$$F_u(x) = \int_{-\infty}^{+\infty} \hat{S}(\sigma) \exp 2\pi i\sigma x \; d\sigma \qquad [\text{W}]. \qquad (6.169)$$

This relation and (6.138) form the basic Fourier transform pair when a two-sided interferogram is employed. It is not necessary, of course, for $F_u(x)$ to be actually unsymmetrical for the method to be used. If a symmetrical interferogram is observed, then $\phi(\sigma) = 0$ and

$$p(\sigma) = B_e(\sigma) \quad \text{and} \quad q(\sigma) = 0. \qquad (6.170)$$

However, the recording of an exactly symmetrical interferogram will be a most unusual occurrence and in general one will have an unsymmetrical interferogram $F_u(x)$ and a complex spectrum $\hat{S}(\sigma)$.

6.14. THE APPARATUS FUNCTION

If the inevitable practical truncation of the interferogram is taken into account, the complex spectrum is not calculated according to the exact relation (6.138) but is given by

$$\hat{S}^c(\sigma) = \int_{-D}^{+D} F_u(x) \exp\left(-2\pi i \sigma x\right) dx \qquad [\text{W m}]. \qquad (6.171a)$$

We call the interval $(-D) \leftrightarrow (+D) = 2D$ the *span* of the interferogram and the centre point of the span the *origin* of the span. Ideally this origin coincides with $x = 0$, the origin of path difference. In practice there will usually be a small separation of the two and the effects this has will be treated more fully later. For the moment we will ignore them and, writing (6.171a) in the form

$$\hat{S}^c(\sigma) = \int_{-\infty}^{+\infty} F_u(x) \sqcap(x/2D) \exp\left(-2\pi i \sigma x\right) dx \qquad [\text{W m}], \quad (6.171b)$$

the convolution theorem gives

$$\hat{S}^c(\sigma) = \hat{S}(\sigma) * A_0(\sigma D) \qquad [\text{W m}]. \qquad (6.172)$$

Now, since $\hat{S}(\sigma)$ is complex, we see that the real and imaginary parts of it, $p(\sigma)$ and $-iq(\sigma)$, are each separately convolved with the unapodized apparatus function to give

$$p^c(\sigma) = p(\sigma) * A_0(\sigma D) \qquad (6.173a)$$

and

$$q^c(\sigma) = q(\sigma) * A_0(\sigma D). \qquad (6.173b)$$

The modulus spectrum is therefore

$$B_e^c(\sigma) = \{[p^c(\sigma)]^2 + [q^c(\sigma)]^2\}^{1/2}, \qquad (6.174)$$

but, because of the non-linear combination of the two transforms, the apparatus function corresponding to the modulus power spectrum is, in general, not obvious.

To investigate the matter in more depth, one considers, as usual, the case of strictly monochromatic radiation of wavenumber σ_0 studied with an interferometer having residual asymmetry described by $\phi(\sigma_0)$. The interferogram

$$F_u(x) = 2\tilde{b}_e(\sigma_0) \cos\left[2\pi\sigma_0 x - \phi(\sigma_0)\right] \qquad [\text{W}] \qquad (6.175)$$

is then the sum of even and odd parts:

$$F_e(x) = 2\tilde{b}_e(\sigma_0) \cos\phi(\sigma_0) \cos 2\pi\sigma_0 x \qquad [\text{W}] \qquad (6.176a)$$

and

$$F_o(x) = 2\tilde{b}_e(\sigma_0) \sin\phi(\sigma_0) \sin 2\pi\sigma_0 x \qquad [\text{W}]. \qquad (6.176b)$$

We therefore have

$$p^c(\sigma) = 2D\bar{6}_e(\sigma_0) \cos \phi(\sigma_0)[\text{sinc } 2(\sigma + \sigma_0)D + \text{sinc } 2(\sigma - \sigma_0)D] \quad \text{[W m]}$$

(6.177a)

and

$$q^c(\sigma) = 2D\bar{6}_e(\sigma_0) \sin \phi(\sigma_0)[\text{sinc } 2(\sigma + \sigma_0)D - \text{sinc } 2(\sigma - \sigma_0)D] \quad \text{[W m]}$$

(6.177b)

The two sinc functions rapidly fall to zero when their arguments are more than a few times $(2D)^{-1}$ from the values corresponding to the peaks. If σ_0 is considerably larger than zero, therefore, it follows that on squaring and adding (6.177a) and (6.177b) to give the modulus, the cross terms will be everywhere small in addition to being of opposite sign and we may write

$$|\hat{S}^c(\sigma)| = 2D\bar{6}_e(\sigma_0)[\text{sinc}^2 2(\sigma + \sigma_0)D + \text{sinc}^2 2(\sigma - \sigma_0)D]^{1/2} \quad \text{[W m]}.$$

(6.178)

Continuing the same arguments, it follows that the first term in the modulus will be very small where the second is large, and vice versa, and we may therefore take the square root of each term separately to give as a second close approximation:

$$B_e^c(\sigma) = |\hat{S}^c(\sigma)| \simeq 2D\bar{6}_e(\sigma_0)[|\text{sinc } 2(\sigma + \sigma_0)D| + |\text{sinc } 2(\sigma - \sigma_0)D|]. \quad (6.179)$$

This result is applicable at all values of σ except $\sigma \simeq 0$. Near the origin the approximations are weakest, but if $\sigma_0 \gg 0$ then (6.179) will hold, for all practical purposes, everywhere. One is therefore led to consider the function

$$B_e^c(\sigma)/\bar{6}_e(\sigma_0) = 2D |\text{sinc } 2(\sigma - \sigma_0)D| \quad (6.180)$$

for the normalized spectrum, and the quantity

$$M_0(\sigma D) \triangleq 2D |\text{sinc } 2\sigma D| \quad (6.181)$$

for the unapodized apparatus function appropriate to the modulus spectrum. This apparatus function (6.181), shown in Figure 6.18, is numerically equal to the modulus of the normal unapodized apparatus function $|A_0(\sigma D)|$. It therefore has the same 'width' as $A_0(\sigma D)$ and the resolution is the same—however, twice as much interferogram is needed to achieve this compared to the one-sided case.

6.15. THE EFFECT OF WEIGHTING TWO-SIDED INTERFEROGRAMS

Just as in the one-sided case, one may seek to reduce the troublesome side lobes of the apparatus function by weighting the interferogram. The

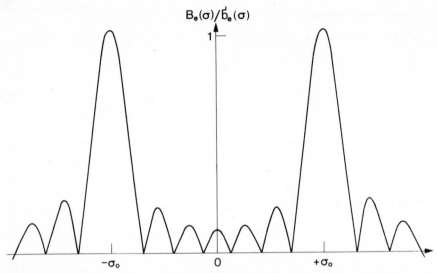

$B_e(\sigma)/b_e'(\sigma)$

Figure 6.18. Spectral window resulting from the observation of a monochromatic line over a finite range of mirror travel. This is the two-sided analogue of the spectral window shown in Figure 6.2

calculated spectrum is then given by

$$\hat{S}^c(\sigma) = \int_{-\infty}^{+\infty} F_u(x)\,\sqcap(x/2D)\,\Omega\!\left(\frac{x}{D}\right)\exp\left(-2\pi i\sigma x\right)\mathrm{d}x \qquad [\text{W m}]$$

$$= \hat{S}(\sigma) * A_0(\sigma D) * \Omega(\sigma D)$$

$$= \hat{S}(\sigma) * A(\sigma D) \tag{6.182}$$

or

$$\hat{S}^c(\sigma) = p^c(\sigma) - iq^c(\sigma), \tag{6.183}$$

where

$$p^c(\sigma) = p(\sigma) * A(\sigma D)$$

and

$$q^c(\sigma) = q(\sigma) * A(\sigma D).$$

As in the one-sided case, $A(\sigma D)$ is the apodized apparatus function, but just as previously, $|A(\sigma D)|$ is not strictly the apodized spectral window of the power spectrum because of the non-linear nature of the modulus-taking operation. However, with the same assumptions as used previously, it turns out to be an excellent approximation in practice and one may write

$$M(\sigma D) = |A(\sigma D)| \tag{6.184}$$

with confidence. We conclude, therefore, that in general the apparatus function appropriate to the modulus spectrum obtained from a full two-sided transform is identical with the modulus of that obtained from the comparable one-sided cosine transformation.

6.16. EFFECTS OF OFFSET IN THE CENTRE OF THE SPAN

We now return to the problem of using an offset span in which the centre of the span and the zero of geometrical path difference do not coincide. The conclusions will be general, so we can simplify the treatment by considering an inherently symmetrical interferogram. The spectrum we would calculate, in the ideal case, that is, with the span centred on $x = 0$, is

$$B_e^c(\sigma) = \int_{-\infty}^{+\infty} F_e(x) \, \Pi(x/2D) \exp(-2\pi i \sigma x) \, dx \qquad \text{[W m]}, \qquad (6.185)$$

which is real and even. If there is a displacement Δ in the location of the centre of the span, the calculated spectrum is not even but complex. Assuming that the *only* error is a misidentification of the origin, (6.185) becomes

$$\hat{S}^c(\sigma) = \int_{-D}^{+D} F_e(x' + \Delta) \exp[-2\pi i \sigma (x' + \Delta)] \, dx' \qquad \text{[W m]}, \qquad (6.186)$$

which by simple change of variable becomes

$$\hat{S}^c(\sigma) = \int_{-D+\Delta}^{D+\Delta} F_e(x) \exp[-2\pi i \sigma x] \, dx,$$

i.e.

$$\hat{S}^c(\sigma) = \int_{-\infty}^{+\infty} F_e(x) \, \Pi\!\left(\frac{x - \Delta}{2D}\right) \exp[-2\pi i \sigma x] \, dx$$

$$= \int_{-\infty}^{+\infty} F_e(x) \left[\Pi\!\left(\frac{x}{2D}\right) * \delta(x - \Delta)\right] \exp[-2\pi i \sigma x] \, dx \qquad \text{[W m]}. \qquad (6.187)$$

Application of the convolution theorem to (6.187) gives

$$\hat{S}^c(\sigma) = B_e(\sigma) * [A_0(\sigma D) \exp(-2\pi i \sigma \Delta)] \qquad \text{[W m]}, \qquad (6.188)$$

which shows that the apparatus function is modified by the inclusion of a periodic factor $\exp(-2\pi i \sigma \Delta)$. Splitting (6.188) into its real and imaginary components gives

$$p^c(\sigma) = p(\sigma) * [A_0(\sigma D) \cos 2\pi \sigma \Delta] \qquad \text{[W m]} \qquad (6.189a)$$

and

$$q^c(\sigma) = q(\sigma) * [A_0(\sigma D) \sin 2\pi \sigma \Delta] \qquad \text{[W m]}. \qquad (6.189b)$$

These relations show that there are periodic terms introduced into the apparatus functions and that these are $\pi/2$ out of phase in the cosine and sine transforms. If the offset Δ is very much less than D, $\cos 2\pi \sigma \Delta$ and $\sin 2\pi \sigma \Delta$ are much more slowly varying than is $A_0(\sigma D)$ and may, to a good approximation, be regarded as purely multiplicative factors in (6.189).

Under this approximation, the cosine and sine Fourier transforms are overlaid with periodic structure of period Δ^{-1}, but this vanishes in the formation of the modulus. However, if the period Δ^{-1} is made smaller, the periodic terms in (6.189) cannot be considered apart from the convolutions. Consequently, these do not cancel in the modulus, which is given by

$$|\hat{S}^c(\sigma)|^2 = \{p(\sigma) * [A_0(\sigma D) \cos 2\pi\sigma\Delta]\}^2$$
$$+\{q(\sigma) * [A_0(\sigma D) \sin 2\pi\sigma\Delta]\}^2 \quad [W^2\, m^2]. \quad (6.190)$$

Whilst the terms in braces in (6.190) show perturbations which are roughly periodic even when Δ/D is not small, there is no well-defined phase relation between these perturbations, with the result that the modulus, in (6.190), shows fluctuations which increasingly have the appearance of 'noise' as Δ/D increases. The analysis can be carried through analogously for the case where a weighting function is included, leading to the expression

$$\hat{S}^c(\sigma) = B_e(\sigma) * [A(\sigma D) \exp(-2\pi i \sigma\Delta)] \quad (6.191)$$

with analogues comparable to (6.189a) and (6.189b).

6.16.1. The Hilbert transform—symmetrical interferogram

An alternative approach enables us to develop (6.187) in terms of an even spectrum closely related to that required (6.185) and a complex spectrum which we can regard as an 'error' arising from the existence of the offset Δ. It will be seen that the presence of offset inevitably leads to perturbations in the modulus spectrum that involve Hilbert transforms.

We can divide the integral (6.187) into two components

$$\hat{S}^c(\sigma) = \int_{-(D+\Delta)}^{+(D+\Delta)} F_e(x) \exp(-2\pi i \sigma x)\, dx - \int_{-D-\Delta}^{-D+\Delta} F_e(x) \exp(-2\pi i \sigma x)\, dx. \quad (6.192)$$

The first is calculated over a symmetrical range, it is therefore real and very similar to the ideal spectrum apart from having a slightly better resolution, $[2(D+\Delta)]^{-1}$ instead of $[2D]^{-1}$. The second term is complex and may be written as the difference

$$\int_{-D-\Delta}^{-D+\Delta} F_e(x) \exp(-2\pi i \sigma x)\, dx = \int_{-D_+}^{0} F_e(x) \exp(-2\pi i \sigma x)\, dx$$

$$- \int_{-D_-}^{0} F_e(x) \exp(-2\pi i \sigma x)\, dx$$

$$= \int_{-\infty}^{+\infty} F_e(x) \sqcap\left(\frac{x}{2D_+}\right) H(-x) \exp(-2\pi i \sigma x)\, dx$$

$$- \int_{-\infty}^{+\infty} F_e(x) \sqcap\left(\frac{x}{2D}\right) H(-x) \exp(-2\pi i \sigma x)\, dx, \quad (6.193)$$

where $H(x)$ is the Heaviside step function (see section 2.9.5) and

$$D_+ \triangleq D + \Delta \quad \text{and} \quad D_- \triangleq D - \Delta.$$

Using the convolution theorem, this difference becomes

$$B_e(\sigma) * A_0(\sigma D_+) * \left[\tfrac{1}{2}\delta(\sigma) + \frac{i}{2\pi\sigma}\right] - B_e(\sigma) * A_0(\sigma D_-) * \left[\tfrac{1}{2}\delta(\sigma) + \frac{i}{2\pi\sigma}\right] \tag{6.194}$$

by virtue of (2.102) after the appropriate change of sign. Now, using the properties of the Hilbert transform (2.9.7), it follows that

$$A_0(\sigma D_\pm) * i/2\pi\sigma = -\tfrac{1}{2}i\mathcal{H}\{A_0(\sigma D_\pm)\}, \tag{6.195}$$

so that the complex integral (6.193) becomes

$$\tfrac{1}{2}B_e(\sigma) * [A_0(\sigma D_+) - A_0(\sigma D_-)] - \tfrac{1}{2}iB_e(\sigma) * \mathcal{H}\{A_0(\sigma D_+) - A_0(\sigma D_-)\}. \tag{6.196a}$$

Substitution of this into (6.192) leads to

$$\hat{S}^c(\sigma) = \tfrac{1}{2}B_e(\sigma) * [A_0(\sigma D_+) + A_0(\sigma D_-)] + \tfrac{1}{2}iB_e(\sigma) * \mathcal{H}\{A_0(\sigma D_+) - A_0(\sigma D_-)\}, \tag{6.196b}$$

which can also be written as

$$\begin{aligned}
\hat{S}^c(\sigma) &= B_e(\sigma) * \{[2D \text{ sinc }(2\sigma D) \cos 2\pi\sigma\Delta] \\
&\quad - i\mathcal{H}\{2\Delta \text{ sinc }(2\sigma\Delta) \cos 2\pi\sigma D\}\} \tag{6.197a} \\
&= B_e(\sigma) * \{[2D \text{ sinc }(2\sigma D) \cos 2\pi\sigma\Delta] \\
&\quad - i[2D \text{ sinc }(2\sigma D) \sin 2\pi\sigma\Delta]\} \tag{6.197b} \\
&= B_e(\sigma) * [2D \text{ sinc }(2\sigma D) \exp(-2\pi i\sigma\Delta)], \tag{6.197c}
\end{aligned}$$

(6.197b) follows because

$$\begin{aligned}
\mathcal{H}\{2\Delta \text{ sinc }(2\sigma\Delta) \cos 2\pi\sigma D\} &= \mathcal{H}\{D_+ \text{ sinc }(2\sigma D_+) - D_- \text{ sinc }(2\sigma D_-)\} \\
&= 2D \text{ sinc }(2\sigma D) \sin 2\pi\sigma\Delta. \tag{6.198}
\end{aligned}$$

The calculated spectrum, therefore, consists of the required spectrum convolved with an instrument function $2D \text{ sinc }(2\sigma D) \cos 2\pi\sigma\Delta$, to which is added the convolution of the required spectrum with the Hilbert transform of the related function $2\Delta \text{ sinc }(2\sigma\Delta) \cos 2\pi\sigma D$. Evaluation of this transform leads to (6.197c), which is identical with (6.188).

Vanasse and Sakai[44] have described a different approach which also highlights the part played by the Hilbert transform when the span is offset. They define two partial spectra

$$\hat{S}^c_-(\sigma) = \int_{-D_-}^{0} F_e(x) \exp(-2\pi i\sigma x)\, dx \quad \text{[W m]} \tag{6.199a}$$

and

$$\hat{S}^c_+(\sigma) \int_0^{D_+} F_e(x) \exp{(-2\pi i \sigma x)}\, dx \qquad \text{[W m]} \qquad (6.199b)$$

evaluated over the negative and positive half-axes, respectively. These integrals are easily rearranged as

$$\hat{S}^c_-(\sigma) = \int_{-\infty}^{+\infty} F_e(x) \sqcap\!\left(\frac{x}{2D_-}\right) H(-x) \exp{(-2\pi i \sigma x)}\, dx \qquad \text{[W m]}$$
$$(6.200a)$$

and

$$\hat{S}^c_+(\sigma) = \int_{-\infty}^{+\infty} F_e(x) \sqcap\!\left(\frac{x}{2D_+}\right) H(x) \exp{(-2\pi i \sigma x)}\, dx \qquad \text{[W m]}$$
$$(6.200b)$$

to give

$$\hat{S}^c_-(\sigma) = \tfrac{1}{2} B_e(\sigma) * [A_0(\sigma D_-) + i\mathscr{H}\{A_0(\sigma D_-)\}]$$
$$= \tfrac{1}{2}[B_e(\sigma) + i\mathscr{H}\{B_e(\sigma)\}] * A_0(\sigma D_-) \qquad \text{[W m]} \qquad (6.201a)$$

and

$$\hat{S}^c_+(\sigma) = \tfrac{1}{2} B_e(\sigma) * [A_0(\sigma D_+) - i\mathscr{H}\{A_0(\sigma D_+)\}]$$
$$= \tfrac{1}{2}[B_e(\sigma) - i\mathscr{H}\{B_e(\sigma)\}] * A_0(\sigma D_+) \qquad \text{[W m]}. \qquad (6.201b)$$

These results show that it is possible to interpret the spectra $\hat{S}^c_-(\sigma)$ and $\hat{S}^c_+(\sigma)$ in two different ways: either as the convolution of the true spectrum with the sum of two different spectral windows $A_0(\sigma D_\mp)$ and $\pm i\mathscr{H}\{A_0(\sigma D_\mp)\}$, which form a Hilbert transform pair, or else as the convolution of the *symmetrical* spectral window $A_0(\sigma D_\mp)$ with the sum of the true spectrum $B_e(\sigma)$ and its Hilbert transform $\pm i\mathscr{H}\{B_e(\sigma)\}$. Either representation therefore gives the spectrum in terms of the sum of a Hilbert transform pair. This is to be contrasted with (6.196b), where the two terms are not Hilbert transforms of one another.

This treatment can be used to derive a general result which applies even when Δ is identically zero. If we again consider monochromatic irradiation, then the calculated complex spectra are

$$\hat{S}^c_\mp(\sigma) = \tfrac{1}{2}6_e(\sigma_0)[A_0(\sigma D_\mp) \pm i\mathscr{H}\{A_0(\sigma D_\mp)\}] \qquad \text{[W m]}$$
$$(6.202a)$$

$$= 6_e(\sigma_0) D_\mp \left[\operatorname{sinc}(2\sigma D_\mp) + i\!\left(\frac{\cos 2\pi\sigma D_\mp - 1}{2\pi\sigma D_\mp}\right) \right] \qquad \text{[W m]}.$$
$$(6.202b)$$

The modulus spectra are

$$|S^c_\mp(\sigma)| = 6_e(\sigma_0) D_\mp\, |\operatorname{sinc}(\sigma D_\mp)| \qquad \text{[W m]}, \qquad (6.203)$$

which have apparatus functions

$$M_0(\sigma D_{\mp}) = |D_{\mp} \operatorname{sinc}(\sigma D_{\mp})| \qquad [\text{m}]. \qquad (6.204)$$

It will be seen that the effect of the Hilbert transform is to double the width of the spectral window. This means that the width of the spectral window in two-sided interferometric spectroscopy is determined by half of the span, regardless of the location of the centre. This result agrees with the conclusions given earlier, where it was stated that the resolution in a single-sided observation running from $x = 0$ to $x = D$ equals exactly that in a two-sided observation running from $x = -D$ to $x = +D$.

6.16.2. The Hilbert transform—unsymmetrical interferogram

When the interferogram is inherently unsymmetrical, the quantity calculated,

$$\hat{S}^c(\sigma) = \int_{D_-}^{D_+} F_u(x) \exp(-2\pi i\sigma x)\, dx \qquad [\text{W m}], \qquad (6.205)$$

can, as before, be resolved into two components

$$\begin{aligned}
\hat{S}^c_-(\sigma) &= \int_{-\infty}^{+\infty} F_u(x) \sqcap\!\left(\frac{x}{2D_-}\right) H(-x) \exp(-2\pi i\sigma x)\, dx \\
&= \tfrac{1}{2}[p(\sigma) - iq(\sigma)] * [A_0(\sigma D_-) + i\mathcal{H}\{A_0(\sigma D_-)\}] \qquad (6.206a)
\end{aligned}$$

and, by similar reasoning,

$$\hat{S}^c_+(\sigma) = \tfrac{1}{2}[p(\sigma) - iq(\sigma)] * [A_0(\sigma D_+) - i\mathcal{H}\{A_0(\sigma D_+)\}]. \qquad (6.206b)$$

If the interferogram origin coincides with the centre of the span, this reduces, as expected, to a total spectrum

$$\hat{S}^c(\sigma) = [p(\sigma) - iq(\sigma)] * A_0(\sigma D), \qquad (6.207)$$

but when there is a finite offset, a very complicated expression

$$\begin{aligned}
\hat{S}^c(\sigma) = &\{p(\sigma) * [A_0(\sigma D)\cos 2\pi\sigma\Delta] - q(\sigma) * [A_0(\sigma D)\sin 2\pi\sigma\Delta]\} \\
&- i\{p(\sigma) * [A_0(\sigma D)\cos 2\pi\sigma\Delta] + q(\sigma) * [A_0(\sigma D)\sin 2\pi\sigma\Delta]\}
\end{aligned}$$
$$(6.208)$$

results. However, provided that the usual condition $D \gg \Delta$ applies, the modulus spectrum will closely approximate the required ideal spectrum.

6.17. EFFECTS OF AN ERROR IN THE ORIGIN OF THE INTERFEROGRAM

Equation (6.208) describes the form of the computed spectrum when there is an offset Δ between the origin $x = 0$ of geometrical path-difference and the centre of the span. If the interferogram is symmetrical, the origin of path difference coincides with the place where all the cosinusoidal interference fringes are in phase. At this place the interference is constructive for all wavenumbers and the power shows a pronounced grand maximum. If the interferogram is unsymmetrical, the grand maximum does not necessarily coincide with the zero of geometrical path difference. The grand maximum is, however, a place of *stationary phase* and, since this is an easily recognizable place, we shall refer to the location of the grand maximum of an interferogram as the *centre* of the interferogram.

If the interferogram is rather unsymmetrical, making recognition of stationary phase difficult, or if there is simply an error in the labelling of the interferogram abscissae, the interferogram recorded is not $F_u(x)$, related to the true origin $x = 0$, but $F_u(x' + \delta)$, related to the incorrect origin at $x = \delta$. The spectrum we calculate is then

$$\hat{S}'^c(\sigma) = \int_{-\infty}^{\infty} F_u(x' + \delta) W\left(\frac{x'}{D}\right) \sqcap \left(\frac{x'}{2D}\right) \exp\left(-2\pi i \sigma x'\right) dx'$$

$$= \int_{-\infty}^{\infty} [F_u(x') * \delta(x' + \delta)] W\left(\frac{x'}{D}\right) \sqcap \left(\frac{x'}{2D}\right) \exp\left(-2\pi i \sigma x'\right) dx', \quad [\text{W m}]$$

(6.209)

which can be transformed to

$$\hat{S}'^c(\sigma) = [p(\sigma) \cos 2\pi\sigma\delta + q(\sigma) \sin 2\pi\sigma\delta] * A(\sigma D)$$
$$- i[q(\sigma) \cos 2\pi\sigma\delta - p(\sigma) \sin 2\pi\sigma\delta] * A(\sigma D) \quad [\text{W m}].$$

The similarities and differences between this expression and (6.208) should be carefully noted, for here it is the spectral terms $p(\sigma)$ and $q(\sigma)$ that are multiplied by the periodic factors.

From the cosine Fourier transform

$$p'^c(\sigma) = [p(\sigma) \cos 2\pi\sigma\delta + q(\sigma) \sin 2\pi\sigma\delta] * A(\sigma D) \quad [\text{W m}] \quad (6.211)$$

and the sine Fourier transform

$$q'^c(\sigma) = [q(\sigma) \cos 2\pi\sigma\delta - p(\sigma) \sin 2\pi\sigma\delta] * A(\sigma D) \quad [\text{W m}], \quad (6.212)$$

we find the modulus spectrum

$$|B_e'^c(\sigma)|^2 = \{[p(\sigma) \cos 2\pi\sigma\delta] * A(\sigma D)\}^2 + \{[q(\sigma) \cos 2\pi\sigma\delta] * A(\delta D)\}^2$$
$$+ \{[p(\sigma) \sin 2\pi\sigma\delta] * A(\sigma D)\}^2 + \{[q(\sigma) \sin 2\pi\sigma\delta] * A(\sigma D)\}^2$$
$$+ 2\{[p(\sigma) \cos 2\pi\sigma\delta] * A(\sigma D)\}\{[q(\sigma) \sin 2\pi\sigma\delta] * A(\sigma D)\}$$
$$- 2\{[q(\sigma) \cos 2\pi\sigma\delta] * A(\sigma D)\}$$
$$\times \{[p(\sigma) \sin 2\pi\sigma\delta] * A(\sigma D)\} \quad [\text{W}^2 \text{ m}^2],$$

(6.213)

which reduces to

$$[B_e'^c(\sigma)]^2 \simeq [p(\sigma) * A(\sigma D)]^2 + [q(\sigma) * A(\sigma D)]^2 \quad [\text{W}^2 \text{ m}^2] \quad (6.214)$$

if δ is small so that $\cos 2\pi\sigma\delta$ and $\sin 2\pi\sigma\delta$ are slowly varying.

6.18. EFFECT OF SAMPLING THE INTERFEROGRAM

As in the one-sided case, the interferogram is generally sampled at finite intervals β for the purpose of digital computation. The computed spectrum is then

$$\hat{S}^c(\sigma) = \int_{-\infty}^{\infty} F_u(x) \sqcap\left(\frac{x}{2D}\right) \sqcup\left(\frac{x}{\beta}\right) W\left(\frac{x}{D}\right) \exp\left(-2\pi i\sigma x\right) dx \tag{6.215}$$

$$= \beta\hat{S}(\sigma) * A(\sigma N\beta) * \sqcup(\sigma\beta) = \hat{S}(\sigma) * \wedge(\sigma N\beta) \qquad [\text{W m}]. \tag{6.216}$$

We find, just as in the one-sided case, that the computed spectrum is replicated at intervals $1/\beta$. However, a point of distinction to be noted is that it is the *complex spectrum* that is replicated, not merely the required power spectrum. However, if $|\sigma| < \sigma_\beta$ as formerly, $\hat{S}^c(\sigma)$ is given by $\hat{S}(\sigma) * A(\sigma D)$ and there is no overlap of the aliases, thus

$$\hat{S}^c(\sigma) = S(\sigma) * A(\sigma N\beta), \qquad |\sigma| < \sigma_\beta \qquad [\text{W m}]. \tag{6.217}$$

We can express (6.215) in the alternative form

$$\hat{S}^c(\sigma) = \beta \sum_{r=-N}^{N} F_u(r\beta)[1 - \tfrac{1}{2}\delta(r \pm N)]W\left(\frac{r}{N}\right) \exp\left(-2\pi i\sigma r\beta\right) \qquad [\text{W m}]. \tag{6.218}$$

In practice, the weighting function is generally zero at $x = \pm N\beta$, so the half-weight terms in (6.218) are zero and we can write

$$\hat{S}^c(\sigma) = \beta \sum_{r=-N}^{N} F_u(r\beta)W\left(\frac{r}{N}\right) \exp\left(-2\pi i\sigma r\beta\right) \qquad [\text{W m}], \tag{6.219}$$

which is the analogue of (6.70).

The practical results of this, and of much that has gone before, can be illustrated by a simple numerical example and, for purposes of comparison with section 6.11.1, we consider the reconstruction of the triangle function by the Fourier transformation of the sinc^2 function. We have already calculated the Fourier transforms (cosine and sine) of the positive ordinates of the offset ($\delta = 0.2$) sinc^2 function in Table 6.1, so we may, remembering that $a_7 = a_{-1}$, $a_6 = a_{-2}$, $a_5 = a_{-3}$, immediately calculate the modulus function $\sqrt{p^2 + q^2}$. The results for $\wedge_m$ are 0.9478, 0.7572, 0.4970, 0.2511, 0.0316 etc. It will be seen that these agree excellently with the results calculated either from the $\delta = 0$ ordinates or from the $\delta = 0.2$ ordinates with the Gibbs phase correction. The two-sided transform therefore not only gives the positions of bands correctly but gives absolute power levels correctly as well. Absorption spectra calculated by the two-sided method will therefore only be limited in accuracy by the limitations imposed by the finite resolution. Against this, however, must be set the considerations that, when compared with single-sided methods, twice as much interferogram needs to be observed. The calculation time is quadrupled because, in addition, two Fourier transforms, rather than one, need to be evaluated.

The advent of FFT methods and the continued advance in computer power make the latter point less of a consideration nowadays, but, when one is working near the limit of resolution, the fact that twice as high a resolution is possible in single-sided work often militates against the two-sided approach. This, coupled with the availability of sophisticated phase-correction procedures of the Forman, Steele, and Vanasse type, makes single-sided interferometry the preferred approach for work at the limits of experimental technique. However, there is still a major role for two-sided interometry in routine laboratory work where accurate intensity results are required.

CHAPTER 7

The determination of phase spectra from two-sided interferograms

7.1. THE PHASE OF THE COMPLEX SPECTRUM

The phase ph $\hat{S}(\sigma)$, being a measure of the asymmetry of the interferogram, has been discussed so far only in connection with the method of Forman *et al.*[40] for the correction of phase error (section 6.11.2). It will be recalled that the phase $\phi(\sigma)$ of $\hat{S}(\sigma)$ is given by

$$\tan \phi(\sigma) = q(\sigma)/p(\sigma), \qquad (7.1)$$

but unfortunately inversion of (7.1) gives only the principal value of the arctan of $q(\sigma)/p(\sigma)$ which lies in the range $-\pi/2$ to $\pi/2$, so that the phase we deduce is uncertain by an arbitrary integral number of π radians, i.e.

$$\phi(\sigma) = \arctan\left[q(\sigma)/p(\sigma)\right] + l\pi \qquad \text{[rad]}, \qquad (7.2)$$

where l is an integer. The situation can be improved, so that the phase is defined over the interval $-\pi$ to $+\pi$, by noting that $p(\sigma)$ is positive in the first and fourth quadrants whilst $-q(\sigma)$ is positive in the third and fourth quadrants. We can therefore write

$$\text{ph } \hat{S}(\sigma) = -\arctan\left[\frac{q(\sigma)}{p(\sigma)}\right] - \frac{1}{2}\left[1 - \frac{p(\sigma)}{|p(\sigma)|}\right]\frac{q(\sigma)}{|q(\sigma)|}\,\pi \qquad \text{[rad]}. \qquad (7.3)$$

In normal interferometric spectroscopy, it is extremely unlikely that the background phase would ever exceed $\pi/2$ radians; however, as we shall see, large phase differences are often introduced deliberately and under these conditions equation (7.3) must be modified to include an integral number of 2π radians and we write

$$-\phi(\sigma) = \text{ph } \hat{S}(\sigma) + 2m\pi \qquad \text{[rad]}. \qquad (7.4)$$

It should be noted that the phase is always defined everywhere except where $p(\sigma)$ is zero. In normal spectroscopy, where $q(\sigma)$ would either be zero or very small, this means that the phase could become indeterminate in regions where the power spectrum becomes zero. It must also be borne in mind that the transforms $q(\sigma)$ and $p(\sigma)$ are derived from a truncated and weighted interferogram so the quantities appearing in equations (7.1) to

(7.4) should bear a superscript c, e.g. $\phi^c(\sigma)$ to denote the limitations imposed in their calculation.

7.2. INTERFEROMETER WITH DISPERSIVE PATH DIFFERENCE

In normal interferometric spectroscopy, the absorbing specimen† (which is also usually dispersive) is placed in front of the detector, or, and more rarely, between the source and the beam divider: that is, it is in one of the two passive arms of the interferometer where the two partial beams are not separated. If, however, it is placed in one of the active arms, one partial beam is traversing a dispersive medium and the other is not, so the symmetry of the interferometer is disturbed. The difference between the phase shifts of the partial beams will then show significant variations and the interferogram will be markedly asymmetric. A similar conclusion is drawn if one of the partial beams is reflected from an imperfect reflector (for example a dispersive dielectric) whilst the other is reflected from a perfect mirror. It is the main point of this chapter to show how the variation of the complex refractive index of the specimen with frequency may be calculated from the Fourier transform of such an asymmetric interferogram.

In the ideal non-dispersive case, there are no residual phase errors and the zero path difference position corresponds to a position of stationary (and in fact zero) phase. This is because all the cosinusoidal waves which add to give the observed interferogram are exactly in phase at the zero path difference position, and this position is therefore called the *centre* of the interferogram. With monochromatic irradiation, the interference function is a simple cosine function in which all the bright fringes have identical power, all the dark fringes have zero power, and the visibility is constant (see section 1.3.1). If, however, a plane-parallel, absorbing, but non-opaque, specimen is placed normal to the radiation in (say) the fixed mirror arm, these conclusions are modified. The increment of path difference in that arm due to the specimen may be compensated by a displacement of the movable mirror, but the fringes are uniformly reduced in visibility. The optical thickness of a specimen of physical thickness d is $n_0 d$, where $n_0 = n(\sigma_0)$ is precisely and uniquely the phase refractive index at wavenumber σ_0. The increase in optical path length over the vacuum case is therefore $\Delta_0 = 2(n_0 - 1)d$. However (see Figure 7.1) this displacement is equal to the fringe shift x_0 and therefore

$$x_0 = \Delta_0 = 2(n_0 - 1)d. \tag{7.5}$$

If the monochromatic source is replaced by a quasi-monochromatic source of bandwidth, $\sigma_0 - \Delta\sigma$ to $\sigma_0 + \Delta\sigma$, small compared with the central

† Assumed to be an isotropic plane-parallel solid plate; with liquid or gaseous specimens suitably confined there are certain differences of detail but the general conclusions are unaffected.

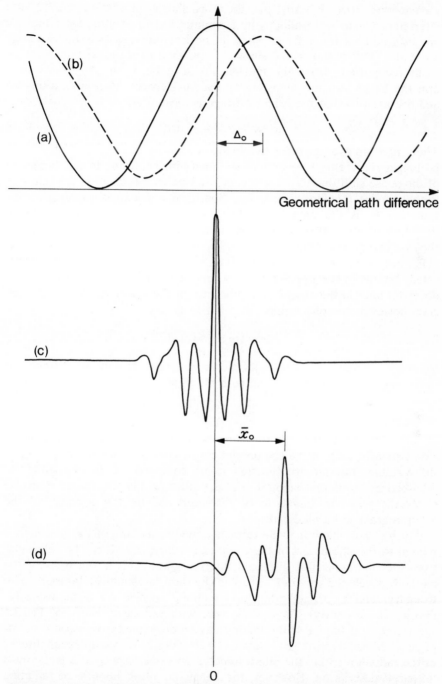

Figure 7.1. Non-dispersive interferograms for (a) monochromatic and (c) broadband irradiation. The resulting dispersive equivalents produced when the specimen is placed in the fixed mirror arm are shown in (b) and (d), respectively

wavenumber (that is, $\Delta\sigma \ll \sigma_0$) then, as shown in section 4.2.1.2, the interference function produced by a symmetrical interferometer exhibits a unique grand maximum fringe when the optical path lengths of the two arms are equal. This bright fringe allows the position of stationary phase to be easily recognized. When the specimen is introduced into the fixed mirror arm the fringe structure depicted by the interference function is displaced and distorted so that the brighest fringe is located at

$$\bar{x}_0 = 2(\bar{n}_0 - 1)d \qquad [\text{m}], \qquad (7.6)$$

where $\bar{n}_0$ is an average refractive index defined by this relation. This fringe no longer represents a position of stationary phase, owing to the dispersion of the phase shift due to the specimen, and we call its location the *centre* of the *dispersive interference function*. Specifically, for quasi-monochromatic radiation $\bar{n}_0$ is the group refractive index provided that the dispersion $dn(\sigma)/d\sigma$ is constant over the interval $\sigma_0 - \Delta\sigma$ to $\sigma_0 + \Delta\sigma$; this is generally the case in a narrow region of normal (as opposed to anomalous) dispersion.

For polychromatic (or broad-band) radiation the situation is more complicated. Just as in the previous case, a unique bright ('white light') fringe is observed both in the absence and presence of the specimen. In this general case, however, the phase shift

$$2\pi\sigma\Delta(\sigma) \qquad [\text{rad}]$$

following introduction of the specimen is different for each wavenumber σ in the spectral band because the path difference

$$\Delta(\sigma) = 2[n(\sigma) - 1]d \qquad [\text{m}] \qquad (7.7)$$

is dependent on σ through the dependence of the refractive index $n(\sigma)$. Consequently, each of the component cosinusoids of which we may imagine the resultant interference function to be composed is displaced and diminished by a different amount. The net effect on the interference function is for the grand maximum to be displaced and for the structure of the interferogram to be modified.

We shall now show that the refractive index spectrum of the specimen is related to the phase part of the complex spectrum which can be calculated from the unsymmetrical displaced interferogram. This is the basis of the method developed by Bell[45] (called by him asymmetric Fourier spectrometry) and by Chamberlain and Gebbie[46] (called by them dispersive Fourier spectrometry). As was the case with the basic theory of Fourier spectrometry (Chapter 4) the theory may be developed from various points of view. We shall derive the relations by considering the complex amplitudes of the radiation within the interferometer. In some respects this is the most 'physical' description. However, the behaviour of an impulse of radiation launched into the interferometer is revealing, particularly from the point of view of causality.

7.3. RELATION BETWEEN REFRACTION SPECTRUM AND UNSYMMETRICAL INTERFEROGRAM

7.3.1. Elementary wave theory

As before (section 4.2), let the disturbance incident at the beam divider of the Michelson interferometer, in which the moving mirror arm (PP_1) is longer than the fixed mirror arm (PP_2), be

$$\hat{E}(t) = \int_0^\infty \hat{a}(\sigma) \exp(-2\pi i\sigma ct)\, d\sigma \qquad [\text{V m}^{-1}] \qquad (7.8)$$

and let a plane parallel specimen of thickness d and complex refractive index

$$\hat{n}(\sigma) = n(\sigma) - i\frac{\alpha(\sigma)}{4\pi\sigma} \qquad (7.9)$$

be placed normal to the radiation in the fixed mirror arm.

The radiation in the fixed mirror arm traverses the specimen twice. For each passage there is an attenuation $\exp[-\tfrac{1}{2}\alpha(\sigma)d]$ and a phase shift $2\pi\sigma[n(\sigma)-1]d$ due to the path difference $[n(\sigma)-1]d$ introduced when a vacuum path of length d is replaced by an equal thickness of specimen; moreover, there are reflection losses and phase shifts at the surfaces bounding the specimen, the effects of which may be collectively described by $\hat{t}_R(\sigma) = t_R(\sigma) \exp[-i\phi_R(\sigma)]$. For each passage of the specimen, the complex amplitude of the radiation is modified by the factor[47]

$$\hat{\mathcal{L}}(\sigma) = t_R(\sigma) \exp[-\tfrac{1}{2}\alpha(\sigma)d] \exp\{-2\pi i\sigma[n(\sigma)-1]d\} \exp[-i\phi_R(\sigma)], \quad (7.10)$$

which is the complex insertion loss† $\hat{\Gamma}_{12}(\sigma)/\hat{\Gamma}_1(\sigma)$. Thus, for double passage, the complex amplitude $\hat{a}(\sigma)$ is modified by the factor

$$[\hat{\mathcal{L}}(\sigma)]^2 = \tau_R(\sigma)\tau_A(\sigma) \exp[-2\pi i\sigma\Delta(\sigma) - 2i\phi_R(\sigma)], \qquad (7.11)$$

where

$$\tau_R(\sigma) = [t_R(\sigma)]^2, \qquad (7.12)$$

$$\tau_A(\sigma) = \exp[-\alpha(\sigma)d], \qquad (7.13)$$

and

$$\Delta(\sigma) = 2[n(\sigma)-1]d \qquad [\text{m}]. \qquad (7.14)$$

† Here $\hat{\Gamma}$, which must not be confused with the mutual coherence of the field, denotes the complex propagation factor of the medium. If subscript 2 denotes the medium and subscript 1 the surrounding vacuum, then

$$\hat{\Gamma}_{12}(\sigma) = \hat{t}_{12}\hat{t}_{21}\hat{\gamma}_2 \sum_0^\infty (\hat{\gamma}_2\hat{r}_{21})^{2m},$$

where $\hat{\gamma}_2$, the complex propagation factor within the medium, is given by $\hat{\gamma}_2 = \exp-[2\pi i\hat{n}(\sigma)d]$. $\hat{\Gamma}_{12}(\sigma)$ therefore accounts for all interface and multiple beam effects.

The disturbance approaching the detector from the moving mirror arm PP_1 is

$$\hat{E}_1(t-t_1) = \int_0^\infty \hat{r}_0(\sigma)\hat{t}_0(\sigma)\hat{a}(\sigma) \exp\left[-i\phi_1(\sigma) - 2\pi i\sigma c(t-t_1)\right] d\sigma \qquad [\text{V m}^{-1}]$$

(7.15)

as before (4.20), except for the inclusion of a phase shift term $\phi_1(\sigma)$, while that from the fixed mirror arm is

$$\hat{E}_2(t-t_2) = \int_0^\infty \hat{t}_0(\sigma)\hat{r}_0(\sigma)\tau(\sigma)\hat{a}(\sigma) \exp\left[-2\pi i\sigma\Delta(\sigma) - 2i\phi_R(\sigma) - i\phi_2(\sigma)\right.$$

$$\left. - 2\pi i\sigma c(t-t_2)\right] d\sigma, \qquad (7.16)$$

where

$$\tau(\sigma) = \tau_R(\sigma)\tau_A(\sigma) \qquad (7.17)$$

is the *power* transmission factor for *single* passage of the specimen and $\phi_2(\sigma)$ is analogous to $\phi_1(\sigma)$. The resultant disturbance travelling to the detector is, therefore,

$$\hat{E}_D(t_1-t_2) = \hat{E}_1(t-t_1) + \hat{E}_2(t-t_2) \qquad (7.18)$$

$$= \int_0^\infty \hat{d}(\sigma) \exp(-2\pi i\sigma ct) d\sigma \qquad [\text{V m}^{-1}], \qquad (7.19)$$

where

$$\hat{d}(\sigma) = \hat{r}_0(\sigma)\hat{t}_0(\sigma)\hat{a}(\sigma)\{\exp 2\pi i\sigma ct_1 + \tau(\sigma) \exp\left[-i\phi(\sigma)\right] \exp 2\pi i\sigma ct_2\} \quad (7.20)$$

is the detected complex spectral amplitude. In this expression,

$$\phi(\sigma) = 2\pi\sigma\Delta(\sigma) + 2\phi_R(\sigma) + \phi_0(\sigma) \qquad [\text{rad}] \qquad (7.21)$$

is the phase difference resulting from the summation of the phase shift $2\pi\sigma\Delta(\sigma)$ due to the optical path difference $\Delta(\sigma)$, the shift $\phi_R(\sigma)$ due to surface and internal reflections and the small residual difference

$$\phi_0(\sigma) = \phi_2(\sigma) - \phi_1(\sigma) \qquad [\text{rad}]. \qquad (7.22)$$

This small phase difference between the two arms of the interferometer may be due, for example, to an inherent lack of symmetry arising from imperfect instrumental adjustment. The detected power (see section 4.2.1) is

$$I_D(\tau) = \lim_{T \to \infty} \frac{A\varepsilon_0 c}{4T} \int_0^\infty |\hat{d}_D(\sigma, t_1, t_2)|^2 d\sigma \qquad (7.23a)$$

$$= \tfrac{1}{2} \int_0^\infty \{1 + [\tau(\sigma)]^2\} B(\sigma) d\sigma$$

$$+ \int_0^\infty \tau(\sigma) B(\sigma) \cos\left[2\pi\sigma c\tau - \phi(\sigma)\right] d\sigma \qquad [\text{W}], \qquad (7.23b)$$

where $B(\sigma)$ is the spectrum that would be observed in the absence of a specimen. We call this the *background spectrum*. (To maintain continuity with the earlier chapters we retain the symbol τ for the time delay. It must not be confused with the transmissivity of the specimen.) The first term in the right member of (7.23b) is the constant part of the transmitted power and is composed of two unequal parts

$$I_1 = \tfrac{1}{2} \int_0^\infty B(\sigma)\, d\sigma \qquad [\text{W}] \qquad (7.24a)$$

and

$$I_2 = \int_0^\infty [\tau(\sigma)]^2 B(\sigma)\, d\sigma \qquad [\text{W}] \qquad (7.24b)$$

contributed by the partial beams which traverse the arms that are without and with the specimen, respectively. The dependence on $[\tau(\sigma)]^2$ rather than $\tau(\sigma)$ in (7.24b) should be noted. The second term in (7.23b) is the interferogram

$$F_u(x) = \int_0^\infty B_s(\sigma) \cos[2\pi\sigma x - \phi(\sigma)]\, d\sigma \qquad [\text{W}], \qquad (7.25)$$

where

$$B_s(\sigma) = \tau(\sigma) B(\sigma) \qquad [\text{W m}] \qquad (7.26)$$

is the spectrum that would be detected in the presence of the specimen. When the specimen is perfectly plane-parallel the relationship found here between $B(\sigma)$ and $B_s(\sigma)$ is identical to that (section 6.6) found when the specimen is placed before the detector. This is because both partial beams traverse the specimen once when it is placed before the detector, whereas one partial beam traverses the specimen twice in the present case; the net result in the modulus spectrum is the same.

The cosine Fourier transform of (7.25) is

$$p(\sigma) = \int_{-\infty}^\infty F_u(x) \cos 2\pi\sigma x\, dx = B_{e(s)}(\sigma) \cos \phi(\sigma) \qquad [\text{W m}] \quad (7.27a)$$

and the sine Fourier transform is

$$q(\sigma) = \int_{-\infty}^\infty F_u(x) \sin 2\pi\sigma x\, dx = B_{e(s)}(\sigma) \sin \phi(\sigma) \qquad [\text{W m}]. \quad (7.27b)$$

The complex spectrum

$$\hat{S}(\sigma) = \int_{-\infty}^{+\infty} F_u(x) \exp(-2\pi i \sigma x)\, dx \qquad [\text{W m}] \qquad (7.28)$$

therefore has a modulus

$$|\hat{S}(\sigma)| = \{[p(\sigma)]^2 + [q(\sigma)]^2\}^{1/2} = B_{e(s)}(\sigma) \qquad [\text{W m}] \qquad (7.29)$$

and a phase

$$-\phi(\sigma) = \text{ph } \hat{S}(\sigma) + 2m\pi \qquad \text{[rad]}. \qquad (7.30)$$

From (7.14 and 7.21) one therefore has

$$n(\sigma) = 1 - \frac{1}{4\pi\sigma d} [\text{ph } \hat{S}(\sigma) + 2m\pi + 2\phi_R(\sigma) + \phi_0(\sigma)]. \qquad (7.31)$$

The refractive index spectrum of the specimen may hence be found from the full Fourier transform (7.28) of the unsymmetrical interferogram, provided the phase shift $\phi_R(\sigma)$ and the background phase $\phi_0(\sigma)$ are known. The latter may readily be found from the complex spectrum obtained from the Fourier transformation of a background interferogram, that is, one observed in the absence of a specimen. One therefore has

$$n(\sigma) = 1 - \frac{1}{4\pi\sigma d} [\text{ph } \hat{S}(\sigma) - \text{ph } \hat{S}_0(\sigma) + 2(m - m_0)\pi + 2\phi_R(\sigma)] \quad (7.32)$$

where the zero subscripts denote background quantities. It likewise follows of course that the transmission spectrum may be obtained from

$$\tau(\sigma) = \frac{|\hat{S}(\sigma)|}{|\hat{S}_0(\sigma)|}. \qquad (7.33)$$

In this simple theory, it is assumed that the specimen is exactly plane parallel, completely homogeneous, and free from surface defects. If these conditions are not satisfied, the wave-front in the specimen arm will be distorted relative to that in the other arm and a complicated interferogram will be produced, the transform of which will not be simply related to the optical constants of the specimen. This means that much higher quality is required of a specimen for dispersive FTS than for absorptive FTS.

7.3.1.1. Restrictions on the magnitude of the dispersive path difference

The calculation indicated by (7.32) is most easily executed if the integers m and m_0 do not deviate from zero or at least assume only small values, say ± 1. We see from (7.30) that the phase differs from the calculated principal value when $m \neq 0$, that is, whenever $|-\phi(\sigma)| > \pi$. Thus the true value and the principal value are identical if

$$|-\phi(\sigma)| < \pi \qquad \text{[rad]} \qquad (7.34a)$$

or

$$2 |n(\sigma) - 1| d < \lambda/2 \qquad \text{[m]}, \qquad (7.34b)$$

which states that the magnitude of the optical path difference $2 |n(\sigma) - 1| d$ introduced must be nowhere greater than one half of the wavelength $\lambda (= 1/\sigma)$ of the radiation. (For the present discussion, we assume $\phi_0(\sigma)$ and $\phi_R(\sigma)$ to be negligible.) If the condition is violated, $m = 0$ only over that wavenumber region where (7.34) holds and m is finite integral over the remaining regions. Consequently, the calculated phase has a branched appearance due to the discontinuities where m changes. When the specimen has only normal dispersion over the detected spectral range, the refractive index is monotonic and the condition (7.34) needs to be satisfied for the smallest wavelength (largest wavenumber) if m is to be zero everywhere. When there is anomalous dispersion, the value of d necessary to ensure that (7.34) is satisfied everywhere is less easily found.

7.3.1.2. Displacement of the origin of the Fourier transformation

The stringency of this condition can be relaxed somewhat if we displace the origin about which the Fourier transform is carried out. Instead of placing the centre of the span at the true origin ($x = 0$) of path difference, we place it at the centre of the displaced dispersive interferogram. The position of this is usually easily recognized, being located at $x = \bar{x}$, where the displaced grand maximum occurs. The variable x is therefore replaced by $y = x - \bar{x}$ and the interferogram becomes

$$F_u(y + \bar{x}) = \int_{-\infty}^{\infty} \hat{S}(\sigma) \exp 2\pi i \sigma \bar{x} \exp 2\pi i \sigma y \, d\sigma \qquad [\text{W}]. \qquad (7.35)$$

This new description of the interferogram may be expressed as the identical function

$$G_u(y) = \int_{-\infty}^{\infty} \hat{S}_y(\sigma) \exp 2\pi i \sigma y \, d\sigma \qquad [\text{W}], \qquad (7.36a)$$

where

$$G_u(y) \triangleq F_u(y + \bar{x}) \qquad [\text{W}] \qquad (7.36b)$$

and

$$\hat{S}_y(\sigma) = \hat{S}(\sigma) \exp 2\pi i \sigma \bar{x} \qquad [\text{W m}] \qquad (7.36c)$$

is the complex spectrum evaluated from the inversion

$$\hat{S}_y(\sigma) = \int_{-\infty}^{\infty} G_u(y) \exp (-2\pi i \sigma y) \, dy \qquad [\text{W m}] \qquad (7.37)$$

of (7.36a). The spectrum (7.36c) may also be written as

$$\hat{S}_y(\sigma) = B_{e(s)}(\sigma) \exp [-i\Psi(\sigma)] \qquad [\text{W m}], \qquad (7.38a)$$

where

$$\Psi(\sigma) = \phi(\sigma) - 2\pi\sigma\bar{x} \qquad [\text{rad}]. \qquad (7.38b)$$

Both (7.36c) and (7.38b) show that the phase of the new complex spectrum $\hat{S}_y(\sigma)$ differs from that of the spectrum $\hat{S}(\sigma)$ by the term $-2\pi\sigma\bar{x}$, but the moduli of the two spectra are identical. It is of use in practice to notice that the phase $\Psi(\sigma)$ consists of a linear term with a superimposed fluctuation and that both terms vanish when $\sigma = 0$.

The spectrum $\hat{S}_y(\sigma)$ can be expressed in terms of its real and imaginary parts,

$$\hat{S}_y(\sigma) = P(\sigma) - iQ(\sigma) \qquad [\text{W m}], \qquad (7.39)$$

which are the cosine Fourier transform

$$P(\sigma) = \int_{-\infty}^{\infty} G_u(y) \cos 2\pi\sigma y \, dy \qquad [\text{W m}] \qquad (7.40\text{a})$$

and the sine Fourier transform

$$Q(\sigma) = \int_{-\infty}^{\infty} G_u(y) \sin 2\pi\sigma y \, dy \qquad [\text{W m}] \qquad (7.40\text{b})$$

of the interferogram $G_u(y)$. The modulus and phase are calculated from these according to

$$|\hat{S}_y(\sigma)| = \{[P(\sigma)]^2 + [Q(\sigma)]^2\}^{1/2} = B_{e(s)}(\sigma) \qquad [\text{W m}] \qquad (7.41\text{a})$$

and

$$-\Psi(\sigma) = \text{ph}\,\hat{S}_y(\sigma) + 2M\pi \qquad [\text{rad}]. \qquad (7.41\text{b})$$

The refraction spectrum is now given by

$$n(\sigma) = 1 + \frac{\bar{x}}{2d} - \frac{1}{4\pi\sigma d}\,|\text{ph}\,\hat{S}_y(\sigma) - \text{ph}\,\hat{S}_0(\sigma) + 2(M - m_0)\pi + 2\phi_R(\sigma)| \qquad (7.42)$$

in terms of the complex spectra $\hat{S}_y(\sigma)$ and $\hat{S}_0(\sigma)$. It should be noted that it is necessary to know the magnitude of $\bar{x}$.

7.3.1.3. *Relaxed restriction on the magnitude of the dispersive path difference*

The integer M in (7.41b) is zero wherever

$$|-\Psi(\sigma)| < \pi \qquad [\text{rad}], \qquad (7.43\text{a})$$

that is, where

$$2\,|n(\sigma) - \bar{n}|\,d < \lambda/2, \qquad (7.43\text{b})$$

since $\bar{x} = 2(\bar{n} - 1)d$. This condition is less stringent than (7.34b) for, in general, $|n(\sigma) - \bar{n}| < |n(\sigma) - 1|$ as $\bar{n} - 1 > 0$. Thus, for a given thickness of specimen, greater refractive index variations may be investigated if the calculation of the Fourier transformation is executed about $y = 0$ rather than $x = 0$.

7.3.2. Nature of the dispersive interferogram

The dispersive interferogram has, in practice, a striking appearance. The brightest fringe is shifted from the position of true zero path difference

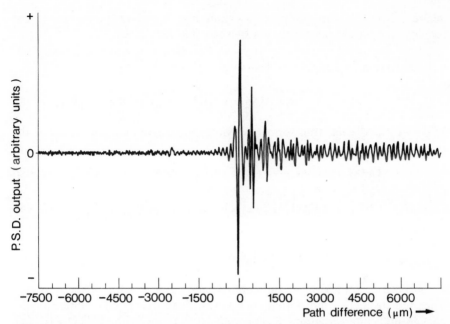

Figure 7.2. Phase-modulated dispersive interferogram obtained with ammonia gas in the fixed mirror arm (J. R. Birch and C. E. Bulleid, unpublished). The spectral range involved is approximately 10–$200\ \text{cm}^{-1}$ and the structure therefore arises from the pure rotation spectrum

by an amount equal to the average optical thickness $2(\bar{n}-1)d$ and the displaced brightest fringe may become markedly asymmetric depending on the dispersion $dn(\sigma)/d\sigma$ of the specimen. However, the new structure giving rise to the asymmetry is imported only into one side of the interference function; if the specimen is placed in the static arm, the new structure appears at positive values of x. An argument based on causality will show why this is, but first we must consider the complex insertion loss.

7.3.2.1. Complex insertion loss and the interferogram

Since, from (7.38a), (7.38b), and (7.21)

$$\hat{S}_y(\sigma) = B_{e(s)}(\sigma) \exp\left[-2\pi i\sigma\Delta(\sigma) - 2i\phi_R(\sigma) - i\phi_0(\sigma)\right] \exp 2\pi i\sigma\bar{x} \qquad [\text{W m}]$$
$$(7.44)$$

and from (7.26), (7.12), and (7.13)

$$\begin{aligned}
B_{e(s)}(\sigma) &= \tau(\sigma)B_e(\sigma) \\
&= \tau_R(\sigma) \exp\left[-\alpha(\sigma)d\right]\hat{S}_0(\sigma) \exp\left[i\phi_0(\sigma)\right] \qquad [\text{W m}],
\end{aligned} \qquad (7.45)$$

we find, from (7.11), that

$$\hat{S}_y(\sigma) = [\hat{\mathscr{L}}(\sigma)]^2 \exp 2\pi i\sigma\bar{x}\hat{S}_0(\sigma) \qquad [\text{W m}], \qquad (7.46)$$

where $[\hat{\mathscr{L}}(\sigma)]^2$ is the complex insertion loss for double passage of the specimen. Fourier transformation of each side now gives

$$\int_{-\infty}^{\infty} \hat{S}_y(\sigma) \exp(-2\pi i\sigma\bar{x}) \exp 2\pi i\sigma x \, d\sigma$$

$$= \int_{-\infty}^{\infty} [\hat{\mathscr{L}}(\sigma)]^2 \hat{S}_0(\sigma) \exp 2\pi i\sigma x \, d\sigma \qquad \text{[W]}. \quad (7.47)$$

If we evaluate each transform, applying the convolution theorem for the purpose, we find

$$G_u(x) * \delta(x - \bar{x}) \equiv G_u(x - \bar{x}) = F_{0u}(x) * R^{(2)}(x) \qquad \text{[W]} \qquad (7.48a)$$

or

$$F_s(x) = F_{0u}(x) * R^{(2)}(x) \qquad \text{[W]}, \qquad (7.48b)$$

where

$$R^{(2)}(x) = \int_{-\infty}^{\infty} [\hat{\mathscr{L}}(\sigma)]^2 \exp 2\pi i\sigma x \, d\sigma \qquad [\text{m}^{-1}] \qquad (7.49)$$

is the Fourier transform of the complex insertion loss for *double* passage. Thus the dispersive interferogram $F_s(x)$ is given by the convolution of the background interferogram $F_{0u}(x)$ with $R^{(2)}(x)$ or with $R^{(1)}(x) * R^{(1)}(x)$, where

$$R^{(1)}(x) = \int_{-\infty}^{\infty} \hat{\mathscr{L}}(\sigma) \exp 2\pi i\sigma x \, d\sigma \qquad [\text{m}^{-1}] \qquad (7.50)$$

is the Fourier transform of the complex insertion loss for *single* passage.

We shall show in the next section that the form of $R^{(1)}(x)$ is determined by the operation within the interferometer of physical restraints which stem from the principle of causality.

7.3.2.2. The impulse response function

A consideration of the operation of the interferometer in terms of an impulse of radiation launched into it from the source leads to an expression for the impulse response function of the specimen.

Let the disturbance incident at the beam divider be $\hat{E}(t)$ given by (7.8). The disturbance reaching the detector from the arm PP_1 (see section 7.3.1) is $\hat{E}_1(t - t_1)$ while that from the fixed mirror arm is

$$\hat{E}_2(t - t_2) = \mathscr{R}(t - t_2) * \hat{E}_1(t - t_2) \qquad [\text{V m}^{-1}], \qquad (7.51)$$

where $\mathscr{R}(t)$ is the impulse response of the specimen when placed in the interferometer. The total signal at the detector is the sum (7.18) of the partial signals and the power detected is

$$I(c\tau) = \lim_{T \to \infty} \frac{\varepsilon_0 cA}{4T} \int_{-\infty}^{\infty} \overline{|\hat{E}_D(t_1, t_2)|^2} \, dt \qquad (7.52a)$$

$$= I_1 + I_2 + F(c\tau) \qquad \text{[W]}, \qquad (7.52b)$$

where

$$I_1 = \lim_{T \to \infty} \frac{\varepsilon_0 cA}{4T} \int_{-\infty}^{\infty} \overbrace{|E_1(t-t_1)|^2}\, dt \quad \text{[W]} \tag{7.53a}$$

and

$$I_2 = \lim_{T \to \infty} \frac{\varepsilon_0 cA}{4T} \int_{-\infty}^{\infty} \overbrace{|E_2(t-t_2)|^2}\, dt \quad \text{[W]} \tag{7.53b}$$

are the powers contributed via the arms 1 and 2 of the interferometer and

$$F(c\tau) = \lim_{T \to \infty} \frac{\varepsilon_0 cA}{4T} \int_{-\infty}^{\infty} \overbrace{2\,|\hat{E}_1(t-t_1)\hat{E}_2(t-t_2)|}\, dt$$

$$= \lim_{T \to \infty} \frac{\varepsilon_0 cA}{4T} \int_{-\infty}^{\infty} \overbrace{2\,|E_1(t-\tau)E_2(t)|}\, dt \tag{7.54}$$

is the interferogram. (As before, τ is the delay time of one partial beam with respect to the other.) Using (7.51) this becomes

$$F(c\tau) = \lim_{T \to \infty} \frac{\varepsilon_0 cA}{4T} \int_{-\infty}^{\infty} \left| \overbrace{2\hat{E}_1(t-\tau) \left[\int_{-\infty}^{\infty} \mathcal{R}(t')\hat{E}_1(t-t')\, dt' \right]} \right| dt \tag{7.55a}$$

$$= \int_{-\infty}^{\infty} \mathcal{R}(t') \left[\lim_{T \to \infty} \frac{\varepsilon_0 cA}{4T} \int_{-\infty}^{\infty} \overbrace{|2\hat{E}_1(t)\hat{E}_1(t+\tau-t')|}\, dt \right] dt'$$

$$= \int_{-\infty}^{\infty} \mathcal{R}(t')F_0(c\tau - ct')\, dt' \tag{7.55b}$$

$$= \mathcal{R}(\tau) * F_0(c\tau) \quad \text{[W]}, \tag{7.55c}$$

where

$$F_0(c\tau) = \lim_{T \to \infty} \frac{\varepsilon_0 cA}{4T} \int_{-\infty}^{\infty} \overbrace{|2\hat{E}_1(t)\hat{E}_1(t+\tau)|}\, dt \quad \text{[W]} \tag{7.56}$$

is the interferogram observed in the absence of the specimen; this is the autocorrelation of the signals $\hat{E}_1(t)$ and $\hat{E}_1(t+\tau)$. A comparison of (7.55c) and (7.48b) shows that $\mathcal{R}(\tau)$ is equivalent to $R^{(2)}(x)$; thus, $[\hat{\mathcal{L}}(\sigma)]^2 = \hat{T}(\sigma)$ is equivalent to the transfer function of the specimen when placed in the interferometer. Of necessity, (see 2.9.7.1), $\mathcal{R}(\tau) = 0$ for $\tau < 0$ (no response before the application of the signal) and therefore the real and imaginary parts of $\hat{T}(\sigma)$ are Hilbert transforms of each other. These transforms are the basis of the Kramers–Kronig relations frequently used to derive either component of $\hat{n}$ when the other is known over a wide frequency range.

The general relation (7.55c) shows that when a distorting mechanism, whatever it is, is introduced into one arm of the interferometer, the new interferogram is the convolution of the old interferogram with the impulse response of the distortion. As the impulse response $R(x)$ ($\equiv \mathcal{R}(c\tau)$) is zero

202

$R^2(x)$

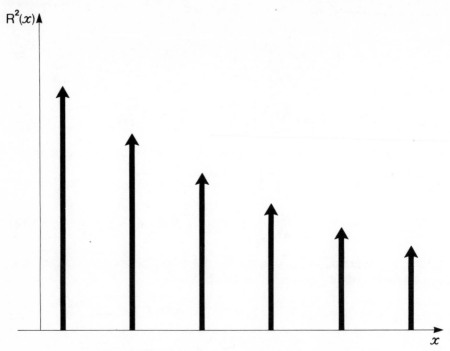

Figure 7.3. Impulse response of a non-dispersive specimen

for $x < 0$, it follows from (7.48b) in the form

$$F_u(x) = \int_{-\infty}^{\infty} F_{0u}(x') R(x - x') \, dx' \qquad [\text{W}] \qquad (7.57)$$

that $F_u(x)$ differs from $F_{0u}(x)$ only where $x > 0$, which verifies that the new structure is developed in the interferogram only at positive path-differences.

7.3.2.3. The impulse response for a non-dispersive specimen

Although the impulse response is quite complicated in detail in the general case, the basic form is related to the impulse response we evaluate in a specially simple case, namely that of a non-dispersive specimen.

If we make the crude assumption that the refractive index $n(\sigma)$ and the absorption coefficient $\alpha(\sigma)$ are constant over the spectral range studied, the insertion loss for single passage is

$$\hat{\mathscr{L}}(\sigma) = (1 - r^2) \exp(2\pi i \sigma d)$$

$$\times \sum_{m=0}^{\infty} r^{2m} \exp\left[-\tfrac{1}{2}(2m + 1)\alpha d\right] \exp\left[-2(2m + 1)\pi i \sigma n d\right], \qquad (7.58)$$

where, for near transparent media, π reflective is assumed. We then obtain

$$R^{(1)}(x) = (1 - r^2)\delta(x + d) * \sum_{m=0}^{\infty} r^{2m} \exp\left[-\tfrac{1}{2}(2m + 1)\alpha d\right]\delta(x - (2m + 1)nd)$$

$$[\text{m}^{-1}] \qquad (7.59)$$

and

$$R^{(2)}(x) = (1 - r^2)^2\delta(x + 2d) * \sum_{m=0}^{\infty} r^{4m} \exp\left[-(2m + 1)\alpha d\right]\delta(x - 2(2m + 1)nd)$$

$$= (1 - r^2)^2\delta(x + 2d) * [\exp(-\alpha d)\delta(x - 2nd) + r^4 \exp(-3\alpha d)\delta(x - 6nd)$$

$$+ r^8 \exp(-5\alpha d)\delta(x - 10nd) + \ldots] \qquad [\text{m}^{-1}]. \qquad (7.60)$$

This impulse response consists of a series of successively smaller Dirac functions corresponding to single passage, triple passage, quintuple passage, etc., of the specimen. The first is at $x = 2(n-1)d$, the second at $2(n-1)d + 4nd$, the third at $2(n-1)d + 8nd$, and so on. This series represents the 'echoes' of the impulse response function: it is illustrated schematically in Figure 7.3, from which it will be seen that the series of echoes has some analogies with the series of signatures sometimes observed in an interferogram. In the general case, of course, where both n and α are varying over the range studied, the echoes will not be simple Dirac functions.

7.3.3. Multiple reflections within the specimen

According to (7.42) the calculation of the refractive index spectrum requires a knowledge of $\phi_R(\sigma)$, that is, the phase shift at the boundaries due to single and multiple reflections. In the general case, the problem of determining $\phi_R(\sigma)$ is formidable, but at the two extremes, i.e. when $\alpha(\sigma)$ is small and when $\alpha(\sigma)$ is large, it becomes relatively straightforward. In the latter case, the effects of multiple reflections are negligible and one merely has the problem of determining the phase shift at a single reflection. This can be determined experimentally either by exact physical substitution of a perfect reflector or else by the use of two specimens of differing thickness. In the former case, one can assume that the phase shift at the boundary with the now slightly lossy medium is the same, namely π, that it would be if the medium were in fact transparent. By summing the infinite series for the resulting amplitude, one has

$$\hat{t}_R = \left[\frac{(1 - r^2)^2 a^2}{1 - 2r^2 a^2 \cos\delta + r^4 a^4}\right]^{1/2} \exp i \arctan\left[\frac{r^2 a^2 \sin\delta}{1 - r^2 a^2 \cos\delta}\right], \quad (7.61a)$$

where $r = r(\sigma)$, $a = \exp\left[-\tfrac{1}{2}\alpha(\sigma)d\right]$, and $\delta = 4\pi n(\sigma)\sigma d$, and hence

$$\phi_R(\sigma) = \arctan\left[\frac{r^2 a^2 \sin\delta}{1 - r^2 a^2 \cos\delta}\right]. \qquad (7.61b)$$

The multiple reflection contribution to the phase $\phi_R(\sigma)$ therefore depends on both $n(\sigma)$ and $\alpha(\sigma)$ and one has a non-linear equation to solve. However, an approximate value of $n(\sigma)$ is quite good enough to produce a sufficiently accurate value of $r(\sigma)$, so that equation (7.33) may be used to calculate $\alpha(\sigma)$. The values of $r(\sigma)$ and $\alpha(\sigma)$ are then fed into equation (7.61b) to produce $\phi_R(\sigma)$, which is used with (7.42) to give a new value of $n(\sigma)$. The process can be repeated, if necessary a few times, till the values for $n(\sigma)$ converge.

7.3.4. Comparison of dispersive and non-dispersive measurements

In non-dispersive FTS, the specimen is usually placed immediately before the detector and each partial beam traverses it before entering the detector window. The complex spectral amplitude in the detector plane is modified from $\hat{d}(\sigma, t_1, t_2)$ to $\hat{\mathcal{L}}(\sigma)\hat{d}(\sigma, t_1, t_2)$, where $\hat{\mathcal{L}}(\sigma)$ is, as before, the complex insertion loss for single passage. The interferogram is therefore changed from

$$F_0(x) = \int_{-\infty}^{+\infty} B_{0e}(\sigma) \exp\left[2\pi i \sigma x - i\phi_0(\sigma)\right] d\sigma \tag{7.62}$$

to

$$F(x) = \int_{-\infty}^{\infty} B_{0e}(\sigma) |\hat{\mathcal{L}}(\sigma)|^2 \exp\left[2\pi i \sigma x - i\phi_0(\sigma)\right] d\sigma \qquad [\text{W}], \tag{7.63}$$

from which it appears that

$$F(x) = [R^{(1)}(x) * R^{(1)}(-x)] * F_0(x) \qquad [\text{W}], \tag{7.64}$$

where we have noted that, since $\hat{n}(\sigma)$ is Hermitian, so is $\hat{\mathcal{L}}(\sigma)$ and therefore

$$\int_{-\infty}^{\infty} \hat{\mathcal{L}}^*(\sigma) \exp 2\pi i \sigma x \, d\sigma = R(-x) \qquad [\text{m}^{-1}]. \tag{7.65}$$

The relation (7.64) is to be compared with

$$F(x) = [R^{(1)}(x) * R^{(1)}(x)] * F_0(x) \qquad [\text{W}], \tag{7.66}$$

which holds when the same specimen is placed in one arm. The impulse response $R^{(1)}(x) * R^{(1)}(-x)$, which is equivalent to the autocorrelation $R^{(1)}(x) \star R^{(1)}(x)$, is symmetrical about $x = 0$, whereas the convolution $R^{(1)}(x) * R^{(1)}(x)$ exists only for $x > 0$. In conventional Fourier spectrometry, it is $R^{(1)}(x) \star R^{(1)}(x)$, or the real $|\hat{\mathcal{L}}(\sigma)|^2$, that is measured—which is essentially a power measurement; in dispersive Fourier spectrometry, it is $R^{(1)}(x) * R^{(1)}(x)$, or the complex $[\hat{\mathcal{L}}(\sigma)]^2$, that is measured—which is an amplitude measurement.†

† In some interferometers, the Mach–Zehnder for example, the radiation traverses the specimen, placed in one arm, once only; in which case $R^{(1)}(x)$ rather than $R^{(1)}(x) * R^{(1)}(x)$ is measured. Under these circumstances, the contrast between the power measurement and the amplitude measurement is more striking. This is because power measurements intuitively involve 'real squared' quantities, whereas amplitude measurements involve 'complex unsquared' quantities.

7.3.5. The quality of the interference function

The magnitude of the interference fluctuations (the interferogram) relative to that of the background level is less in a dispersive interference function than in a corresponding non-dispersive interference function. This is because the attenuation of the specimen causes the interfering partial beams to be dissimilar, with the result that interference is reduced and, perhaps, even eliminated for the more strongly attenuated frequency components. The corresponding frequency components traversing the other arm are un-attenuated and contribute to the background level but not to the inter-ference. It is shown later, when noise is considered (Chapter 9), that this effect can be troublesome in practice and can degrade signal-to-noise ratio.

In the non-dispersive case with either d.c. detection or, more usually, amplitude modulation the detected power is

$$I(x) = \bar{I} + F(x) \qquad [\text{W}], \tag{7.67a}$$

where the mean level is

$$\bar{I} = \int_{-\infty}^{\infty} B_e(\sigma)\tau(\sigma)\,d\sigma \qquad [\text{W}] \tag{7.67b}$$

and the interferogram is

$$F(x) = \int_{-\infty}^{\infty} B_e(\sigma)\tau(\sigma)\cos 2\pi\sigma x\,d\sigma \qquad [\text{W}]. \tag{7.67c}$$

A measure of the magnitude of the interference relative to the mean level is the ratio

$$Q_I = [F(0)/\bar{I}] \tag{7.68}$$

of the peak interferogram power $F(0)$ to the mean level $\bar{I}$. We may call this the *quality* of the interferogram for it represents the magnitude of the interferometrically modulated and therefore useful radiation, relative to the unmodulated and therefore useless radiation. With an ideal instrument, this ratio, from (7.67a) and (7.67b), is unity, but in practice it is often appreci-ably less.

In the case of a dispersive interference function

$$J(y) = \bar{J} + G(y) \qquad [\text{W}], \tag{7.69a}$$

where

$$\bar{J} = \tfrac{1}{2}\int_{-\infty}^{\infty} \{1 + [\tau(\sigma)]^2\}B_e(\sigma)\,d\sigma \qquad [\text{W}] \tag{7.69b}$$

and

$$G(y) = \int_{-\infty}^{\infty} B_e(\sigma)\tau(\sigma)\cos\left[2\pi\sigma y - \Psi(\sigma)\right]d\sigma \qquad [\text{W}], \tag{7.69c}$$

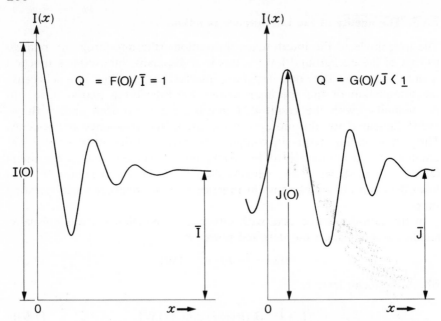

Figure 7.4. Quality functions for ideal amplitude modulated non-dispersive interferometry (left) and dispersive interferometry (right)

the quality is

$$Q_J = [G(0)/\bar{J}]. \tag{7.70}$$

The general relations for Q_J, found by substituting the integrals (7.69b) and (7.69c) into (7.70), are hard to examine unless specific forms are assumed for the spectrally variable functions. If we assume $\Psi(\sigma)$ to be small and $\tau(\sigma)$ to be constant, then

$$Q_J \sim \frac{2\tau}{(1+\tau^2)}, \tag{7.71}$$

which is always less than 1 and becomes small as τ decreases. For a given specimen, the quality of a dispersive interference function is less than that of the corresponding non-dispersive function ($Q_J < Q_I$). It is possible to develop a definition for Q_J/Q_I in the phase-modulated case, but this is not so easy to use and one usually confines oneself to qualitative statements. The use of phase modulation can usually more than make up for the loss of quality entailed by the use of dispersive FTS.

7.3.6. Truncation of the dispersive interferogram

The results obtained above are strictly valid only if the interferograms are infinite in extent. In practice, they are truncated to possess a finite span and they are also weighted. The spectrum we calculate is, therefore,

$$\hat{S}^c(\sigma) = \int_{-\infty}^{\infty} G_u(y) W\left(\frac{y}{D}\right) \sqcap\left(\frac{y}{2D}\right) \exp\left(-2\pi i \sigma y\right) dy \qquad (7.72a)$$

$$= \hat{S}(\sigma) * A(\sigma D) \qquad [\text{W m}] \qquad (7.72b)$$

when the span $2D$ is centred on $y = 0$. The phase of the calculated complex spectrum is given by

$$\tan \Psi^c(\sigma) = \frac{Q(\sigma) * A(\sigma D)}{P(\sigma) * A(\sigma D)}. \qquad (7.73)$$

The deviation of this from the true phase is least when the effects of the convolutions with $A(\sigma D)$ are least, that is, when the transforms $Q(\sigma)$ and $P(\sigma)$ are relatively slowly varying by comparison with $A(\sigma D)$. The background interferogram is truncated and weighted as described in Chapter 6, and so the refraction spectrum is given by (7.42) with the calculated transforms replacing the ideal ones; i.e.

$$n^c(\sigma) = 1 + \frac{\bar{x}}{2d} - \frac{1}{4\pi\sigma d} [\text{ph } \hat{S}^c(\sigma) - \text{ph } \hat{S}_0^c(\sigma) + 2(M - m_0)\pi + 2\phi_R(\sigma)]. \quad (7.74)$$

7.3.6.1. Calculation with offset span

It was shown in section 7.3.1.2 that it is desirable to take $x = \bar{x}$ rather than $x = 0$ as the origin for the Fourier transformation of the dispersive interferogram if the condition to be satisfied by the specimen is not to be too demanding. When the interferogram is truncated, there is an additional reason, for computational artefacts arise if the place of stationary phase $(x = \bar{x})$ does not lie near the centre of the span.

The spectrum we calculate from the Fourier transform of the truncated weighted interferogram having the centre of its span at $x = 0$ is

$$S'^c(\sigma) = \int_{-\infty}^{\infty} F_u(x) W\left(\frac{x}{D}\right) \sqcap\left(\frac{x}{2D}\right) \exp\left(-2\pi i \sigma x\right) dx \qquad [\text{W m}]. \quad (7.75)$$

Although the span is symmetrically disposed about $x = 0$, we should remember that the main structure of the interferogram in the region of its centre is near $x = \bar{x}$. Consequently, (7.75) may also be written as

$$S'^c(\sigma) = \exp\left(-2\pi i \sigma \bar{x}\right) \int_{-\infty}^{\infty} G_u(y) W\left(\frac{y + \bar{x}}{D}\right) \sqcap\left(\frac{y + \bar{x}}{2D}\right) \exp\left(-2\pi i \sigma y\right) dy \quad (7.76a)$$

$$= \exp\left(-2\pi i \sigma \bar{x}\right) \{\hat{S}_y(\sigma) * [A(\sigma D) \exp 2\pi i \sigma \bar{x}]\}$$

$$= \exp\left(-2\pi i \sigma \bar{x}\right) \{[\hat{S}(\sigma) \exp 2\pi i \sigma \bar{x}] * [A(\sigma D) \exp 2\pi i \sigma \bar{x}]\} \qquad [\text{W m}].$$
$$(7.76b)$$

Using the arguments of section 6.16, we may express this as the sum of real and imaginary parts, from which we deduce the phase. Writing

$$P'^c(\sigma) = \tfrac{1}{2}\{P(\sigma) * [A(\sigma\bar{D}_+) + A(\sigma\bar{D}_-)] - Q(\sigma)$$
$$* [A(\sigma\bar{D}_+) - A(\sigma\bar{D}_-)]\} \qquad [\text{W m}], \quad (7.77a)$$

$$Q'^c(\sigma) = \tfrac{1}{2}\{Q(\sigma) * [A(\sigma\bar{D}_+) + A(\sigma\bar{D}_-)] - P(\sigma)$$
$$* [A(\sigma\bar{D}_+) - A(\sigma\bar{D}_-)]\} \qquad [\text{W m}], \quad (7.77b)$$

where

$$\bar{D}_+ = D + \bar{x} \qquad [\text{m}] \qquad (7.78a)$$

and

$$\bar{D}_- = D - \bar{x} \qquad [\text{m}], \qquad (7.78b)$$

we have

$$\hat{S}'^c(\sigma) = \hat{S}(\sigma) * [A(\sigma D) \exp 2\pi i \sigma \bar{x}] = P'^c(\sigma) - iQ'^c(\sigma) \qquad [\text{W m}], \quad (7.79)$$

and hence the computed phase $\Psi'^c(\sigma)$ of the complex spectrum (7.76) is given by

$$\tan \Psi'^c(\sigma) = \frac{Q'^c(\sigma)}{P'^c(\sigma)}, \qquad (7.80)$$

which should be compared with (7.73) and the ideal relation

$$\tan \Psi = \frac{Q(\sigma)}{P(\sigma)}. \qquad (7.81)$$

The consequence of the dependences of $Q'^c(\sigma)$ and $P'^c(\sigma)$ on the Hilbert transforms (7.77) is to make $\Psi'^c(\sigma)$ deviate considerably from the required value through the appearance of 'noise' and rapid branching. The effects vanish if and only if the calculation is executed about $y = 0$ as origin as opposed to $y = -\bar{x}$. There are, therefore, two important reasons for calculating the Fourier transform of the observed interferogram about $x = \bar{x}$ ($y = 0$):

    I. to relax the constraint on $n(\sigma)d$;
    II. to minimize the computational artefacts associated with transforming a truncated function.

7.3.7. Effects of sampling the interferogram

For the purposes of carrying out the numerical evaluation of the Fourier transformation of the interferogram, it is common practice for the interferogram function to be sampled at equal intervals β of path difference so that the computed complex spectrum is given by

$$\hat{S}^c(\sigma) = \int_{-\infty}^{\infty} G_u(y) W\left(\frac{y}{D}\right) \sqcap\left(\frac{y}{2D}\right) \sqcup\left(\frac{y}{\beta}\right) \exp\left(-i2\pi\sigma y\right) dy \qquad [\text{W m}]. \quad (7.82)$$

In this section, we shall consider the effect that this discrete path difference sampling will have on the computed phase shift spectrum of a specimen or, more correctly, on the computed phase spectrum of its complex insertion loss. In particular, we shall be concerned with the consequences of the non-coincidence of the position of zero optical path difference with any of the points of the sampling comb of the interferogram. The effects of this on the computed modulus spectrum have already been shown to be insignificant if the difference between the origin of computation and the position of zero optical path difference is small. However, the positions chosen for the origins of computation of the specimen and reference interferograms in a dispersive experiment will critically affect the computed phase spectra and this has led to some confusion in the literature over the correct procedure to follow in the evaluation of phase spectra, largely due to the introduction of the unnecessary phase correction procedure of Chamberlain et al.[46] The correct procedure to follow in such a situation has been outlined by Birch and Bulleid[48] and recently considered more fully by Birch and Parker.[48a] This is discussed in the following.

The object of a dispersive Fourier transform experiment is to determine the complex insertion loss of a specimen. This is the complex factor by which an incident electric field is changed by its interaction with the specimen and is, therefore, the ratio of the complex spectra obtained by transformation of the specimen and reference interferograms using the same position of geometric path difference as the computational origin. This origin can be the position of zero geometric path difference, but need not necessarily be so; the definition of complex insertion loss only requires that the same position be used. In an ideal experiment, the complex insertion loss would therefore be determined from

$$\hat{\mathscr{L}}(\sigma) = \hat{S}_x^S(\sigma)/\hat{S}_x^R(\sigma), \qquad (7.83)$$

where the superscripts S and R refer to the specimen and reference measurements, respectively, and the subscript x indicates the position of path difference used as the origin for the computation of each complex spectrum. Thus, the phase of the complex insertion loss would be found from

$$\phi_L = \text{ph } \hat{S}_x^S(\sigma) - \text{ph } \hat{S}_x^R(\sigma) + 2(m - m_0)\pi. \qquad (7.84)$$

In practice the optical thickness of a specimen causes the specimen interferogram to shift to positive path differences and severe phase branching would result, as discussed in sections 7.3.1.1 and 7.3.1.2, if the position of zero geometric path difference were used for its origin of computation. It is generally more convenient to use a sampling point, x_s', near to the centre of the displaced fringe as origin instead. Similarly, a sampling point will not, in general, coincide exactly with the centre of the reference interferogram, so a point at x_0', near to zero geometric path difference is used. With these arbitrarily chosen points used as the computational origins, the Fourier transform shift theorem shows that the phase of the complex insertion loss is now given by

$$\phi_L = \text{ph } \hat{S}^S_{x'_s}(\sigma) - \text{ph } \hat{S}^R_{x_0}(\sigma) + 2\pi\sigma(x'_s - x'_0) + 2(m' - m'_0)\pi \qquad (7.84a)$$

in terms of the experimentally observable quantities. This is a perfectly general result for any specimen in a reflection or transmission experiment. The third term in the right-hand member of equation (7.84) is known as the fringe shift term, and is readily determined since the difference

$$x'_s - x'_0 = N\beta \qquad (7.85)$$

is a whole number, N, of sampling intervals. Thus, in a double-pass transmission experiment, the real refractive index spectrum is correctly calculated from

$$n^c(\sigma) = 1 + \frac{x'_s - x'_0}{2d} + \frac{1}{4\pi\sigma d} [\text{ph } \hat{S}^S_{x'_s}(\sigma) - \text{ph } \hat{S}^R_{x_0}(\sigma) + 2(m' - m'_0)\pi], \quad (7.86)$$

which ignores the reflection phase shifts as being small compared with $4\pi\sigma(n-1)d$. In the analysis of Chamberlain et al.,[46] the precise separation, $\bar{x}$, of the positions of the centres of the specimen and reference interferograms is determined and used to calculate a phase spectrum

$$\phi' = \text{ph } \hat{S}^S_{x'_s}(\sigma) - \text{ph } \hat{S}^R_{x_0}(\sigma) + 2\pi\sigma\bar{x} + 2(m' - m'_0)\pi, \qquad (7.87)$$

which is not, in general, equal to the true phase of the complex insertion loss given by (7.84). In a transmission experiment (the situation considered by these authors), an erroneous refractive index is then calculated from this ϕ' value as

$$n'^c(\sigma) = 1 + \frac{1}{4\pi\sigma d} [\phi'], \qquad (7.88)$$

which is subsequently corrected by addition of the term

$$\frac{(H - H_0)\beta}{2d} \qquad (7.89)$$

obtained by estimating values of the small shifts, $H_0\beta$ and $H\beta$ of the computational origins of the reference and specimen interferograms from their respective grand maxima. This is an unnecessary and time-consuming process for the following reasons.

(i) The first step of the procedure, the calculation of ϕ' from equation (7.87), introduces the error that the rest of the procedure sets out to remove.

(ii) Neither $\bar{x}$ or $(H - H_0)\beta$ is a whole number of sampling intervals, so their necessary derivation by interpolation in the interferogram will introduce systematic errors. Moreover, Chamberlain et al.[46] suggest finding $\bar{x}$ from a subsidiary experiment using more finely sampled interferograms. Thus, additional measurements are required.

These problems are all avoided if the interferograms are recorded in a systematic way, so that the shift $N\beta$ between their grand maxima is known and the phase spectrum determined from equation (7.84).

Birch and Bulleid[48] have pointed out that the arguments leading to equation (7.84) relating the phase of the complex insertion loss to the experimentally determined quantities do not necessarily require the points x'_s and x'_0 used as the computational origins to be particularly close to the grand maxima of the interferograms. Thus, if these were chosen so that

$$2\pi\sigma(x'_s - x'_0) \tag{7.90}$$

was quite large, the difference

$$\mathrm{ph}\,\hat{S}^S_{x'_s}(\sigma) - \mathrm{ph}\,\hat{S}^R_{x'_0}(\sigma) \tag{7.91}$$

would be found to have changed by an equal and opposite amount, so that the total computed phase, ϕ_L, would remain constant. This is strictly only true if both phase spectra remain within their principal value regions. In practice this is usually the case. This compensating effect is illustrated in Figure 7.5.

7.3.8. Linking the calculated phase

Even if the condition (7.43b) has been met, for the normal case of slowly varying dispersion of the specimen across the spectral range concerned, by suitable choice of its thickness, there still remains the possibility that the required value of the phase may deviate from the calculated principal value in regions of strong anomalous dispersion.[†] In determining whether or not an integral number of 2π has to be added to, or subtracted from, the phase function when its principal value becomes discontinuous, it is necessary to examine the signs of the functions $P^c(\sigma)$ and $Q^c(\sigma)$ and relate these to the form of $B^c_e(\sigma)$. If the condition (7.43b) is satisfied, all the values of $\mathrm{ph}\,\hat{S}^c(\sigma)$ calculated as a function of wavenumber lie on the $M=0$ branch; otherwise some values may lie in branches characterized by $M=\pm 1, \pm 2$, etc. Non-zero values of M may be determined provided that there are sufficient adjacent data in the computed principal-value spectrum requiring the same M values. Special consideration is required when a discontinuity exactly coincides with the centre of a narrow absorption feature, as may be the case when strong and sharp (anomalous) absorption features are being studied and the turning points in the refraction spectrum on either side of the centre of the feature are not observed. This is the case when, for example, narrow features characteristic of a gas spectrum are studied with insufficient resolution to resolve the detail—the refraction feature then has the appearance of

† 'Anomalous' dispersion is a term that has been handed down to us from the early days of optics. It was then believed that the refractive index always increased as the wavelength decreased. When examples of the opposite behaviour were encountered, these were labelled 'anomalous'. In infrared spectroscopy, anomalous dispersion is associated with the line-centre region.

212

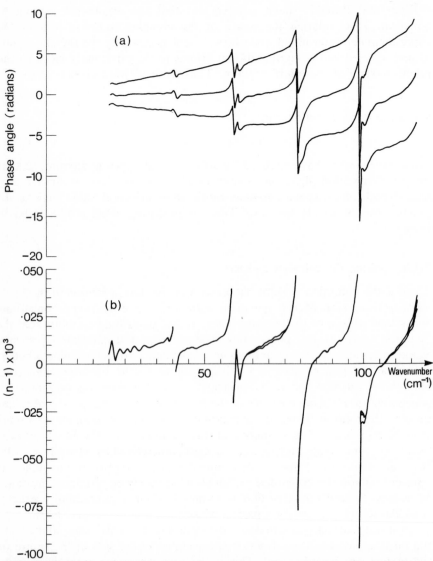

Figure 7.5. Phase angle curves (a) resulting from three different choices of inter-ferogram origin (the middle one being 'correct') give virtually identical refraction spectra (b). The data come from the work of Birch and Bulleid[48] and show the dispersion through the pure rotation lines of ammonia

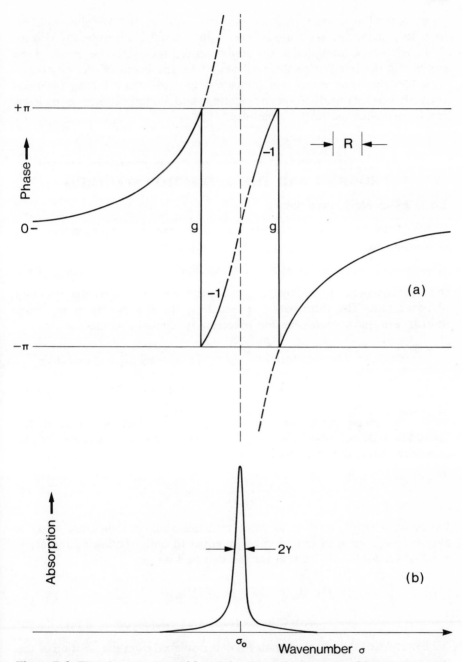

Figure 7.6. The phase spectrum (a) and the absorption spectrum (b) for an unresolved feature centred at σ_0. The full lines in (a) show the principal value. The discontinuities at g may be eliminated by addition (or subtraction) of 2π but the final discontinuity at σ_0 must be retained since $R \ll 2\gamma$ and the phase spectrum therefore resembles that for an undamped resonance

an undamped resonance. Under these circumstances, the M-values must be used to provide the computed phase with a suitable discontinuity (Figure 7.6). Whenever ambiguities or doubts arise, these can be resolved by comparing the refraction values calculated for specimens of differing thickness. Clearly, these values will coincide only if the phase joining has been correctly carried out. Real spectral features and multiple interference effects can be distinguished in the same way.

7.4. RELATION BETWEEN THE UNSYMMETRICAL INTERFEROGRAM AND THE REFLECTION SPECTRUM

7.4.1. Elementary wave theory

If one of the mirrors of the two-beam interferometer is exactly replaced by the surface of a medium having a frequency-dependent reflection factor

$$\hat{r}(\sigma) = r(\sigma) \exp i\theta(\sigma), \tag{7.92}$$

the interferometer is no longer symmetrical and neither is the resultant interferogram. The disturbance approaching the detector from the fixed 'mirror' arm (now containing the reflecting specimen) becomes

$$\hat{E}_2(t - t_2) = \int_{-\infty}^{\infty} \hat{t}_0(\sigma)\hat{r}_0(\sigma)\hat{d}(\sigma) \frac{r(\sigma)}{r_p(\sigma)} \exp\{i[\theta(\sigma) - \pi]\} \exp[-2\pi i\sigma c(t - t_2)$$
$$- i\phi_2(\sigma)] \, d\sigma \qquad [\mathrm{W}^{1/2}\,\mathrm{m}^{-1}], \quad (7.93)$$

where the phase term $\theta(\sigma) - \pi$ represents the effect of removing the reference reflector with $\hat{r}_p(\sigma) = r_p(\sigma)e^{i\pi}$ and replacing it exactly by the specimen. Thus, the insertion loss is

$$\hat{\mathscr{L}}(\sigma) = \frac{\hat{r}(\sigma)}{r_p(\sigma)} \simeq \hat{r}(\sigma)e^{-i\pi}, \tag{7.94}$$

the approximation holding in most practical situations, because $r_p(\sigma)$ is almost independent of σ and virtually equal to unity. Following the argument of section 7.3.1, we find for the detected power

$$I(x) = \tfrac{1}{2}\int_{-\infty}^{\infty} \{1 + [r(\sigma)]^2\}B_{0e}(\sigma)\, d\sigma + \int_{-\infty}^{\infty} r(\sigma)B_{0e}(\sigma)$$
$$\times \cos[2\pi\sigma x + (\theta(\sigma) - \pi) - \phi_0(\sigma)]\, d\sigma \qquad [\mathrm{W}]. \quad (7.95)$$

The centre of the interferogram $F_u(x)$ is displaced from the position of the true zero path difference ($x = 0$) by a small amount $\bar{x}$ given by the average value of $[\theta(\sigma) - \pi]/2\pi\sigma$, thus

$$\bar{x} = \left(\frac{\overline{\theta(\sigma) - \pi}}{2\pi\sigma}\right) \qquad [\mathrm{m}]. \tag{7.96}$$

This displacement is much smaller than the $\bar{x}$ encountered in dispersive transmission spectrometry and usually amounts to less than one sampling point. In general, therefore, for a dispersive reflection measurement, it is not necessary to use a displaced origin of computation for the specimen interferogram. Thus, from equation (7.84), the phase of the complex insertion loss would be determined as

$$\phi_L = \text{ph } \hat{S}^S_{x_0}(\sigma) - \text{ph } \hat{S}^R_{x_0}(\sigma) + 2(m - m_0)\pi \qquad (7.97)$$

and the phase of the complex reflectivity of the specimen as

$$\theta(\sigma) = \phi_L + \pi. \qquad (7.98)$$

7.4.2. The impulse response function

The relation between the specimen interferogram $F(c\tau)$ and the background interferogram has already been given as the convolution (7.55c) involving $\mathcal{R}(\tau)$, the impulse response function of the specimen. Since

$$\int_{-\infty}^{\infty} F(c\tau) \exp\left(-2\pi i\sigma c\tau\right) d(c\tau) = \int_{-\infty}^{\infty} [\mathcal{R}(\tau) * F_0(c\tau)] \exp\left(-2\pi i\sigma c\tau\right) d(c\tau),$$

$$(7.99)$$

it follows that

$$\hat{S}(\sigma) = \mathcal{F}\{\mathcal{R}(\tau)\}\hat{S}_0(\sigma) \qquad [\text{W m}], \qquad (7.100)$$

where $\mathcal{F}\{\mathcal{R}(\tau)\}$ is the transfer function of the specimen in the interferometer. The transfer function is therefore equivalent to the complex insertion loss, which in turn is equal to the complex reflectivity of the specimen,

$$\hat{\mathcal{L}}(\sigma) = \hat{r}(\sigma)e^{-i\pi}, \qquad (7.101)$$

so

$$\mathcal{R}(\tau) = \int_{-\infty}^{\infty} \hat{r}(\sigma) \exp\left(-i\pi\right) \exp 2\pi i\sigma c\tau \, d\sigma \qquad [\text{m}^{-1}]. \qquad (7.102)$$

CHAPTER 8

The generation and recording of the interferogram in a practical interferometer

8.1. DEFECTS IN PRACTICAL INTERFEROMETRIC SYSTEMS

In the foregoing sections we have assumed the interference to be produced in an idealized interferometer. The various non-ideal aspects so far considered have all concerned the difficulties that can be associated with recording and evaluating the data once the signal reaches the detector. When these have been dealt with in an appropriate manner there still remains the question as to whether the signal delivered from the detector is related to the required spectral distribution in the way described.

It is easy to think of many shortcomings in our assumptions. The source we use in practice cannot be a point; if it were no radiation would be detectable. Because it is not a point, non-parallel rays are present within the interferometer even when the radiation is collimated. Second, the beam splitter is not a perfect divider with equal reflection and transmission coefficients and its transmission will vary with wavelength. In addition, the mirrors may not be accurately aligned, the mirror drive may not be uniform, the source may fluctuate in intensity, and noise will always accompany the recorded electric signal.

The consequences of recorded noise are so extensive and important that they are dealt with separately in Chapter 9. Here we deal with the effects concerning the interferometer itself. It is important to note that the elimination of some of these is quite beyond the control of the operator, since they are an essential part of any practical instrument; for example, the use of an extended source. Those remaining can be classified as imperfections due to insufficient care on the part of the maker and user of the interferometer, for example, imperfect mirror alignment.

8.2. USE OF A FINITE LIMITING APERTURE

The source used in the practical interferometer has to be finite in extent. Ideally, we suppose the source to be very small so that the rays passing through the interferometer all travel parallel to the direction of propagation. The solid angle of the beam traversing the interferometer is, therefore, effectively zero. We record the radiant flux $F(x)$ passing through the (small) exit aperture for each value x of path difference in the interferometer (measured along the direction of propagation) and then perform a harmonic analysis on the resultant record in order to evaluate the spectral distribution. In practice, however, we record the flux transported in a *finite* solid angle disposed symmetrically about the direction of the mirror movement. We therefore record an interferogram function that is more or less different from $F(x)$ depending on whether the flux angle is great or small. The important parameter in this discussion is the diameter of the limiting stop, which we generally arrange to be the detector window. Figure 8.1 shows two possible paths that may be followed by rays from an off-axis point S in the source aperture. They meet at D on the detector window. The half-angle subtended by the source aperture at the collimating lens L_1 is α_S and the half-angle subtended by the detector aperture (window) at the condensing lens is α_D. The inclination of the full ray passing through the collimating lens is α ($<\alpha_S$) and this is equal to the inclination of the broken rays at the lens L_2. If the point D lies at the edge of the detector aperture, $\alpha = \alpha_D$ and α_D represents the maximum value that α may have. If the point S lies at the edge of the source aperture, $\alpha = \alpha_S$ and α_S represents the maximum value of α. There is, therefore, a *limiting stop* or *field stop* in the source plane which subtends a semi-angle α_F. This stop may be the source aperture itself ($\alpha_F = \alpha_S$) or the image in the source plane of the detector window ($\alpha_F = \alpha_D$). For the situation shown in Figure 8.1, $\alpha_F = \alpha_D$.

8.2.1. Dependence of resolving power on limiting stop

In order to gain an impression of the effects of a finite source aperture we need consider only the rudimentary parts of the interferometer as shown in Figure 8.2. The interferometer itself can be simplified to two planes: the moving mirror M_1 and the image M_2' of the fixed mirror in the moving mirror arm. The separation of the two planes is $\frac{1}{2}x$, where x is measured in terms of axial path difference. We consider a ray from some point S in the effective source aperture inclined at angle α to the principal axis. It is then rays such as those reflected at A and B that interfere when they are recombined at the detector. The path difference x_α between the two rays in question is

$$x_\alpha = (AB + BC) - AD \tag{8.1a}$$

$$= x/\cos \alpha - x \sin \alpha \tan \alpha$$

$$= x \cos \alpha \quad \text{[m]} \tag{8.1b}$$

and this should replace x in the expressions already used. (Note that $\alpha = 0$ for $x_\alpha \equiv x$.) The off-axis path difference is *less* than the axial path difference. The phase corresponding to (8.1) for some wavenumber σ is

$$\phi_\alpha = 2\pi\sigma x_\alpha = 2\pi\sigma x \cos \alpha \quad \text{[rad]}. \tag{8.2}$$

218

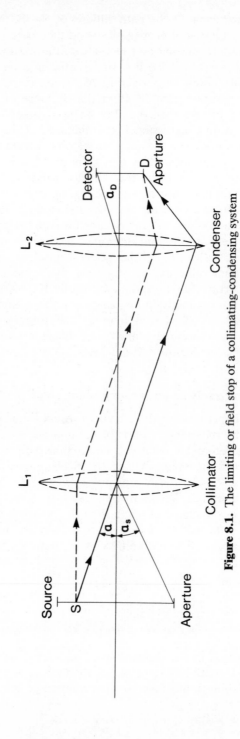

Figure 8.1. The limiting or field stop of a collimating-condensing system

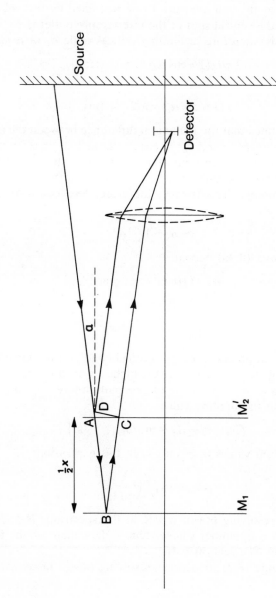

Figure 8.2. Equivalent optical system of a two-beam interferometer

The difference between the on-axis and off-axis phases,

$$\phi - \phi_\alpha = 2\pi\sigma x(1 - \cos\alpha) \quad \text{[rad]}, \tag{8.3}$$

therefore increases as α increases and becomes a maximum when $\alpha = \pi/2$.

If we arrange for the exit aperture to be just filled by the radiation cone corresponding to the central spot of the interference pattern, $\phi - \phi_\alpha$ will not exceed π. There is, therefore, a limiting critical value α_c of α satisfying

$$2\pi\sigma x(1 - \cos\alpha_c) = \pi \quad \text{[rad]} \tag{8.4a}$$

or

$$x(1 - \cos\alpha_c) = \lambda/2 \quad \text{[m]}. \tag{8.4b}$$

Equation (8.4b) states that the maximum difference between the on-axis and off-axis path differences, x and $x_c = x\cos\alpha_c$, is one half-wavelength. If $x - x_c$ exceeds this value, destructive interference reduces the amount of power entering the detector aperture at maxima of the interference function and increases it at minima. The fringes, therefore, lose contrast. Since α_c is sufficiently small for

$$\cos\alpha_c \sim 1 - \tfrac{1}{2}\alpha_c^2 \tag{8.5}$$

and, since $\Omega_c = \pi\alpha_c^2$, (8.4a) becomes

$$\alpha_c = (1/\sigma x)^{1/2} \quad \text{[rad]} \tag{8.6a}$$

or

$$\Omega_c = \pi/\sigma x \quad \text{[sr]}. \tag{8.6b}$$

We must arrange, in practice, for this critical condition for the containment of the central fringe to be satisfied by the highest wavenumber present in the detected spectrum as well as for the largest path difference, that is, Ω_c must be no greater than the limiting value

$$\Omega_M = \pi/\sigma_M D = m\pi/\mathcal{R}_M \quad \text{[sr]} \tag{8.7}$$

set by the maximum values of σ and x. In this expression

$$\mathcal{R}_M = \sigma_M \Big/ \left(\frac{1}{mD}\right) \tag{8.8}$$

is the maximum resolving power σ_M/R in the spectrum, $R = 1/mD$ is the resolution, and m is a number whose value is dependent on the form of the weighting function (see section 6.4).

We can rearrange (8.7) to express resolving power in terms of limiting solid angle:

$$\mathcal{R}_M = m\pi/\Omega_M. \tag{8.9a}$$

We find, therefore, an important result concerning the resolving power in the spectrum. It appeared from our elementary discussion that the only limitation imposed on the spectral resolving power was that due to the need to record a finite length of interferogram. Thus, any finite resolving power would be achieved simply by recording sufficient interferogram. We now find that a further restriction is imposed when the limiting stop is finite. The resolving power can be no greater than

$$\mathcal{R} = m\pi/\Omega_F \qquad (8.9b)$$

for a limiting stop Ω_F. If Ω_F is chosen to have the value Ω_M, (8.9a) and (8.9b) are identical. Thus, the resolving power can be increased to the value given by (8.9b) by increasing the maximum value of x, but the value of D must satisfy the relation

$$D \leqslant \pi/\sigma\Omega_F. \qquad (8.10)$$

If greater resolving power is required, Ω_F must be reduced as appropriate, thereby reducing the throughput. We therefore recover the general result which applies to all spectrometers, that resolution is bought at the price of reduced throughput and therefore reduced signal-to-noise ratio. Of course for an interferometric spectrometer one gets the best bargain and an increase in resolution of at least an order of magnitude is available compared with a grating spectrometer—for the same final signal-to-noise ratio— but an ultimate limitation is, nevertheless, present.

8.2.2. Effect of finite field stop on monochromatic interferogram

Consider a small bundle of rays passing through the interferometer at inclinations to the axis lying in the interval α to $\alpha + d\alpha$. The radiant power collected from the source of spectral radiance $l(\sigma)$ is

$$l(\sigma)\, d\sigma\Omega_c\, dA_s \qquad [\text{W}], \qquad (8.11a)$$

where Ω_c is the solid angle subtended at the source by the collimator of focal length f_c and dA_s is the area of the annulus giving rise to the bundle of rays. If the radius of the annulus is r and its width is dr,

$$dA_s = 2\pi r\, dr = f_c^2\, d\Omega \qquad [\text{m}^2], \qquad (8.11b)$$

where we have introduced the increment $d\Omega$ of the solid angle

$$\Omega = \pi r^2/f_c^2 = \pi\alpha^2 \qquad (8.11c)$$

included by the cone of rays inclined at the angle α. Thus the power carried by the bundle of rays is

$$l(\sigma)\, d\sigma A_c\, d\Omega \qquad [\text{W}], \qquad (8.11d)$$

since $\Omega_c f_c^2 = A_c$. A convenient representation for (8.11d) in terms of $B(\sigma)$ is obtained by using (4.5) and noting that the *étendue* $E = \Omega_c A_s$ is also equal to $A_c\Omega_s$; the power can then be written

$$B(\sigma)\, d\sigma\, d\Omega/\Omega_F \qquad [\text{W}], \qquad (8.11e)$$

where Ω_s is replaced by Ω_F, the solid angle subtended by the field stop at the collimator. For monochromatic radiation of wavenumber σ_i and total power 6_i, the power carried by the rays is

$$6_i \, d\Omega/\Omega_F \quad [\text{W}] \tag{8.11f}$$

and the interference function produced is

$$dK_i(x) = \frac{6_i}{\Omega_F} d\Omega + \frac{6_i}{\Omega_F} \cos 2\pi \left(1 - \frac{\Omega}{2\pi}\right) \sigma_i x \, d\Omega \quad [\text{W}], \tag{8.12}$$

where, since α is generally small, we have used

$$\cos \alpha \sim 1 - \tfrac{1}{2}\alpha^2 = 1 - \Omega/2\pi. \tag{8.13}$$

Consequently, the flux of a given wavenumber that reaches the detector from the entire aperture Ω_F is

$$K_i(x) = \int_{\Omega=0}^{\Omega=\Omega_F} dK_i(x) \tag{8.14a}$$

$$= 6_i + 6_i \operatorname{sinc}(\sigma_i x \Omega_F/2\pi) \cos[2\pi\sigma_i x(1 - \Omega_F/4\pi)] \tag{8.14b}$$

$$\triangleq 6_i + G_i(x) \quad [\text{W}]. \tag{8.14c}$$

The constant part of $K_i(x)$, i.e. the average value of the interference record, is directly proportional to the solid angle since 6_i is proportional to Ω_F. The varying part is also proportional to Ω_F but has its amplitude modulated by the sinc function. Therefore as x increases the fringe contrast gets less and less. The fringe contrast first becomes zero where

$$\sigma_i x \Omega_F = 2\pi, \tag{8.15a}$$

i.e. where

$$\sigma_i x = 2\pi/\Omega_F. \tag{8.15b}$$

The condition (8.15b) shows that for a wavenumber σ_i, the interference fluctuations first die away when a path difference

$$x_e = 2\pi/\Omega_F\sigma_i \tag{8.16}$$

is attained; this is called the *extinction path difference*. Combining (8.16) with (8.7) we have

$$x_e = 2D \tag{8.17}$$

This lies, of course, beyond the permissible range of the mirror travel and will not, therefore, be reached. At the maximum path difference that is reached (i.e. $x = D$), the damping sinc function has reached 64% of its value at $x = 0$. Because the damping is so slow it can be ignored when wavelengths or wavenumbers are being measured, for it is only the periodicity of the interferogram which is of interest. The wavelength we measure,

$$\lambda_i^c = \frac{1}{\sigma_i(1 - \Omega_F/4\pi)} \quad [\text{m}], \tag{8.18a}$$

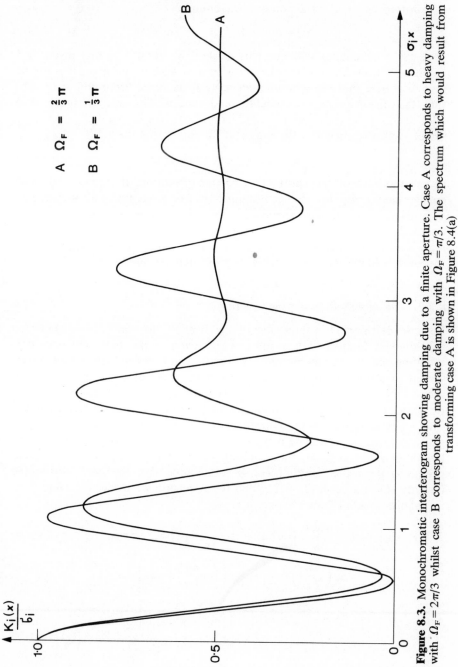

Figure 8.3. Monochromatic interferogram showing damping due to a finite aperture. Case A corresponds to heavy damping with $\Omega_F = 2\pi/3$ whilst case B corresponds to moderate damping with $\Omega_F = \pi/3$. The spectrum which would result from transforming case A is shown in Figure 8.4(a)

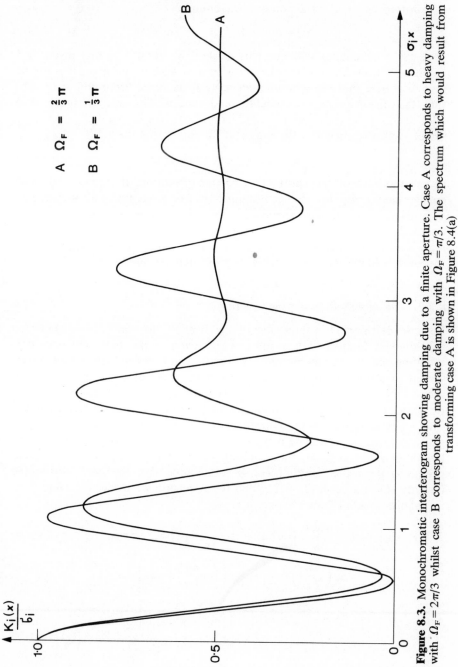

A $\quad \Omega_F = \frac{2}{3}\pi$

B $\quad \Omega_F = \frac{1}{3}\pi$

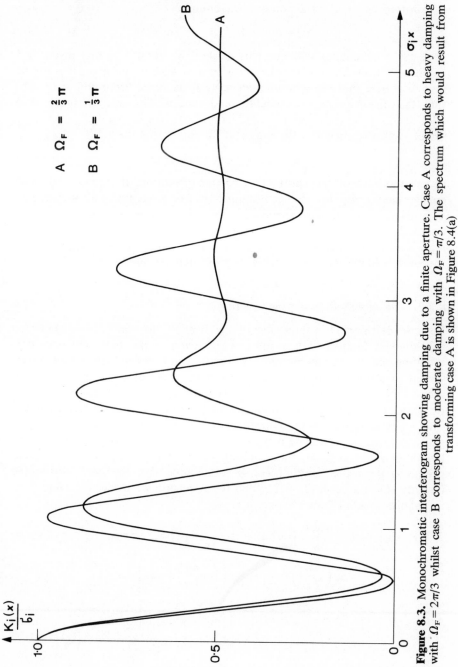

223

is distorted and the corresponding wavenumber

$$\sigma_i^c = \sigma_i(1 - \Omega_F/4\pi) \qquad [\text{m}^{-1}] \tag{8.18b}$$

appears to be smaller than the true value. This result was first given by J. Connes.[16] In finding the Fourier transform of an interferogram it is important to realize that the wavenumber scale is distorted according to (8.18b), that the distortion is unavoidable, and that in the most accurate work allowance for the contraction should be made by correcting the wavenumber scale. In this connection, the important parameter is the ratio

$$|\sigma_i^c - \sigma_i|/R = (m/4\pi)\Omega_F\sigma_i D \tag{8.19}$$

of the modulus of the contraction to the resolution. If this ratio is much smaller than unity, the shift is insignificant. For a fixed resolution (fixed D) in a given interferometer (fixed Ω_F), this ratio increases with wavenumber and indicates that for high σ_i (mid- and near infrared) the shift can be significant even for modest resolutions. Alternatively, at low σ_i (far infrared) the shift is likely to be significant only in high resolution work.[61]

8.2.3. The modified apparatus function

Because of the slow attenuation of the fringes, the transformed spectrum will show a feature with a finite width even if the extended source is radiating strictly monochromatic radiation, and the interferogram is observed to infinite path difference. It follows, therefore, that the interferograms (and consequently the resulting spectra) of an extended monochromatic source and a point quasi-monochromatic source are qualitatively similar. This analysis applies to the consideration of an infinite length of interferogram, but one can conclude just as soundly that if one is transforming a *finite* length of interferogram, the resulting resolution will be worse than that predicted from the length used. If we assume that the field stop is the limiting stop (i.e. that $\Omega_F = \Omega$), then the spectrum we would calculate from an infinite length of interferogram is

$$B'^c(\sigma) = \mathfrak{b}_i \int_{-\infty}^{\infty} \text{sinc}\left(\frac{\sigma_i x\Omega}{2\pi}\right) \cos\left[2\pi\sigma_i x\left(1 - \frac{\Omega}{4\pi}\right)\right] \cos 2\pi\sigma x \, dx \tag{8.20a}$$

$$= \tfrac{1}{2}\mathfrak{b}_i \int_{-\infty}^{\infty} \text{sinc}\left(\frac{\sigma_i x\Omega}{2\pi}\right) \cos\left\{2\pi\left[\sigma_i\left(1 - \frac{\Omega}{4\pi}\right) + \sigma\right]x\right\} dx$$

$$+ \tfrac{1}{2}\mathfrak{b}_i \int_{-\infty}^{\infty} \text{sinc}\left(\frac{\sigma_i x\Omega}{2\pi}\right) \cos\left\{2\pi\left[\sigma_i\left(1 - \frac{\Omega}{4\pi}\right) - \sigma\right]x\right\} dx$$

$$= \tfrac{1}{2}\mathfrak{b}_i(2\pi/\Omega\sigma_i) \sqcap (2\pi\sigma/\sigma_i\Omega) * [\delta(\sigma_i^c + \sigma) + \delta(\sigma_i^c - \sigma)] \qquad [\text{W m}]. \tag{8.20b}$$

This consists of two features located (see Figure 8.4(a)), respectively, at

$$\sigma = \pm\sigma_i^c = \pm(1 - \Omega/4\pi)\sigma_i \qquad [\text{m}^{-1}]; \tag{8.21}$$

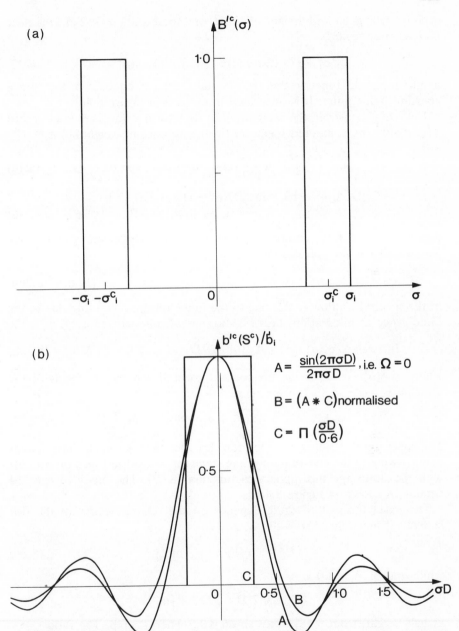

Figure 8.4. In (a) is shown the apparatus function which results from the study of monochromatic radiation using an interferometer of finite aperture. For convenience the heights of the box-cars are shown as unity. Their true heights are $\pi \eth_i/\sigma_i \Omega_F$ and the area under the two features therefore comes out, as it should, to simply $\eth_i$. In (b) is shown the spectral window which results when the interferometer scan has also a finite range. Again for convenience the curves are shown normalized, but the true height of curve C (resulting when the box-car and sinc function have the same width) is 0.82

each is rectangular with height $\pi \bar{b}_i/(\sigma_i \Omega)$ and total width $\sigma_i \Omega/(2\pi)$. It is clear that we may regard

$$A_\Omega(\sigma_i^c - \sigma) = (2\pi/\sigma_i \Omega) \sqcap (2\pi\sigma/\sigma_i \Omega) * \delta(\sigma_i^c - \sigma) \qquad (8.22)$$

as the apparatus function that arises as a direct consequence of the source aperture being finite. It is illustrated in essence in Figure 8.4(a).

If we now constrain the integration in (8.20a) to a finite range to accord with reality the computed spectrum (in the absence of weighting) is

$$B'^c(\sigma) = \bar{b}_i \int_{-\infty}^{\infty} \text{sinc} \left(\frac{\sigma_i x \Omega}{2\pi} \right) \sqcap \left(\frac{x}{2D} \right) \cos 2\pi\sigma_i^c x \cos 2\pi\sigma x \, dx \qquad (8.23a)$$

$$= 2\bar{b}_i D[A_\Omega(\sigma_i^c + \sigma) * \text{sinc } 2D(\sigma_i^c + \sigma) + A_\Omega(\sigma_i^c - \sigma)$$
$$* \text{sinc } 2D(\sigma_i^c - \sigma)] \qquad [\text{W m}]. \qquad (8.23b)$$

Again we have two features: that at the physically meaningful positive wavenumber is the convolution

$$B'^c(\sigma) = 2\bar{b}_i DA_\Omega(\sigma_i^c - \sigma) * \text{sinc } 2D(\sigma_i^c - \sigma), \qquad \sigma > 0 \qquad [\text{W m}] \qquad (8.24a)$$

of the spectral window (8.22) due to the finite aperture with that due to the finite range of integration; thus (8.24a) may also be written as

$$B'^c(\sigma) = \bar{b}_i A_\Omega(\sigma_i^c - \sigma) * A_0((\sigma_i^c - \sigma)D), \qquad \sigma > 0 \qquad [\text{W m}], \qquad (8.24b)$$

where $A_0[(\sigma_i^c - \sigma)D]$ is the unapodized spectral window. In terms of the variable $s^c = \sigma_i^c - \sigma$, (8.24b) becomes

$$b'^c(s^c)/\bar{b}_i = A_\Omega^i(s^c) * A_0(s^c D) \qquad [\text{m}]. \qquad (8.25)$$

The spectrum we evaluate is therefore broader than we might have anticipated, owing to the convolution of the standard unapodized spectral window with the finite aperture apparatus function (8.22). The modified spectral window is shown in Figure 8.4(b).

The exact form of $b'^c(s^c)/\bar{b}_i$ depends on the relative widths of the two convolved functions. The width of $A_\Omega^i(s^c)$ is

$$\delta A_\Omega^i = \sigma_i \Omega/2\pi \qquad [\text{m}^{-1}] \qquad (8.26)$$

while that of $A_0(s^c D)$ is

$$\delta A_0 = 0.7/D \simeq \tfrac{3}{4}/D \qquad [\text{m}^{-1}] \qquad (8.27)$$

at half-height (from 6.58), since sinc $\theta = 0.5$ when $\theta \sim 0.6$. The ratio

$$\frac{\delta A_\Omega^i}{\delta A_0} = \frac{\sigma_i \Omega}{2\pi} \cdot \frac{4D}{3} = \frac{2}{3} \frac{\sigma_i D\Omega}{\pi} \qquad (8.28)$$

is the significant factor. If $\delta A_\Omega^i \gg \delta A_0$, the aperture term dominates the spectral window, while, if $\delta A_0 \gg \delta A_\Omega^i$, the truncation term is dominant. Now since from (8.7) there exists a maximum value of Ω set by the maximum wavenumber present and the maximum path difference introduced, it follows at once that

$$\frac{\delta A'_\Omega}{\delta A_0} = \frac{2}{3} \cdot \frac{\Omega}{\Omega_M} \qquad (8.29)$$

and the ratio is therefore less than unity under normal circumstances.

A detailed consideration of the broadening effect of the aperture term has been given by J. Connes.[16,27] She shows that for $\delta A_\Omega = \delta A_0$, the peak height is reduced to 82% of its value where $\delta A_\Omega = 0$ and the width of the feature is increased by 10% (see Figure 8.4(b)).

8.2.4. Modified apparatus function—weighted interferogram

We can now introduce weighting of the interferogram so that the computed spectrum is

$$B'^c(\sigma) = 2\int_0^\infty G_i(x) W\left(\frac{x}{D}\right) \sqcap\left(\frac{x}{2D}\right) \cos 2\pi\sigma x \, dx \qquad (8.30a)$$

$$= \mathfrak{G}_i[A_\Omega(\sigma_i^c + \sigma) * A((\sigma_i^c + \sigma)D) + A_\Omega(\sigma_i^c - \sigma)$$
$$* A((\sigma_i^c - \sigma)D)] \qquad \text{(W m]}, \quad (8.30b)$$

where $A(\sigma D)$ is the apodized spectral window. Clearly, the spectral distribution may be written

$$b'^c(s^c)/\mathfrak{G}_i = A'_\Omega(s^c) * A(s^c D) \qquad \text{[m]}. \qquad (8.31)$$

The explicit form of (8.31) needs to be specially calculated for each particular weighting function. However, the general conclusion will in all cases be the same: the effect of a finite aperture for the source will be to broaden the spectral window in the calculated spectrum by comparison with the spectrum calculated for a point source.

8.2.5. Effect of finite field stop for quasi-monochromatic radiation

So far we have only considered the highly artificial situation in which we have assumed the interferometer to be irradiated by a finite source that is strictly monochromatic. The next step is to investigate the situation in which the source has a finite but narrow spectral extent, that is a quasi-monochromatic source.

Suppose the source is isotropic and radiates spectral power

$$B(\sigma) \, d\sigma = b(s) \, ds \qquad \text{[W]} \qquad (8.32)$$

at wavenumber $\sigma = \sigma_i + s$ and suppose that this power is finite over a small range $\Delta\sigma \ll \sigma_i$. The radiant flux that is detected along a given direction α is given by the integral

$$dK(x) = \frac{d\Omega}{\Omega_F} \int_0^\infty B(\sigma) \, d\sigma + \frac{d\Omega}{\Omega_F} \int_0^\infty B(\sigma) \cos 2\pi\sigma x_\alpha \, d\sigma \qquad \text{[W]} \quad (8.33)$$

representing the sum of all terms such as (8.12) given for just a single wavenumber σ_i.

Since the spectral band is narrow, the second integral in (8.33), representing the interferogram observed for a ray along a direction α, may be written

$$\int_0^\infty B(\sigma) \cos 2\pi\sigma x_\alpha \, d\sigma = \int_0^\infty b(s) \cos 2\pi(\sigma_i + s)x_\alpha \, ds$$

$$= \cos 2\pi\sigma_i x_\alpha \int_0^\infty b(s) \cos 2\pi s x_\alpha \, ds \qquad [\text{W}] \qquad (8.34)$$

if we assume that $b(s)$ is symmetrical about $s = 0$ (thus making the sine transform of $b(s)$ equal to zero). Equation (8.33) may then be written

$$dK(x) = \bar{I}\frac{d\Omega}{\Omega_F} + f(x_\alpha) \cos 2\pi\sigma_i x_\alpha \frac{d\Omega}{\Omega_F} \qquad [\text{W}], \qquad (8.35)$$

where

$$f(x_\alpha) = \int_0^\infty b(s) \cos 2\pi s x_\alpha \, ds \qquad [\text{W}] \qquad (8.36)$$

is the cosine Fourier transform of the source function and

$$\bar{I} = \int_0^\infty B(\sigma) \, d\sigma.$$

The flux that reaches the detector from the entire solid angle Ω_F is then given by

$$K_i(x) = \bar{I} + \int_0^{\Omega_F} f(x_\alpha) \cos 2\pi\sigma_i x_\alpha \frac{d\Omega}{\Omega_F} \qquad [\text{W}]. \qquad (8.37a)$$

When the spectral distribution is very narrow, $f(x_\alpha) \sim f(x)$ and

$$K_i(x) \simeq \bar{I} + f(x) \, \mathrm{sinc}\,(\sigma_i x \Omega_F/2\pi) \cos\,[2\pi\sigma_i x(1 - \Omega_F/4\pi)] \qquad [\text{W}]. \qquad (8.37b)$$

The recorded interferogram is, therefore (see (8.14b)), distorted in two ways from its ideal form

$$K_i(x) \simeq \bar{I} + f(x) \cos 2\pi\sigma_i x \qquad [\text{W}]; \qquad (8.38)$$

since there is an additional amplitude modulation, the argument of the cosine function is altered, and both are dependent on σ_i. The spectrum we obtain from (8.37b) is

$$B_i^c(\sigma^c) = 2\int_0^\infty f(x) \, \mathrm{sinc}\left(\frac{\sigma_i x \Omega_F}{2\pi}\right) \cos\left[2\pi\sigma_i x\left(1 - \frac{\Omega_F}{4\pi}\right)\right] \cos 2\pi\sigma x \, dx$$

$$= B(\sigma) * A_\Omega(\sigma_i^c - \sigma) \qquad [\text{W m}], \qquad (8.39)$$

the convolution of the required spectrum with the rectangular scanning function (see (8.22)) $A_{\Omega}(\sigma_i^c - \sigma)$. The spectral feature at positive wavenumbers (see (8.22)) is

$$
\begin{aligned}
B_i^c(\sigma^c) &= B(\sigma) * (2\pi/\sigma_i \Omega) \sqcap (2\pi\sigma/\sigma_i \Omega) * \delta(\sigma_i^c - \sigma) \\
&= b(s^c) * (2\pi/\sigma_i \Omega) \sqcap (2\pi s^c/\sigma_i \Omega) \\
&= b(s^c) * A_{\Omega}^i(s^c) = b_i^c(s^c) \qquad [\text{W m}].
\end{aligned} \tag{8.40}
$$

This result may be extended, with no need for further detailed working. When a finite range of interferogram is used the spectrum becomes

$$
b_i^c(s^c) = b(s^c) * A_{\Omega}^i(s^c) * A_0(s^c D) \qquad [\text{W m}] \tag{8.41}
$$

and, when weighting is employed,

$$
b_i^c(s^c) = b(s^c) * A_{\Omega}^i(s^c) * A(s^c D) \qquad [\text{W m}]. \tag{8.42}
$$

The finding is, as before, that the spectral window is modified to the convolution

$$
A_{\Omega}^i(s^c) * A(s^c D). \tag{8.43}
$$

This window is dependent on σ_i, as we have denoted by carrying the superscript on $A_{\Omega}^i(s^c)$. The spectral window will, therefore, be different when the feature of interest $b(s)$ lies at a different centre σ_i in the available spectral range. The height of $A_{\Omega}^i(s^c)$ is $\pi/(\sigma_i \Omega)$ and the width is $\sigma_i \Omega/(2\pi)$. If the aperture effect is significant, that is, if $A_{\Omega}^i(s^c)$ makes a significant contribution to the convolution (8.43), then considerable care needs to be exercised in comparing line shapes found for spectra having features at various places throughout a spectral range (see Figure 8.5).

If it is possible, this difficulty is best avoided by arranging for $A_{\Omega}^i(s^c)$ to have little influence in the convolution (8.43), that is, by arranging for $\delta A_{\Omega}^i \ll \delta A$. We should use an aperture Ω that is considerably smaller than Ω_M to achieve this. This will be the preferred way in which an interferometer is used, but a compromise is necessary in practice since a very small aperture approximates to the ideal of a point source but passes little radiation, whereas a large aperture provides more radiation but causes distortion in the calculated spectrum. The experimentalist must, as always, in interferometric spectroscopy, choose his operating parameters to suit the result he is trying to achieve.

8.2.6. Effect of finite field stop for broad-band radiation

With broad-band radiation, the approximations that were made to derive Equation (8.39) become inappropriate and (see (8.14b)) we must write

$$
B'^c(\sigma) = 2 \int_0^{\infty} \left[\int_0^{\infty} B(\sigma') \operatorname{sinc}\left(\frac{\sigma' x \Omega_F}{2\pi}\right) \cos\left[2\pi\sigma' x \left(1 - \frac{\Omega_F}{4\pi}\right)\right] d\sigma' \right] \cos 2\pi\sigma x \, dx. \tag{8.44}
$$

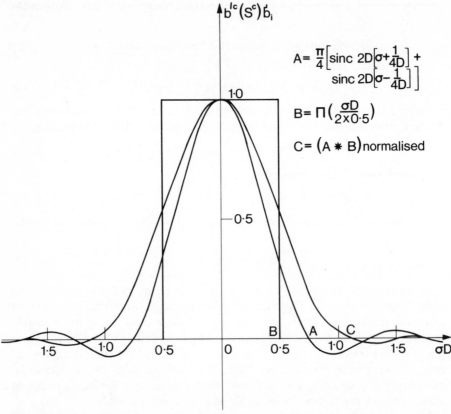

Figure 8.5. Effect of finite aperture on an apodized spectral window. Curve A arises using the cosine weighting function and it will be seen that an extremely large (and in practice improbable) aperture broadening would be necessary before any additional effects would be noticeable

This can also be written (cf. (8.20b))

$$B'^c(\sigma) = \int_0^\infty \left[\frac{2\pi}{\sigma'\Omega_F} \Pi\left(\frac{2\pi\sigma}{\sigma'\Omega_F}\right) * [\delta(\sigma'^c + \sigma) + \delta(\sigma'^c - \sigma)] \right] B(\sigma') \, d\sigma'. \quad (8.45)$$

It is not at all easy to visualize in a simple physical way the meaning of these two equations. The quantity in brackets in (8.45) represents a pair of box-cars centred at $\sigma = \mp\sigma'(1 - \Omega_F/4\pi)$ having width $\sigma'\Omega_F/2\pi$ and height $2\pi/\sigma'\Omega_F$. For a particular value of σ, the box-car scans across the range of σ'^c changing both in height and in width as it does. However Kiselev and Parshin[49] have been able to show that the integrals (8.44) and (8.45) may be written in the remarkably simple form

$$B'^c(\sigma) = \frac{2\pi}{\Omega_F} \int_\sigma^{\sigma(1 - \Omega_F/2\pi)^{-1}} \frac{B(\sigma')}{\sigma'} \, d\sigma', \quad (8.46)$$

which holds provided $B(\sigma')$ is well behaved and that $B(0) = 0$. (By expanding $B(\sigma')$ in a power series it may readily be shown that $B'^c(\sigma) \to B(\sigma)$ as $\Omega_F \to 0$.) The result is therefore quite general and contains no assumptions about the form of the spectral distribution. When a finite interferogram range and a weighting function are introduced, the spectrum that is evaluated is (8.44) with the terms $W(x/D)$ and $\sqcap(x/2D)$ added between the square bracket and the cosine term. By the use of the convolution theorem (section 2.3) and by invoking the definition of a convolution (2.45b), one may write

$$B'^c(\sigma) = \frac{2\pi}{\Omega_F} \int_{-\infty}^{\infty} A((\sigma - \sigma'')D) \left[\int_{\sigma''}^{\sigma''(1 - \Omega_F/2\pi)^{-1}} \frac{B(\sigma')}{\sigma'} \, d\sigma' \right] d\sigma'' \quad (8.47)$$

By interchanging the independent variables in (8.47), we find that the equation may be written

$$B'^c(\sigma) = \frac{2\pi}{\Omega_F} \int_{-\infty}^{\infty} \frac{B(\sigma')}{\sigma'} \left[\int_{\sigma'(1 - \Omega_F/2\pi)}^{\sigma'} A((\sigma - \sigma'')D) \, d\sigma'' \right] d\sigma', \quad (8.48)$$

which we may write as

$$B'^c(\sigma) = \int_{-\infty}^{+\infty} B(\sigma') \mathcal{W}(\sigma, \sigma') \, d\sigma', \quad (8.49a)$$

where

$$\mathcal{W}(\sigma, \sigma') = \frac{2\pi}{\sigma \Omega_F} \int_{\sigma'(1 - \Omega_F/2\pi)}^{\sigma'} A((\sigma - \sigma'')D) \, d\sigma'', \quad (8.49b)$$

summarizes the whole effect in the apodized spectrum of using a finite source. It is not very convenient to use (8.49b) as it stands because of its asymmetry, but a simple change of variable can be used to throw it into a symmetrical form. Writing $u = (\sigma - \sigma'')D$ one has

$$\mathcal{W}(\xi, \mu) = \frac{1}{2\mu} \int_{\xi - \mu}^{\xi + \mu} A(u) \, du, \quad (8.50)$$

where $\xi = (\sigma - \sigma')D + \mu$ and $\mu = \sigma' D \Omega_F/4\pi$: ξ is essentially a frequency variable and μ is essentially a line broadening variable. It will be seen on expanding $A(u)$ in a power series that as $\mu \to 0$ $\mathcal{W}(\xi, 0)$ goes to $A(\xi)$, as it should. Equations (8.49) were first given by Parshin[50] who did, however, use a different notation. Their usefulness lies in their ability to give directly the effect on the calculated spectrum of the use of a finite source even when weighting is used. One merely calculates the spectral window function $A(\sigma D)$ and then substitutes it in (8.49b) to arrive at the aperture distorting function itself and hence to $B'^c(\sigma)$ via (8.49a). Of course one often cannot do this analytically, but the integrations are readily performed numerically on a small computer or even on a programmable calculator.

The process will be illustrated by considering two simple 'analytical' expressions for the weighting function, namely the rectangle function and the triangle function. With rectangular weighting (that is, merely truncation) we have

$$A_0(u) = 2D \text{ sinc } 2u \tag{8.51}$$

(see (6.15)), so that

$$\mathcal{W}(\xi, \mu) = \frac{1}{2\mu} \int_{\xi-\mu}^{\xi+\mu} 2D \text{ sinc } 2u \, du \tag{8.52}$$

becomes a sine integral (section 2.9.2.1) whose value is

$$\mathcal{W}(\xi, \mu) = \frac{D}{2\mu\pi} [\text{Si} (2\pi(\xi+\mu)) - \text{Si} (2\pi(\xi-\mu))]. \tag{8.53}$$

The maximum value, which occurs at $\xi = 0$, is given by

$$\mathcal{W}(0, \mu) = \frac{2D}{(2\mu\pi)} \text{Si} (2\mu\pi). \tag{8.54}$$

It will be seen that this goes to $2D$ (as it ought) when Ω_F, and hence μ, tends to zero. In Figure 8.6 is shown the form of the normalized ratio

$$\mathcal{W}_0(\xi, \mu) = \frac{\mathcal{W}(\xi, \mu)}{\mathcal{W}(0, \mu)} = \frac{1}{2 \text{Si} (2\mu\pi)} [\text{Si} (2\pi(\xi+\mu)) - \text{Si} (2\pi(\xi-\mu))] \tag{8.55}$$

for $\mu = 0$, 0.35, 0.5, and 0.675. The broadening effect of using a large aperture is quite obvious and the width increases as μ increases. However the sidebands of the aperture function do not get progressively smaller as μ increases but pass through a minimum magnitude before increasing again. The minimum occurs for $\mu \approx 0.505$. It will be realized from this or from examination of Figure 8.6 that the finite aperture has the effect of introducing apodization and it may be advantageous in certain experimental situations—especially those that involve the study of sharp lines occupying a relatively narrow frequency interval—to use no additional apodization but rather to choose the aperture such that the optimum value of μ is achieved. For broad-band spectroscopy this procedure is less useful since, for a fixed Ω_F, μ will vary significantly across the band. However some allowance can be made for this.

If apodization is employed, then the effect of the finite aperture is to give a still broader line. This will be illustrated for the case mentioned earlier—that is, triangular apodization. From (6.35b) we have

$$A(u) = D(\text{sinc } u)^2 \tag{8.56}$$

and therefore

$$\mathcal{W}(\xi, \mu) = \frac{D}{2\mu} \int_{\xi-\mu}^{\xi+\mu} (\text{sinc } u)^2 \, du. \tag{8.57}$$

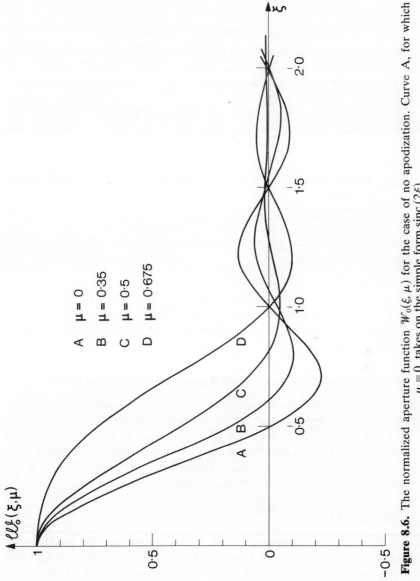

Figure 8.6. The normalized aperture function $\mathcal{W}_0(\xi, \mu)$ for the case of no apodization. Curve A, for which $\mu = 0$, takes on the simple form sinc (2ξ)

Evaluating this integral by means of the method of parts gives

$$
\mathcal{W}(\xi, \mu) = \frac{D}{2\mu\pi} \left[\mathrm{Si}\,(2u) - u \left[\frac{\sin u}{u} \right]^2 \right]_{(\xi-\mu)\pi}^{(\xi+\mu)\pi}
$$

$$
= \frac{D}{2\mu\pi} \left[(\mathrm{Si}\,(2(\xi+\mu)\pi) - \mathrm{Si}\,(2(\xi-\mu)\pi)) + \frac{1}{\pi(\xi^2-\mu^2)} \right.
$$

$$
\times [\mu(1 - \cos 2\xi\pi \cos 2\mu\pi) - \xi \sin 2\xi\pi \sin 2\mu\pi)]. \qquad (8.58)
$$

Likewise, this tends to the limit of D when ξ is zero and μ tends to zero. In Figure 8.7 are shown the normalized forms of $\mathcal{W}(\xi, 0)$ and $\mathcal{W}(\xi, 0.5)$. On comparison with Figure 8.6 it will be seen that the additional apodization (over and above that due to the finite aperture) has led to a broadening and still further attenuation of the side lobes. Further analysis shows that the effect of the finite aperture on the shape of sharp lines is to reduce the power on the high wavenumber side: this effect is greatest when no apodization is used. If Ω_F is known, the effect can be calculated and a correction applied. Parshin[50] has given simple expressions for this correction. Thus for the unapodized case one has

$$
\mathcal{W}(0, \mu)/\mathcal{W}(0, 0) = \mathrm{Si}\,(2\mu\pi)/(2\mu\pi) = 1 - \tfrac{1}{18}(2\mu\pi)^2 + \ldots, \qquad (8.59)
$$

which reveals the high wavenumber attenuation. Therefore if one defines a correction factor

$$
\eta(\mu) = \mathcal{W}(0, 0)/\mathcal{W}(0, \mu), \qquad (8.60)
$$

the true spectral window can be obtained by replacing $\mathcal{W}(\xi, \mu)$ by $\eta(\mu)\mathcal{W}(\xi, \mu)$. The true spectrum can therefore be obtained by correcting the wavenumber scale from σ' to $\sigma(1 - \Omega/4\pi)$ and by calculating the ordinates using $\eta(\mu)\mathcal{W}(\xi, \mu)$.

8.3. IMPERFECTIONS ASSOCIATED WITH MIRRORS

If the mirrors of the interferometer have distorted surfaces or are not correctly adjusted, the resultant interferogram is degraded because of this and its Fourier transform does not give a correct representation of the detected spectral distribution. There is a variation of path difference over the aperture of the mirrors. To take account of this each element of the beam that contributes to the final image must be considered separately.

8.3.1. The defect function

Suppose the power distribution over the beam incident on the interferometer to be uniform and equal to $B(\sigma)\,d\sigma/A$ for radiation of wavenumber σ; A is the area of the beam. For the ideal interferometer the contribution to the total power at the exit aperture from an element $dA = du\,dv$ at (u, v) is

$$
dI(x) = \int_0^\infty d\sigma \frac{B(\sigma)}{A} [1 + \cos 2\pi\sigma x(u, v)]\,dA \qquad [\mathrm{W}] \qquad (8.61)
$$

235

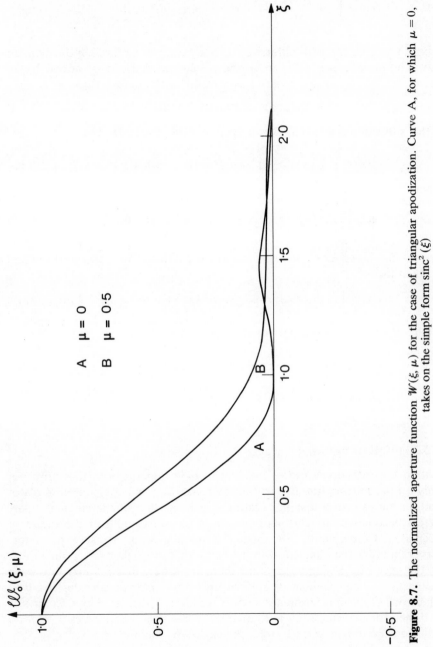

Figure 8.7. The normalized aperture function $\mathscr{W}(\xi, \mu)$ for the case of triangular apodization. Curve A, for which $\mu = 0$, takes on the simple form $\operatorname{sinc}^2(\xi)$

A $\mu = 0$

B $\mu = 0.5$

and the total power is

$$I(x) = \int_0^\infty d\sigma \frac{B(\sigma)}{A} \iint_A [1 + \cos 2\pi\sigma x(u, v)] \, du \, dv \qquad \text{[W]}, \qquad (8.62)$$

where $x(u, v)$ is the path difference for the rays passing through the element dA. The dependence of x on u and v can generally be described by an increment $\delta(u, v)$ arising directly as a result of the defect, thus

$$x(u, v) = x + \delta(u, v) \qquad \text{[m]}. \qquad (8.63)$$

After substituting in (8.62) we find, since $\delta(u, v)$ is small, that

$$I(x) = \int_0^\infty B(\sigma) \, d\sigma + \int_0^\infty d(\sigma) B(\sigma) \cos 2\pi\sigma x \, d\sigma \qquad \text{[W]}, \qquad (8.64)$$

where

$$d(\sigma) = \frac{1}{A} \iint_A \cos [2\pi\sigma\delta(u, v)] \, du \, dv \qquad (8.65)$$

represents the effect on the spectrum of the defect. The new interferogram is, therefore

$$F_D(x) = \int_{-\infty}^\infty B_e(\sigma) \, d(\sigma) \cos 2\pi\sigma x \, d\sigma = F(x) * D(x) \qquad \text{[W]}, \qquad (8.66)$$

where $B_e(\sigma)$ is the even part of the detected spectrum $B(\sigma)$ and

$$D(x) = \int_{-\infty}^\infty d(\sigma) \cos 2\pi\sigma x \, d\sigma \qquad \text{[m}^{-1}\text{]}, \qquad (8.67)$$

the *defect function*, is the Fourier transform of the *spectral distortion factor* $d(\sigma)$.

8.3.2. Effect of distorted surfaces

Within the category 'surfaces' we can include imperfections that arise not only on the mirrors but also within the separating and compensating plates and, in the case of a lamellar grating, on the surface of the elements. Any defect becomes relatively more severe as the wavelength of the radiation decreases. Consequently, the higher frequency components of the interferogram are worst affected, with the result that the higher wavenumber end of the spectrum is degraded more than the lower, that is, $d(\sigma)$ falls as σ increases. As an example of surface distortion, we can consider the case where one of the interferometer mirrors is spherical, discussed by J. Connes.[16]

If the radius of curvature is R_s, and we define polar coordinates (r, θ) for a point on the mirror surface, then the sagitta of the mirror is

$$\frac{r^2}{2R_s} \qquad \text{[m]} \qquad (8.68)$$

and that of the wavefront reflected from it

$$\delta(r) = \frac{r^2}{R_s} \quad \text{[m]}. \tag{8.69}$$

The spectral distortion factor is

$$d(\sigma) = \frac{1}{A} \int_0^{2\pi} d\theta \int_0^R r \, dr \cos\left(2\pi\sigma \frac{r^2}{R_s}\right) = \text{sinc}\left(2\sigma \frac{R^2}{R_s}\right) \tag{8.70}$$

and the defect function

$$D(x) = \int_{-\infty}^{\infty} \text{sinc}\left(2\sigma \frac{R^2}{R_s}\right) \cos 2\pi\sigma x \, d\sigma = \sqcap\left(\frac{xR_s}{2R^2}\right) \quad \text{[m}^{-1}\text{]}. \tag{8.71}$$

The recorded interferogram is, therefore, $F(x) * \sqcap(xR_s/2R^2)$ and the calculated spectrum $B(\sigma) \, \text{sinc}\,(2\sigma R^2/R_s)$. The spectral distortion factor falls to zero first when $\sigma = \sigma_s = R_s/2R^2$. It is clear that the highest wavenumber present in the spectrum must be considerably less than σ_s to avoid excessive attenuation of the calculated intensities. If we make $\sigma_M = 0.1\,\sigma_s$, $d(\sigma)$ falls from 1 at $\sigma = 0$ to only 0.983 at $\sigma = \sigma_M$. Thus the spectral attenuation is less than 1.7% if $\sigma_M \leq R_s/5R^2$.

8.3.3. Effect of maladjustment of the interferometer

The effect of a constant maladjustment is equivalent to a surface fault and is, of course, independent of x. It is also possible to have a maladjustment due to the perturbation of a mirror plane during the motion of the mirror; this effect *is* x-dependent.

8.3.3.1. Constant maladjustment

We assume that during the setting up of the interferometer the mirror tilts were not correctly adjusted to make M_2 and the image of M_1 parallel. As a result, the fringe pattern is degraded.

Suppose the circular, perfectly flat, mirrors are misaligned by an angle e that is small and suppose that the axis of intersection passes through the centre of the aperture. The two reflected wavefronts are inclined at an angle $2e$, thus giving rise to a path difference increment

$$\delta(v) = 2ve \quad \text{[m]}. \tag{8.72}$$

where v is a Cartesian coordinate in the direction of maximum tilt for the plane normal to the beam direction. The spectral distortion factor is

$$\begin{aligned}
d(\sigma) &= \frac{1}{A} \int_{-R}^{R} dv \int_{-(R^2-v^2)^{1/2}}^{(R^2-v^2)^{1/2}} du \cos 4\pi\sigma v e \\
&= \frac{2R}{A} \int_{-R}^{R} \left[1 - \left(\frac{v}{R}\right)^2\right]^{1/2} \cos 4\pi\sigma v e \, dv \\
&= 2\frac{J_1(4\pi\sigma Re)}{4\pi\sigma Re}
\end{aligned} \tag{8.73}$$

and the defect function is

$$D(x) = \int_{-\infty}^{\infty} 2\left(\frac{J_1(4\pi\sigma Re)}{4\pi\sigma Re}\right) \cos 2\pi\sigma x \, d\sigma = \frac{1}{\pi} \left| 1 - \left(\frac{x}{2Re}\right)^2 \right|^{1/2} \quad [m^{-1}].$$

$$(8.74)$$

The calculated spectrum is, therefore $2J_1(4\pi\sigma Re)B(\sigma)/4\pi\sigma Re$. The spectral distortion factor falls to zero first when $4\pi\sigma Re = 3.83$, that is, when $\sigma_s = 0.305/Re$. It is then equal to 98.3% of the value at $4\pi\sigma eR = 0$, the value for perfect alignment ($e = 0$), when $e = 0.041/R\sigma$. Thus, the spectral attenuation is less than 1.7% if

$$\sigma_M \leqslant \frac{0.041}{Re}.$$

8.3.3.2. Variable maladjustment

A variable maladjustment occurs when, for example, the tilt of the movable mirror varies as the mirror is translated. The tilt may be periodic due, for example, to some error in the drive screw or it may be random due, for example, to vibrational effects accompanying the motion. The path difference increment is now x-dependent, so that

$$x(u, v) = x + \delta(u, v; x) \quad [m] \tag{8.75}$$

and the spectral distortion factor is

$$d(\sigma, x) = \frac{1}{A} \iint_A \cos\left[2\pi\sigma\delta(u, v; x)\right] du \, dv. \tag{8.76}$$

The defect function is the integral.

$$D(x) = \int_{-\infty}^{\infty} d(\sigma, x) \cos 2\pi\sigma x \, d\sigma \quad [m], \tag{8.77}$$

which is not, strictly, a Fourier transform, owing to the dependence of $d(\sigma, x)$ on both σ and x.

8.4. THE PRACTICAL PRODUCTION OF THE INTERFEROGRAM

8.4.1. Aperiodic and periodic generation

The variation of path difference (x) within the interferometer takes place, of course, in time (t) so the interferogram, strictly speaking is a function of time. The Fourier transform of a time function is a true frequency function so the spectrum produced has, again strictly speaking, frequency as its abscissa. The frequencies produced in this way are very low, never getting much higher than the audio frequency region, thus reflecting the fact that the recording of the interferogram is a relatively slow procedure. They are, however, strictly proportional to the frequencies of the radiation being measured with the interferometer and one may relabel the abscissa in true radiation frequencies (say in THz) by means of the simple relation

$$Af = \nu, \tag{8.78}$$

where A is a constant, f the audio frequency appearing in the transform and ν the radiation frequency. Different symbols, f and ν are used so as to emphasize the differences between the two kinds of frequency. The movement of the mirror (in time) which produces the interferogram is usually referred to by the descriptive term '*scanning*'. Other terms are sometimes encountered: thus Genzel[51] uses 'interference modulation' but the term 'modulation' is, as will be seen later, best reserved for a rapid fluctuation imposed deliberately on the beam (or beams) to facilitate the experimental problem of translating a radiation power into a quasi-d.c. voltage.

There are two main approaches to the generation of the interferogram and each divides into two subtechniques. The first, main approach is *aperiodic generation*, in which one moves the mirror slowly from $x = -D$ to $x = +D$ (or from $x = 0$ to $x = D$) and records the interferogram ordinates at regular intervals Δx of x. It is essentially a 'one-off' operation. The experimentalist may repeat the observation several times, so that he may check on dubious features, or else so that he may use spectral averaging techniques to improve the signal-to-noise ratio, but these repeats can take place at any random time after the first observation has taken place. With *periodic generation* the mirror moves rapidly from $x = -D$ to $x = +D$ (or sometimes only over half the range) in a regularly repeated fashion. In aperiodic generation the mirror is moving so slowly that the signal level can be assumed to be essentially steady during the observing time of each sample, and some form of artificial modulation of the radiation is desirable so that the problems of d.c. amplification can be avoided. In periodic generation the mirror is moving over the interferogram at a sufficiently fast rate that the fluctuations lie in the audio frequency range and no additional modulation is necessary. This gives periodic generation an advantage, in that there is no interruption of the beam and all the power arrives at the detector all the time, but there are ways (see later) of achieving this advantage in aperiodic scanning.

The two forms of aperiodic generation are the continuous drive and the stepped drive modes. In the former the mirror moves, as its name suggests, continuously and the electronic system records the average value of the interferogram ordinate during a brief time slot disposed symmetrically about the correct sampling time. In the latter the mirror is stationary during the sampling interval and then moves as rapidly as possible to its next position, where it again rests for the time of the observation. The fundamental point which has to be explored in discussing the relative merit of these two techniques and, in fact, in discussing all techniques, is that the observing time has to be *finite* not just because practical electronics cannot respond instantaneously but because a finite observing time is essential so that the noise fluctuations will average down to an acceptable level. The continuous drive method, therefore, becomes open to the objection that the quantity to be measured is changing during the measurement period and that one has no *a priori* reason to believe that the average value of the ordinates over the time slot will equal the true value in the centre of the time slot. It is also open to the objection that it makes poor use of the available time since signal is only being averaged and recorded for a small part of that time. Close analysis of the operation shows that the spectral distortion, brought about by replacing an ordinate by the average of the ordinates throughout a small interval, is not very severe. This is because, if the requirements demanded by the sampling theorem are met, the interferogram will not be varying rapidly between sampling points and the average and the central ordinate will therefore agree quite closely. The time disadvantage does remain, but, since the mirror is moving smoothly, backlash problems can be almost eliminated and there are no sudden jolts to the system—with their attendant problems—such as are almost unavoidable in stepped drive operation. However, in stepped drive operation almost optimum use of the observing time can be achieved since the time for moving the mirror can be made a small fraction of an observing interval. The stepping is realized by driving the mirror, via a linear translator—a micrometer for example—from a stepper motor.

The two forms of periodic generation are the rapid scan and the frequency analysis modes. In the former the interferogram ordinates are digitized as the rapidly moving mirror passes through the appropriate positions. The various values obtained for each value of path difference are then averaged digitally and it is the averaged interferogram which is transformed. In the frequency analysis mode, there is no Fourier transformation of the interferogram. Instead the detector output is fed to a narrow-band tunable filter whose output indicates the power incident in its pass band. Essentially the interferometer scales the frequency from the inaccessible terahertz region down to the highly accessible audio frequency region and direct frequency spectroscopy becomes possible. Frequency analysis spectroscopy of this kind involves, as will be seen on reflection, a loss of the multiplex advantage and in the far infrared this is too high a price to pay for the experimental simplification, although Genzel[51] has described some preliminary experiments using it. In the near infrared, where energy limitation is not a restriction, it is in principle viable, but no one has yet published any results which are superior to those obtainable with a grating spectrometer. The vast majority of near and mid-infrared interferometry has been carried out by the rapid scan techniques which are exemplified by the well-known Digilab spectrometer. This features an air bearing which permits ultra-smooth motion for the mirror and laser monitoring of the mirror position (see later) to give a very reliable operation and one which is, moreover, very suited to computer control. Nevertheless, the most accurate spectroscopy and that featuring the highest resolution is still carried out even in the very near infrared by stepped-drive aperiodic methods.

The question of 'how rapid is rapid-scan?' might meet with the answer that there is theoretically no upper limit apart from that set by inertia of the moving parts and by speed of response of detectors and electronic systems. In practice, however, the speed of the mirror is surprisingly slow. Thus by the usual equation

$$\lambda f = v, \tag{8.79}$$

where λ is the wavelength of the radiation, f the audio frequency, and v the speed of the mirror, it will be seen that for a wavelength of 1 μm and a frequency of 10 kHz, the speed is only 1 cm s^{-1}. There is thus no difficulty in reversing the direction of travel at the end of the full traverse. At 100 μm the corresponding frequency would be 100 Hz so if the spectrometer is to cover the whole infrared region the detector output amplifier has to have a very large bandwidth.

8.4.2. Modulation of the radiant flux in an interferometer

In aperiodic generation it is necessary, as remarked earlier, to employ some form of flux modulation so that a.c. amplification may be used and one may have thereby the additional benefit of synchronous rectification. There are three ways in which an electromagnetic wave may be modulated, its *amplitude*, its *frequency*, or its *phase* may be regularly varied in time. Amplitude modulation (AM) is virtually universal in conventional infrared spectroscopy and, in fact, it was its introduction by American and German workers in the late 1930's and early 1940's which made infrared spectroscopy a practical proposition. In this technique an opaque chopper interrupts the beam regularly. The output of the detector (provided the detector is linear) faithfully mirrors the imposed fluctuations and since the regular waveform may be decomposed into the frequency components which make up its representation in terms of a Fourier series, the steady (i.e. d.c.) radiation intensity has been transcribed into a set of alternating (i.e. a.c.) components, each of whose amplitudes is a measure of the original radiant intensity. There is clearly no advantage in using a broad-band amplifier, since signal power is confined to regularly spaced but very narrow frequency regions. The common practice, therefore, is to use a sharply tuned narrow-band amplifier, in which only a single Fourier component is amplified, all other frequencies being strongly attenuated. There is, therefore, a considerable gain in signal-to-noise ratio, but at the usual price of loss of speed of response. If the pass band of the amplifier has width Δf, then the response time of the amplifier is of the order $(\Delta f)^{-1}$. A practical difficulty and one that bedevilled experimentalists in the early days was drift between the frequency of the chopper and that of the tuned amplifier—this could cause large changes in the interferogram ordinate and lead to drastic effects in the transformed spectrum. Modern equipment is much more stable and the pass band characteristic can usually be arranged to be more or less flat-topped, thus minimizing the effects of residual drifting. The actual frequency of operation is set by the speed of rotation of the chopper and by the number of blades it possesses. The waveform is determined by the blade profile and the choice of this is a most important point experimentally since one wishes to channel as much power as possible into the Fourier component (usually the lowest harmonic) which is to be amplified and detected. The ideal case would be if the modulated waveform were a pure cosinusoid; this is very difficult to arrange exactly, but some close approaches are possible. In Figure 8.8 are shown some waveforms and their resulting Fourier coefficients.

Amplitude modulation was naturally used in the early interferometers since they were constructed by spectroscopists grounded in orthodox infrared technology. The nature of the interferometric process in a two-beam interferometer permits, however, an alternative—phase modulation.[52] Phase modulation (PM) is commonplace in communication systems since it, like its close relative frequency modulation (FM), is more efficient and less prone to interference than AM. The radiation traversing a Michelson interferometer may be phase modulated if one of the mirrors (usually the 'fixed' one) is oscillated (or 'jittered') rapidly back and forth whilst at all times maintaining its surface perpendicular to the axis. The origin of the modulation is shown in Figure 8.9. When the path difference corresponds to a bright or dark fringe of the monochromatic radiation, the oscillation causes little or no change in intensity at the detector and there will be no a.c. output. When, however, the path difference corresponds to a point where the intensity is crossing the average value line, there will be a maximum a.c. signal. A *cosine* intensity wave is therefore translated into a *sine* voltage wave. The argument is readily extended to the broad-band case and one arrives at the general conclusion that a PM interferogram looks very like the first derivative of the corresponding AM interferogram. Some advantages of PM emerge at once from this simple physical visualization of its operation. First, since there is no interruption of the beam by an opaque chopper, the average power reaching the detector is doubled. Second, the average value of the interferogram ordinates is zero and there is no need to use up a large amount of the dynamic range of the digital system with non-interferometrically coded (and therefore useless) information: this wastage is especially serious when dealing with dispersive interferograms of low quality. Third, and for the same reason, the noise carried on the background—due for example to lamp flicker—is eliminated. Thus PM gives a considerable improvement in signal-to-noise ratio and at the same time makes better use of the digital system.

There are one or two minor drawbacks only: thus the aligning of the interferometer, using the detected signal, is a little more tricky since there is

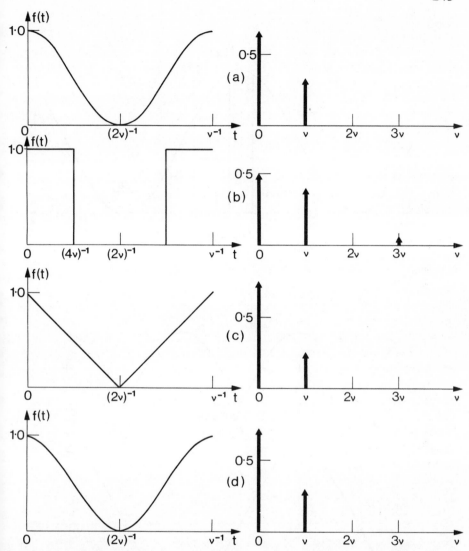

Figure 8.8. Voltage modulation waveforms and their resultant Fourier power coefficients for the cases of (a) pure cosinusoidal, (b) square wave, (c) triangular, and (d) 'straight edged blade over a round source' modulation. In most cases the higher frequency components are too weak to be shown separately

a zero of voltage at ZPD; also the depth of modulation (as will appear later) is wavenumber dependent and the amplitude of mirror jitter has, therefore, to be optimized for any required spectral range. A minor program modification is required so that a sine rather than a cosine transform may be produced. The problem of alignment, using the PM signal, is overcome with a little practice and the operator soon becomes adept at adjusting the

244

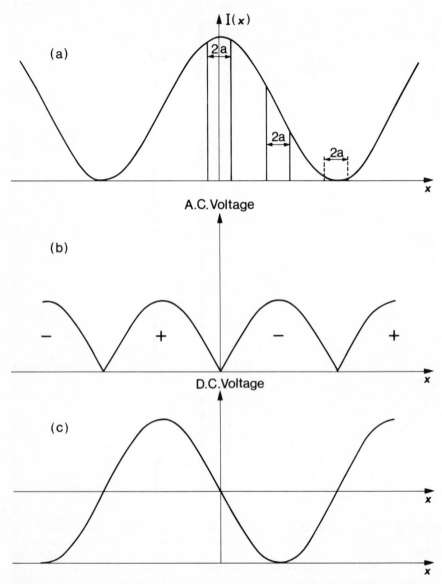

Figure 8.9. Mechanism of phase modulation. The variations of intensity (a) are converted into a varying a.c. voltage (b) by the oscillation of the mirror. The a.c. voltage is either of positive or negative sense (that is, of zero or π phase shift) with respect to the reference and on phase-sensitive rectification a d.c. voltage interferogram (c) is obtained. It is assumed throughout that only the fundamental frequency is passed by the electronic system

instrument to produce the best possible contrast between the maximum and the minimum on either side of ZPD. Alternatively optical techniques such as autocollimation can be used. The wavenumber dependence of the modulation can actually be turned to advantage so that the interferogram carries mostly information about a desired spectral band pass and therefore makes better use of the limited digital range of the encoding system. With plastic film beam division, however, a rather similar limitation is imposed by multiple beam interference in the film and the two phenomena are usually matched to maximize and minimize the same wavenumbers. It will be seen, therefore, that PM represents a considerable advance in interferometric technique and it is becoming widely employed. It, therefore, is of some interest to go into its operation in some detail.

8.4.3. Mathematical theory of phase modulation

There are only two important modulation functions to be considered, square wave modulation and cosinusoidal modulation. In the former the mirror changes discontinuously from $x_0 + a$ to $x_0 - a$, where x_0 is its mean position. In the latter the mirror moves smoothly according to the relation

$$x = x_0 + a \cos 2\pi ft. \tag{8.80}$$

If the interferometer is being irradiated with monochromatic radiation, then the detector signal, at a path difference x, is

$$V(x) = V_0[1 + \cos 2\pi\sigma x]. \tag{8.81}$$

The jittering of the modulating mirror makes x time-dependent, according to the relation $x = x_0 + a\phi(t)$, where $\phi(t)$ is the modulating function (square wave or cosinusoidal as the case may be). We have, therefore, a time-dependent, but periodic, voltage output from the detector:

$$V(t) = V_0 |1 + \cos 2\pi\sigma[x + a\phi(t)]|, \tag{8.82}$$

which we may resolve into its Fourier components.

With square wave modulation, the voltage output $V(t)$ is also a square wave which oscillates between $V_0[1 + \cos 2\pi\sigma(x_0 + a)]$ and $V_0[1 + \cos 2\pi\sigma(x_0 - a)]$ with a frequency f. This is shown in Figure 8.10. The detector output can, therefore, be divided into a d.c. component of magnitude $V_0[1 + \frac{1}{2}\cos 2\pi\sigma(x_0 + a) + \frac{1}{2}\cos 2\pi\sigma(x_0 - a)]$, which is of no further interest, and a purely a.c. part which oscillates between $\pm(V_0/2)[\cos 2\pi\sigma(x_0 + a) - \cos 2\pi\sigma(x_0 - a)]$, i.e. between $\pm V_0 \sin(2\pi\sigma x_0) \sin(2\pi\sigma a)$. The decomposition of a unit amplitude square wave into its Fourier series coefficients is a well-known problem—its solution is given in essence in equation (2.146). Only the odd harmonics appear; their amplitudes go inversely as the odd integers and with the phase change of $\pi/2$ shown in Figure 8.10 as compared with Figure 2.17 change sign alternately. In other words

$$V_{2n+1} = \frac{4}{\pi} \cdot \frac{(-1)^n}{2n+1}, \qquad n = 0, 1, 2, \text{ etc.}, \tag{8.83}$$

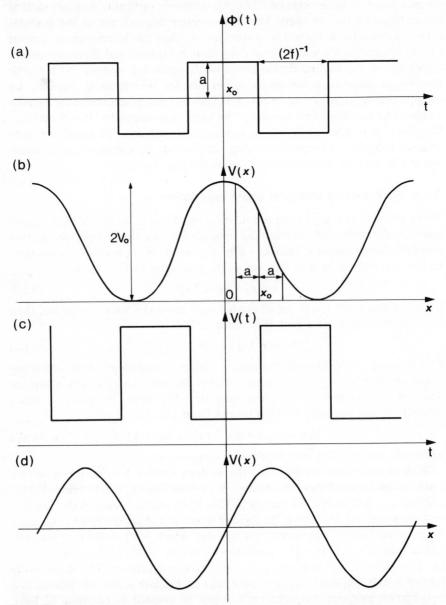

Figure 8.10. Square-wave modulating function (a), intensity (i.e. d.c. voltage) interferogram of monochromatic radiation (b), oscillation of voltage as the phase modulating mirror jitters about x_0 with amplitude a (c), and resulting rectified interferogram (d) corresponding to the fundamental Fourier component

and the harmonic number is given by $m = n + 1$. It follows that since the power goes as the square of the voltage, the powers will be in the proportions 1, $\frac{1}{9}$, $\frac{1}{25}$, etc., and since

$$1 + \frac{1}{9} + \frac{1}{25} + \cdots \frac{1}{(2n+1)^2} \cdots = \frac{\pi^2}{8} = 1.2337, \qquad (8.84)$$

the fundamental ($m = 1$) carries 81% of the power. Thus there is very efficient channelling of the available power into the chosen (i.e. the fundamental) harmonic.† Of course one cannot achieve a square wave in practice since mirrors cannot be accelerated and decelerated instantaneously and the waveform that one could realistically achieve would be more like a 'squared-off' symmetrical saw-tooth. However, the Fourier series representation of the saw-tooth has the coefficients $4(1 - \cos n\pi)/\pi n^2$ which account for equation (8.84) above and the saw-tooth channels the available power even more efficiently. The conclusion, therefore, remains that practical approaches to square-wave modulation are very attractive.

The rectified signal at a given value of x will be proportional to

$$V_0 \sin (2\pi\sigma a) \sin 2\pi\sigma x \qquad (8.85)$$

and this will therefore be the interferogram. The result is easily extended to the broad-band case by the usual infinitesimal arguments and one arrives at

$$I(x) = \int_0^\infty B(\sigma) \sin (2\pi\sigma a) \sin 2\pi\sigma x \, d\sigma. \qquad (8.86)$$

This is antisymmetrical, has no constant component, and its sine Fourier transform will give not $B(\sigma)$, but $B(\sigma) \sin 2\pi\sigma a$. The multiplying factor we may call the *phase modulation characteristic*: it is always wavenumber dependent, but in the present case takes on a relatively simple form. Experimentally, one would have adjusted the spectral band pass, by filters, suitable thickness of beam splitter, etc., to emphasize a particular spectral region. All one has to do additionally is to choose the mirror amplitude a so that $\sigma_{max} a = 0.25$ and then the spectral band pass maximum σ_{max} and the maximum in the phase modulation characteristic will coincide.

For the case of cosinusoidal phase modulation, the equivalent of equation (8.82) is

$$V(t) = V_0[1 + \cos 2\pi\sigma[x + a \cos 2\pi f t]]. \qquad (8.87)$$

It will be recalled from Chapter 6 that the Fourier decomposition of a $\cos (\cos \omega t)$ function will involve Bessel functions and we therefore expect the Fourier components of $V(t)$ to be expressible in terms of the low order

† A criticism is sometimes voiced of PM that it is impossible to get perfect modulation, i.e. 100% channelling using it. In principle, one could with AM but in practice one cannot so the criticism is only academic.

Bessel functions $J_n(f)$, where n is integral. Expanding equation (8.87) we get

$$V(t) = V_0[1 + \cos 2\pi\sigma x \, \cos (2\pi\sigma a \, \cos 2\pi f t) - \sin 2\pi\sigma x \, \sin (2\pi\sigma a \, \cos 2\pi f t)].$$
(8.88)

We can now use the identities

$$\cos (k \cos 2\pi\nu t) = J_0(k) - 2J_2(k) \cos 4\pi\nu t + 2J_4(k) \cos 8\pi\nu t - \dots \text{etc.},$$
(8.89a)

$$\sin (k \cos 2\pi\nu t) = 2J_1(k) \cos 2\pi\nu t - 2J_3(k) \cos 6\pi\nu t + \dots \text{etc.}$$ (8.89b)

to calculate the powers in the various Fourier components. If the pass band of the detector plus amplifier is such as to admit only the fundamental, then the interferogram will be given by

$$V_0[2J_1(2\pi\sigma a)] \sin 2\pi\sigma x.$$
(8.90)

This is similar to that derived above for the case of square wave modulation but the phase modulation characteristic is not elementary (see Figure 6.4). It is interesting to note, however, that if one were to rectify the *second* harmonic one would then obtain a *symmetrical* interferogram. This is different from the square wave case where, of course, the even harmonics are absent. The phase modulation characteristic peaks where $\sigma a = 0.294$ and its first zero occurs at $\sigma a = 0.61$. In these relations, the parameter 'a' is quoted in path difference terms; the actual amplitude is only half this.

Orthodox spectroscopy, that is, with the specimen in one of the passive arms of the interferometer, calls for very little further comment. If two-sided phase modulated interferograms are recorded, the power will be mostly in the sine rather than in the cosine transform but the modulus will be identical with that obtained with AM. If single-sided interferograms are observed, some form of phase error correction is necessary, but the approach of Forman, Steele, and Vanasse[40] is just as readily applied with sine transforms as it is with cosine transforms and no problem is encountered in practice. All the arguments about resolution, apodization and spectral windows apply just the same. When, however, one comes to consider asymmetric or dispersive spectroscopy, in which the specimen is in one of the active arms, a slightly deeper analysis is required.

If one selects a single monochromatic component which contributes

$$\mathfrak{b}_i \cos [2\pi\sigma_i x + \phi_i]$$
(8.91)

to the interferogram, then in the AM case the cosine transform will be $\cos \phi_i$ multiplied by the ideal transform and the sine transform will be $-\sin \phi_i$ multiplied by the ideal transform. The ratio of the two transforms therefore gives $-\tan \phi_i$, as we saw in Chapter 7. In the PM case one has

$$\mathfrak{b}_i \sin [2\pi\sigma_i x + \phi_i]$$
(8.92)

and an identical analysis follows, except that the rôles of the sine and cosine transforms have to be reversed. From this it follows that the phase spectrum, unlike the power spectrum, is independent of the amplitude of the mirror motion. The argument is readily extended to the broad-band case and one would then have the two components of the complex spectrum:

$$P(\sigma) = \mu(\sigma a)B_e(\sigma) \sin \phi(\sigma_i), \qquad (8.93a)$$

$$Q(\sigma) = \mu(\sigma a)B_e(\sigma) \cos \phi(\sigma_i), \qquad (8.93b)$$

where $\mu(\sigma a)$ is the PM characteristic. Their ratio is independent of $\mu(\sigma a)$ but, if $\mu(\sigma a)$ were to take on negative values, an additional π would be needed in the phase angle. In practice, however, $\mu(\sigma a)$ is always positive over the region where there is any spectral energy and this difficulty never obtrudes. The conclusion, therefore, is that all the techniques of interferometric spectroscopy may be applied with PM just as with AM: the experimentalist must, however, choose a suitable value of the amplitude a to make $\mu(\sigma a)$ peak in the region of maximum interest.

8.4.4. Practical realization of phase modulation

Nearly all phase modulation that has been used in practice so far has been of the cosinusoidal type. A very lightweight front surface aluminized mirror is used so that inertial problems are reduced and the tendency to sag and thus put the interferometer out of alignment is avoided as far as possible. Early modulators were construced from slightly modified loudspeakers, the mirror being attached to the diaphragm. The amplitude of oscillation of the mirror was then determined by the audio frequency power supplied to the loudspeaker coil. It was found, however, that the fabric diaphragm of a loudspeaker tends to fatigue and is also subject to the effects of adsorbed water and it is, therefore, usually replaced by a phosphor-bronze 'spider'. Modern modulators tend to use substantial electromagnetic vibrators (essentially very robust loudspeaker coils) and these are found to be very satisfactory. Of course, the amplitude of oscillation has to be very accurately defined in PM as otherwise an additional source of noise will be imported into the spectrum, that due to the time variation of $\mu(\sigma a)$. Moss,[53] at the National Physical Laboratory, has devised a special circuit which keeps the amplitude very steady, but his investigations have revealed that the substantial vibrators tend to be self-compensating due to electromagnetic damping effects and there is little random or systematic variation of amplitude.

In PM interferometry the path difference has to be scanned and simultaneously modulated. With orthodox spectroscopy a sensible solution, and one very attractive from the engineering viewpoint, is to divide the work and let one mirror scan whilst the other modulates. With dispersive interferometry it may not be possible, however, to move the 'fixed' mirror. Thus, for example, when studying liquids, the specimen is in contact with the mirror which, in essence, forms the rear plate of the liquid cell. One has, therefore, to use the variable arm to provide both scanning and modulation. In early applications jittering was actually applied to the moving mirror

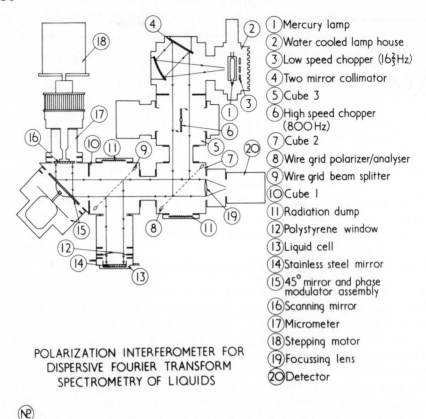

1. Mercury lamp
2. Water cooled lamp house
3. Low speed chopper ($16\frac{2}{3}$Hz)
4. Two mirror collimator
5. Cube 3
6. High speed chopper (800 Hz)
7. Cube 2
8. Wire grid polarizer/analyser
9. Wire grid beam splitter
10. Cube 1
11. Radiation dump
12. Polystyrene window
13. Liquid cell
14. Stainless steel mirror
15. 45° mirror and phase modulator assembly
16. Scanning mirror
17. Micrometer
18. Stepping motor
19. Focussing lens
20. Detector

POLARIZATION INTERFEROMETER FOR
DISPERSIVE FOURIER TRANSFORM
SPECTROMETRY OF LIQUIDS

Figure 8.11. A comprehensive submillimetre interferometer for dispersive Fourier transform spectroscopy. This features a slow ③ and a fast ⑥ amplitude modulating chopper and also an indirect ⑮ phase modulating chopper. (After Afsar, Hasted and Chamberlain (1976) *Infrared Physics*, **16**, 301–310

itself, but present day practice is usually to introduce an additional 45° vibrating reflector into the variable arm. This is illustrated in Figure 8.11.

8.4.5. Synchronous rectification of the amplifier output

The a.c. voltage emerging from the amplifier output could be simply rectified, either half-wave or full-wave, and the resulting signal smoothed using an *LC* or *RC* network to produce the d.c. analogue record needed for digitization. However, this process is inefficient since the noise signals are similarly rectified and one would be relying entirely on the smoothing network to attenuate their random excursions. Universal practice therefore is to use synchronous rectification in which another signal of the same frequency and the same phase as that being investigated is used to switch a synchronous rectifier. The reference signal used to be provided by a lamp and photocell arrangement which was interrupted also by the chopper, but nowadays a reference is usually taken direct from the modulating motor or vibrator. This is the only method available with PM.

The basic principle of a synchronous rectifier is that two input signals are multiplied together and the d.c. resultant (if any) is given as the output. Modern synchronous amplifiers usually take the incoming, and roughly sinusoidal, reference waveform and transform it to a square wave, of invariant amplitude, which oscillates about a true zero of voltage. The phase of this square wave can usually be adjusted by means of a simple rotary control. The operation is revealed in Figure 8.12. If a cosine wave having a phase shift of ϕ relative to the square wave is applied, the resultant average value is

$$V(\phi) \sim \left(\frac{2\pi}{\omega}\right)^{-1} \left[\int_{(\pi/2-\phi)/\omega}^{(\pi/2+\phi)/\omega} A_1 A_2 \cos \omega t \, dt - \int_{(\pi/2+\phi)/\omega}^{(3\pi/2+\phi)/\omega} A_1 A_2 \cos \omega t \, dt \right]$$

$$= \frac{4A_1 A_2}{2\pi} \cos \phi, \tag{8.94}$$

where A_1 is the amplitude of the signal waveform and A_2 that of the rectifying square wave. Thus the output signal is proportional to A_1, which is essential, but its absolute magnitude may swing from a maximum positive value to a maximum negative value through zero as ϕ is varied from 0 to π. It will be realized that noise waveform components whose phases lie near $\pi/2$ relative to the rectifier will be eliminated and, moreover, all except those with zero phase shift will be more or less attenuated. To work out the effect this will have on the signal-to-noise ratio in the interferogram, and hence in the transformed spectrum, it is necessary to make some assumptions about the distribution of noise components with respect to phase. If, as seems likely, the noise components are packed with equal density from $\phi = -\pi$ to $\phi = +\pi$, then the total noise power becomes the integral of $\cos^2 \phi$ averaged over the interval, in other words 0.5. The signal is of course unaffected, so the signal-to-noise ratio is improved by a factor of two. In experimental spectroscopy this is a very significant improvement.

It occasionally happens, in spectroscopy, that the source is not continuous, but gives out a series of short pulses—all regularly spaced. If the pulse length is Δt and the repeat time t, then the ratio $\Delta t/t$ is usually called the duty cycle. The duty cycles actually encountered are often very short and, in such circumstances, synchronous rectification becomes essential. This is because the noise power would be accumulated through the long interpulse period (usually called the 'space') and signal only through the relatively short pulse period (usually called the 'mark'). The signal is, therefore, usually passed to a circuit (called a 'gate') which allows passage when the pulse is on but does not when the pulse is off. The output from this circuit is then integrated on a special integrator, usually based on a high impedance electrometer valve. If the noise is produced *solely* by the detector, this system gives an improvement (over straight detection) in the signal-to-noise

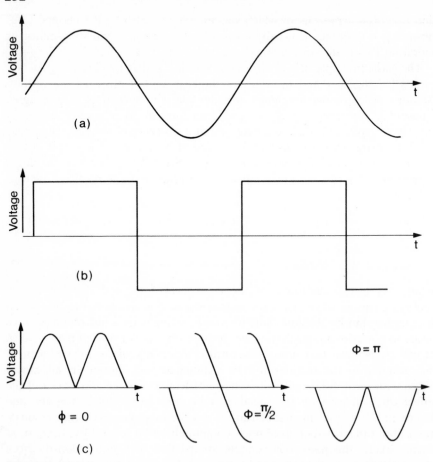

Figure 8.12. The reference voltage waveform (a) is enormously amplified in a saturable amplifier to produce the square wave (b). This is used to operate the phase-sensitive rectifier which is also fed by a signal waveform of the same frequency. The resulting output (as might for example be seen on an oscilloscope display) is shown for various values of the relative phase in (c)

ratio by a factor which is equal to the reciprocal of the duty cycle. If the noise is carried solely by the radiation or is produced solely in the electronics or any combination of these, there is no improvement. In infrared spectroscopy, however, detector noise is always significant and there is always available a substantial improvement in the signal-to-noise ratio by the use of gating circuits. The operation of the gate can also be considered in the frequency domain, where one resolves the regular signal pulses into their Fourier components. Power is again restricted to very narrow frequency regions and one requires of the rectifier that it be sensitive only in those regions and this is tantamount, of course, to saying that the rectifier be 'on' only when there is a pulse to rectify.

8.4.6. Smoothing of the analogue record

After modulation, narrow-band amplification, and phase-synchronous rectification the final output is a voltage whose value should faithfully mirror the original intensity at the detector. As the intensity varies, when, for example, the moving mirror scans, the output voltage will exactly image the variations and for this reason it is usually called the *analogue record*. It is, of course essential that we have an exact analogue record of the intensity interferogram if we are to transform it and derive the spectrum. The only major snag likely to be encountered by the experimentalist in achieving this necessary transliteration is the possibility of non-linear behaviour of the detector or of the electronic chain, but with reasonable care this possibility can be completely eliminated or at least its effects reduced far below the final noise level. The analogue record will normally be accompanied by residual noise and the spectroscopist can often use additional smoothing to advantage.

The simplest smoothing to be applied is a resistance (R)- capacitance (C) network. To see how this works, one imagines a capacitor C to be charged through a resistor R to a final voltage V_0. Solution of the appropriate differential equation gives the time dependence of the voltage on the capacitor as

$$V(t) = V_0[1 - \exp(-t/RC)]; \qquad (8.95)$$

the product RC is thus a characteristic delay time and it is usually called the *time constant* of the circuit. If the voltage V_0 is now allowed to vary sinusoidally in time, we need to determine the output of a potentiometric (i.e. voltage dividing) network which incorporates one or more capacities to provide smoothing. A simple circuit is shown in Figure 8.13. The output of this circuit for an input voltage of $V_0 \sin \omega t$ is

$$V(\omega) = \frac{V_0 R_3 (R_1 + R_2 + R_3) \sin \omega t}{\omega^2 C^2 R_1^2 (R_2 + R_3)^2 + (R_1 + R_2 + R_3)^2}$$
$$- \frac{V_0 \omega C R_1 (R_2 + R_3) \cos \omega t}{\omega^2 C^2 R_1^2 (R_2 + R_3)^2 + (R_1 + R_2 + R_3)^2}. \qquad (8.96)$$

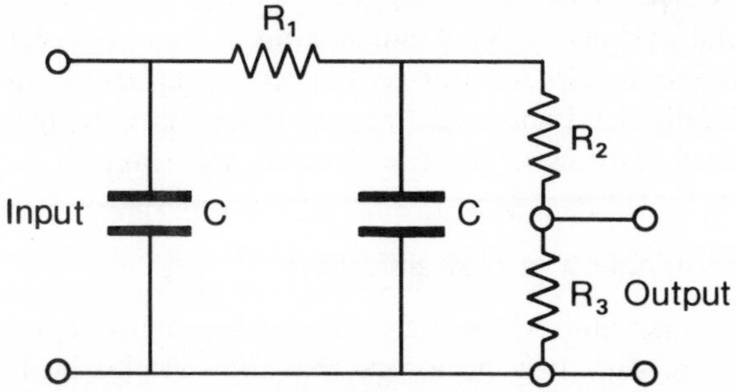

Figure 8.13. Typical resistance/capacitance smoothing network

The first term rapidly becomes negligible unless R_1 is very small, but the second falls off only as ω^{-1} beyond a characteristic value of ω, which we may define, somewhat arbitrarily, as that which makes the two terms in the denominator equal. This value is therefore

$$\omega_c = \frac{R_1 + R_2 + R_3}{CR_1(R_2 + R_3)}. \qquad (8.97)$$

Two limiting cases emerge: first, when R_1 is very small, ω_c becomes $(R_1C)^{-1}$ and, second, when R_1 is very large compared to $(R_2 + R_3)$, when ω_c becomes $[C(R_2 + R_3)]^{-1}$. Thus by a suitable choice of the circuit components one can arrange that frequency components up to a certain value are unaffected but those lying beyond are progressively attenuated. In the continuous scan aperiodic method one would choose the components such that the highest audio frequency produced by the scanning lay below the critical frequency. In the stepped scan mode one would choose the time constant such that it was much smaller than the sampling time. In this way the exponential change from one sampled value to the next would take only a small fraction of the available time and one would be sure of taking a true sample.

A more recent technique of smoothing is the use of integration circuits. In these there is no damping of the response, each instantaneous voltage is added, and a running average taken. The voltages are randomly distributed about the mean value and the running average converges closer and closer to the mean as the integrating time gets longer and longer. This method is theoretically superior to the RC method in all cases, but is particularly apposite to the stepped-scan situation. An extension of this method is to use digital integration in which the instantaneous voltages are digitized and the resulting numbers averaged. This method has in principle the possibility of eliminating random noise from the digitized record altogether, in much the same way that pulse-code-modulated (PCM) telephoning can be made noise free. However, in practice this is only possible if the digitization noise (see Chapter 9) is greater than the random noise, and with modern high reliability, high precision digital systems this would not be a common occurrence.

It is, of course, essential that the overall response time of the electronic/digital system be very much longer than the reciprocal of the modulation frequency, for otherwise the a.c. components in the rectified signal would be digitized and could appear either directly, or else in aliased form, as spurious features in the transformed spectrum.

8.4.7. The determination of path difference

The record which is required for the computer is a digitized sampled version of the analogue record. It is necessary that the samples be taken at equal spacings of path difference and that the absolute value of the spacing be known. There are three methods currently in use for determining the length of the sampling interval and for ensuring that the intervals be closely identical. The simplest, though oddly not the oldest, method is to mount the moving mirror on a micrometer driven by a stepping motor. One then relies on the quality of the micrometer and the reproducibility of the steps provided by the motor. The principal sources of error are periodic variations in the micrometer and random variations in the motor steps. The periodic variations give rise (via the Fourier decomposition of the 'cosine of a cosine' function) to spurious features regularly spaced each side of the genuine feature. By analogy with the artefacts encountered in grating spectroscopy and due to periodic errors in the ruling these are usually called 'ghosts'. They can be reduced by the same method that is used to make better class gratings, namely the use of the 'Merton nut' on the drive screw. The random errors give rise to 'noise' in the transformed spectrum. Both ghosting and the noise due to random errors in the sampling get worse as the wavenumber being studied gets higher, and as the resolution is increased, and it is not at all clear yet whether this method has any future for high resolution work above 1000 cm^{-1}. It is, however, so simple and provides so secure and reliable a digitizing signal that it is very attractive, particularly for the far infrared.

The second method relies on the generation of Moiré fringes. In this two linear transmission gratings are used inclined to one another at a slight angle. One grating spanning the distance of travel is fixed and the other—a short section of grating of indentical ruling—is attached to the carriage which carries the mirror. A beam of light from an incandescent source passes through the gratings and then on to a photocell, and as the carriage moves the Moiré fringe pattern moves across the detector window and produces an alternating voltage output. The frequency of the output is determined by the spacing of the Moiré gratings and by the speed of the carriage and it is, therefore, relatively simple to ensure that the interferogram samples are taken at the correct points. Care has to be taken however to ensure that the system is very robust and completely free from vibration, for unlike the stepper motor system there is always the danger of an extra, supernumerary, sample being taken and this could cause quite severe distortion if it were to occur before the noise-limited portion of the interferogram is reached. This will depend, of course, on the spectral content of the interferogram.

The third method is to use the interferometer, or a small subsidiary interferometer sharing the common moving mirror, to generate fringes in the radiation from a small gas laser. This is an extension of the idea, used by the early workers[54] of monitoring the path difference by observing the fringes of say the mercury green line (546.1 nm) or the cadmium red line (643.8 nm). In principle all random and systematic errors are eliminated. However, if the analysis is pushed to its limit, it will be realized that a laser emission, though extremely close to true monochromaticity and therefore capable of generating fringes over a wide difference of path length, can nevertheless take place anywhere over the spontaneous profile where gain is available. In orther words its *absolute* frequency can vary by many times its frequency spread. In routine spectroscopy at resolutions down to $0.1\,\text{cm}^{-1}$ this source of error is negligible, but at higher resolution one might have to use a stabilized laser, which would be adequate for all possible interferometric spectroscopic purposes. The most popular laser for path difference monitoring is the helium/neon laser operating at 632.8 nm. This is available with tube lengths of less than 300 mm and at very low cost. Sampling once per fringe the cut-off wavelength would be 1.2656 μm, but simple techniques are available for taking two samples per complete fringe and therefore the entire infrared can easily be covered. Similarly, to avoid oversampling, which will occur for example in the far infrared, fringe counting techniques are used to sample every nth fringe. Infrared beam splitters are often opaque in the visible so the same beam splitter often cannot be used for the infrared signal and the laser fringes. In this case, either a small central hole is cut in the beam divider to carry a visible beam splitter, or, more elegantly, the laser fringes are generated from the back surface of the moving mirror in a small subsidiary interferometer. The trigger signals to the digital system are usually generated by enormously amplifying the laser fringe signals in a non-linear saturable amplifier, which has the effect of turning a sine wave into a square wave. The square wave is then differentiated, which produces a series of pulses which are used for triggering the recording system.

The laser method is therefore very appropriate to those cases (continuous or rapid scan) where the carriage is moving steadily, but for the stepped scan mode some modifications are necessary. The most elegant way of doing this has been described by the Connes[55] who use 'cat's eye' retroreflectors as their mirrors. Coarse positioning of the 'cat's eye' is achieved by conventional motor stepping, but then fine adjustment of the path difference is brought about by a piezoelectric element attached to the small mirror of the 'cat's eye'. The signal to the piezoelectric compensator is produced by an electronic system fed from the detector observing the laser fringes via the same 'cat's eye'. Thus the path difference steps can be set accurately equal to the laser wavelength.

8.5. PERIODIC GENERATION OF THE INTERFEROGRAM

As mentioned earlier, there are two important techniques of periodic generation, the frequency analysis and the rapid scan modes. Their analysis has much in common with the aperiodic treatment but there are sufficient points of difference to justify a reasonably detailed account.

8.5.1. The frequency analysis method

So far this method has seen only limited application, principally in the laboratory of Professor Genzel at the University of Frankfurt. The loss of the multiplex advantage is a severe handicap and, although one can partially compensate for this by feeding the detector output to a bank of tuned filters, one then promptly loses the simplicity of operation which was so attractive in the first case! The theory can be developed in considerable depth and has several elegant aspects, but, bearing in mind the limited application of frequency analysis, only the basic principles will be given here.

The scanning function to be considered is the saw-tooth wave, (Figure 8.14) which, in its repeat period $0 \leqslant t \leqslant T_0$ has the form

$$X = tD/T_0. \tag{8.98}$$

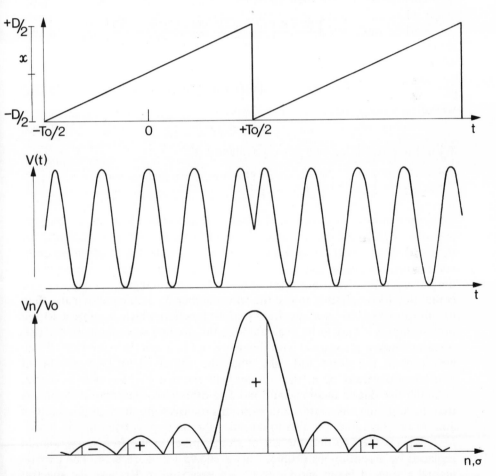

Figure 8.14. The saw-tooth scanning function in the upper inset produces the periodic voltage/time interferogram in the middle inset. This is resolved into its Fourier components in the lower inset, which also shows how the discrete frequency spectrum becomes blurred into a continuous spectral profile. For convenience all the ordinates are drawn above the line, an arrangement which is also readily achieved in practical spectroscopy

Substituting this in the voltage interferogram for monochromatic irradiation,

$$V(x) = V_0[1 + \cos 2\pi\sigma x],$$

gives

$$V(t) = V_0[1 + \cos 2\pi\sigma Dt/T_0], \qquad (8.99)$$

which can be resolved into a Fourier series

$$V(t) = V_0 + \sum_1^\infty V_n \cos 2\pi\left(\frac{n}{T_0}\right)t. \qquad (8.100)$$

The coefficients V_n are then given by

$$V_n = V_0\left[\frac{\sin 2\pi(\sigma D + n/2)}{2\pi(\sigma D + n/2)} + \frac{\sin 2\pi(\sigma D - n/2)}{2\pi(\sigma D - n/2)}\right] \qquad (8.101)$$

which can also be written

$$(V_n/V_0) = (-1)^n \operatorname{sinc}(n\sigma/\sigma_n)[2\sigma^2/(\sigma^2 - \sigma_n^2)], \qquad (8.102)$$

where

$$\sigma_n = n/2D. \qquad (8.103)$$

With this substitution one has equivalently

$$C\left(\frac{\sigma}{\sigma_n}\right) = \left(\frac{V_n}{V_0}\right) = \frac{\sin 2\pi D(\sigma + \sigma_n)}{2\pi D(\sigma + \sigma_n)} + \frac{\sin 2\pi D(\sigma - \sigma_n)}{2\pi D(\sigma - \sigma_n)}. \qquad (8.104)$$

These functions will be seen to degenerate into just two features at $n = +\sigma D$ and $n = -\sigma D$ when σD becomes integral. This is, of course, correct, for then with D an exact integral number of wavelengths the function becomes continuous, without a 'hiccup', at $t = T_0$ and the continuous cosine wave has only a single Fourier component. When σD is not integral, the best way of continuing the analysis is to use the wavenumber σ_n defined above and then to interpret (8.104) as a spectral window function which is subject to the proviso that it is only to be evaluated for the discrete wavenumbers σ_n. This spectral window is identical with that obtained in aperiodic operation. If one now presses the point and asks what the values of (V_n/V_0) will be at non-integral values of n, the answer will be that (V_n/V_0) will be finite, although the simple theory would indicate otherwise. One reason for this is that the scanning has to start and stop; this removes the sharp discontinuities and, in practice, one has also to consider the finite pass band of the filters. One expects then to find the set of sharp spikes in the frequency domain replaced by a continuous profile. This is shown in Figure 8.14. The experimental results of Happ and Genzel[56] are very similar. Because the spectral window function is so similar (identical with ideal filters) to that obtained in aperiodic scanning, the resolution in the spectrum is the same and we have $R = 0.7/D$.

The spectrum can be scanned by changing the filter frequency through the progressive values which correspond to the discrete wavenumbers σ_n or, alternatively, the filter frequency may be kept unchanged and the value of D changed. The former method permits wide spectral coverage, but the latter is restricted to a very narrow region. It is however an apposite way of mapping the spectral window when σ_n can be set close to an available monochromatic input (say a microwave source or a laser).

Apodization can be introduced into periodic scanning in much the same way that it can into aperiodic scanning. The apodizing function, which will be unity at $x = 0$ and zero at $x = D$, will likewise be periodic and may be resolved into its Fourier components. It is quite straightforward to show that the use of apodization (which is tantamount to a form of amplitude modulation) introduces side bands to the frequency spectrum and if the positions of these side bands and their intensities are chosen carefully one can get very efficient apodization in the same way that one does in aperiodic scanning (see Figure 6.3).

In the majority of applications of Fourier transform spectrometry, a broad spectral band of radiation is used and the interferogram has the form typified, for example, by Figure 1.4(c). There are sharp variations near $x = 0$, whereas for large values of x there are few, if any, variations in power. Since the turn-round in the periodic scan can never be perfectly sharp, a turning is best avoided at $x = 0$ and two-sided interferometry where the turnings are at $x = \pm D$ is preferable. If the interferogram is unsymmetrical, then two-sided operation becomes essential. Analysis shows that the a.c. signal from the detector contains two signals which are in quadrature and which may therefore be separated by a phase-sensitive detector. The first of these arises from the symmetrical component of the interferogram and it is identical, apart from numerical factors, with that already derived. It therefore gives a symmetrical spectral window described by equation (8.104). The second arises from the antisymmetrical component and its spectral window function is derived from the coefficients in a *sine* Fourier series. This is, therefore, an odd function and is given by

$$S\left(\frac{\sigma}{\sigma_n}\right) = (-1)^n \ \text{sinc}\left(\frac{n\sigma}{\sigma_n}\right)\left[\frac{2\sigma\sigma_n}{\sigma^2 - \sigma_n^2}\right]. \qquad (8.105)$$

This, like the even spectral window, can be resolved into factors:

$$S\left(\frac{\sigma}{\sigma_n}\right) = -\text{sinc} \ 2(\sigma + \sigma_n)D + \text{sinc} \ 2(\sigma - \sigma_n)D \qquad (8.106)$$

and it will be recalled that essentially this function emerged (equation (6.177b)) in the analysis of two-sided aperiodic interferometry. One can, therefore, carry out identically the same procedure as in Chapter 6 to produce an approximate modulus expression

$$|2D \ \text{sinc} \ 2(\sigma + \sigma_n)D + 2D \ \text{sinc} \ 2(\sigma - \sigma_n)D|, \qquad (8.107)$$

which may be used to derive the calculated power spectrum. The phase of the spectrum may be found from the arc tangent of the ratio of the two components in quadrature.

The commonest case of a phase, different from zero, arises when there is an offset span, for so far no-one has used periodic generation to produce dispersive spectra. The offset span introduces significant difficulties into the theory and these are best avoided by using an offset reference function—rather reminiscently of aperiodic work. Vogel and Genzel[57] analysed this and showed how a Moiré device coupled synchronously to the periodic path difference scanner could produce the necessary offset reference function.

8.5.2. Rapid scan methods

The rapid scan instruments currently on the market feature some of the most advanced technology ever to appear in infrared instrumentation, but, nevertheless, the principles on which they work are basically very simple. Fundamentally the operation is aperiodic since each interferogram is separately observed and recorded each time the mirror completes its scan, but the presence of a zero path difference monitor plus the ability of the built-in computer to carry out immediate phase correction means that the set of interferograms can be added in the memory store and this is formally equivalent to periodic operation. There is no amplitude or phase modulation of the radiation, so the energy throughput is increased and there is no additional imposed spectral characteristic. Also the interferogram oscillates about a true zero of voltage and there may be a consequent gain, just as there is with PM in signal-to-noise ratio. On the other hand, there is no instrumental noise suppression and the achieving of an acceptable noise level depends on the accumulation of a sufficient number of interferograms. This is equivalent to the 'CAT' (computing of average transients) techniques familiar in pulsed magnetic resonance work. It depends on the simple theorem that the signal-to-noise ratio observed in the average of N independent observations will be improved by the factor $N^{1/2}$ compared with that in any single observation.

The rapid scan instruments, of which the Digilab FTS-14 can be taken as typical, follow the detector with a wide-band amplifier. The output voltage of the amplifier is digitized in a high precision analogue-to-digital converter, the output of which proceeds directly to the computer. The digitizing signals are provided from a laser reference interferometer and there is, additionally, a small 'white light' source whose interferometric grand maximum of signal provides a zero path difference fiducial mark. The computer, after phase correction, can store and average a large number of interferograms and can also, if desired, transform a number of averages and average the resulting spectra. The fundamental point is that the signal-to-noise ratio is determined solely by the length of the observing time, but one can either spend the time observing a large number of noisy interferograms or a smaller number of less noisy interferograms. The computer control gives the operator a great deal of flexibility here. A major advantage of interferometric spectroscopy, which is not often stressed, is that the final output being in digital form lends itself very readily to further processing. Thus one can have automatic scale expansion to an arbitrary degree in either axis, one can have subtraction of a known reference spectrum to leave solely the spectral features which differ between the specimen and the reference, one can have spectral intensities automatically estimated by digital integration, and many more things beside.

The broad spectral cover of the rapid scan instruments makes them natural choices for mid- and near infrared work, but here, unlike the far infrared, where interferometry is the only sensible way to do spectroscopy, they encounter severe competition from conventional grating spectrometers. The mid-infrared is where the bulk of infrared spectroscopy is carried out and it is mainly done by semi-skilled spectroscopists. The conventional spectrometers are specially designed to be extremely easy to operate and the designers of interferometric spectrometers have faced this challenge and produced instruments in which all of the internal operations are under computer control and all the operator has to do is to learn to use the teletype keyboard. The sample chamber is designed to be virtually identical with that of a conventional spectrometer and it is quite possible to train staff to use the instrument without their realizing just what is going on inside. Of course, in the hands of skilled spectroscopists these instruments can do wonderful things since they have all the luminosity and multiplex advantages inherent in the method. One can, therefore, study almost opaque specimens, determine weak emission spectra, or on the other hand use the advantages to 'buy' quickness of operation. Thus gas-phase chromatography has been revolutionized by the coupling of the column to an infrared interferometric spectrometer which permits determination of the elutant spectra 'on the fly'. One has to realize, however, that these instruments are very expensive and unlikely to get significantly cheaper and for this reason there will always be a place for grating spectrometers in the lower and middle parts of the market. At the upper end, on the other hand, grating instruments will only be able to compete on price grounds, and this is getting more and more difficult.

The success of the rapid scan interferometers in the mid- and near infrared has led to many publications and also to a few more general accounts and reviews. Thus Bell[58] gives a good general introduction, Hanel and his colleagues[59] describe astronomical applications, and Griffiths, Foskett, and Curbelo[60] review chemical uses. It seems likely that these successes will continue and that new fields of endeavour will be opened up by these remarkable and very sophisticated spectrometers.

CHAPTER 9

Noise effects in Fourier transform spectrometry

9.1. THE TYPES OF NOISE ENCOUNTERED

As we have seen in Chapter 8, the interferogram signal is invariably accompanied by noise, that is, random electrical signals, and the object of the recording system is to 'clean-up' the wanted signal so that the signal-to-noise ratio (S/N) will be high enough to give an acceptable spectrum when the interferogram is transformed. There are many profound questions to be answered here, not least being that of just how one decides the maximum tolerable noise level in the interferogram. However, before tackling these we need to consider the types of noise encountered and their origins.

The noise may arise in the detector (or recording electronics) or else may be carried by the radiation itself. Also, since some form of modulation will almost certainly be being used, this too may inject noise due, for example, to erratic fluctuations of the modulating system (chopper in AM, jittering mirror in PM). Noise generated in the interferometer or its associated electronics will merely be added to the wanted interferometric signal, provided the detector and electronic chain have strictly linear characteristics, and we therefore call this type *additive noise*. Noise associated with the radiation and carried by it will have a power which will vary with the signal level and we therefore call it *multiplicative noise*. Either kind of noise may have a frequency-independent power spectrum, i.e. may be 'white' or else may exhibit some particular form of frequency dependence. In particular, the so-called '$1/f$' noise mentioned in the previous chapter is very common. Multiplicative noise is found in two principal categories, *photon noise* and *signal noise*. Photon noise has its origin in the discrete or 'grainy' nature of electromagnetic radiation, i.e. that it is propagated in the form of packets of energy called photons. The number of photons arriving at a given surface in a specified time interval will fluctuate in an ergodic fashion and, since one is dealing with the problem of random and uncorrelated arrival, the fluctuations will be proportional to the square root of the number of photons arriving. It follows that the noise level will be proportional to the square root of the signal power (I_s) and that the signal-to-noise ratio will likewise be proportional to the square root of the signal power. Photon noise is a fundamental quantity. It is related to the random nature of the spontaneous emission process and is therefore an unavoidable concomitant of the use of broad-band thermal sources. When lasers are being considered the matter becomes much more subtle since one is dealing with *coherent stimulated* radiation. Fortunately it is unnecessary to go into any details of this very difficult field here since interferometers are almost exclusively used with thermal sources.

Signal noise arises from fluctuations in the output of the source or else from fluctuations in the transmissive medium between the source and the detector. The noise level is therefore proportional to the signal power and the signal-to-noise ratio is independent of the signal power. Therefore, in the three cases considered, we have

Noise type	S/N
Detector noise	$\sim I_s$
Photon noise	$\sim I_s^{1/2}$
Signal noise	Constant

We shall see later that the multiplex advantage is gained only in the case where the noise is independent of the signal power—that is, for the case of additive noise. In the photon noise case the multiplex advantage is just balanced by the increased noise and there is no net gain. In the signal noise dominated case multiplexing is transformed into a disadvantage when comparison is made with sequential operation.

In the infrared region of the spectrum, photon energies are of the same order of magnitude as thermal energies at ambient temperatures and, in the far infrared, the photon energies are in fact *less* than kT at room temperature, when $\sigma < 200 \text{ cm}^{-1}$. Infrared detectors will therefore be limited by their own noise characteristics unless they are operated at very low temperatures. The dominant noise source will therefore usually be the fundamental Johnson noise whose power in a bandwidth Δf is given by

$$P_J = 4kT\Delta f. \tag{9.1}$$

This expression is derived classically as part of the argument which leads ultimately to the Rayleigh–Jeans radiation formula. It is therefore an approximation and it only holds under the condition where the Planck formula degenerates into the Rayleigh–Jeans formula, namely $hf < kT$. However, since f will be an audio frequency, the temperatures where equation (9.1) will fail will never be reached in practice, and this equation can be taken, for all intents and purposes, to be exact. It will therefore be seen that it is advantageous to use thermal detectors (bolometers, etc.) at as low a temperature as possible to reduce the Johnson (i.e. thermal) noise to as low a value as possible. However, even at the cryogenic temperatures provided by pumped liquid helium—that is, a few kelvins—the responsivity of most infrared detectors is still not high enough to prevent the Johnson noise dominating, and it follows that with such a detector used on an interferometer the full multiplex advantage will be available. It is assumed, of course, in this that the source is stable. The development of Fourier transform spectroscopy in the infrared, and especially the far infrared region of the spectrum, rather than in the optical regions, came about for several reasons, but the availability of the multiplex advantage was one of the principal of these. In recent years, significant progress has been made in developing very sensitive infrared detectors and with these it is now possible to approach the photon-noise-limited condition. With these there is no multiplex advantage, but the interferometer still enjoys the luminosity or *étendue* advantage over a dispersive instrument. In fact this advantage is so great that interferometers are still the preferred instruments for infrared spectro-astronomy, even though the signal noise due to 'twinkling' and poor 'seeing' gives them a multiplex disadvantage!

In the optical regions there are available some very sensitive detectors and the photon noise tends to dominate. One needs here to distinguish very carefully between two quite different meanings which have become attached to the phrase 'photon noise'. To the optical spectroscopist the term refers, as mentioned earlier, to the fluctuations resulting from the random arrival of photons. To the infrared spectroscopist (and especially to the infrared astronomer) the term refers to the random fluctuations of the arrival at the detector of *background* photons. Detectors, of course, cannot be constrained to 'see' only the source, but this does not matter at optical frequencies where, because the peak of the black body curve is long past, the background is dark. In the infrared, on the other hand, thermal photons from the background may represent the ultimate limitation to detector performance. The most sensitive infrared detectors are photoconductors and this has given rise to a well known acronym—BLIP, i.e. background limited infrared photoconductor! If the limitation is strictly due to the background photon noise, then this will be independent of the signal level, the noise will be purely additive, and the multiplex advantage will obtain. If however one is dealing with signal photon noise, there will be, as stated above, no advantage. The minimum detectable power in the signal-photon-noise-limited state for an ideal detector is given by

$$P_S = 4h\nu\Delta f, \tag{9.2}$$

where Δf is, as before, the receiver bandwidth. This equation is interesting to analyse. First, it states directly that if two photons are arriving on average in a response time interval (i.e. $(\Delta f)^{-1}$), then one stands a 50% chance of detecting the fact. This explains the emergence of photon counting techniques in recent years, especially in connection with low luminosity phenomena such as Raman scattering. The second point to emerge from the analysis is that photon noise on the signal becomes more significant as the frequency rises. Physically this is because the photon energy is rising too, and the fluctuation due to the arrival or non-arrival of a single photon becomes more important. At far infrared frequencies an enormous number of photons is required to give detectable power levels, whereas at optical frequencies a much smaller number is required. It follows therefore that at optical frequencies the combination of high sensitivity and increasing photon noise will mean that the multiplex advantage will seldom apply. The background and intrinsic photon noise limitations are illustrated in Figure 9.1, which also shows the performance of some modern detectors in relation to them. The background limitation is defined in terms of a full hemispherical field of view (FOV) with a background temperature of 300 K. For cooled detectors within a cold cavity, the FOV can be restricted by proper design and detectivities far higher than the ambient temperature background limit can now be achieved, as indicated in the figure for a screened InSb bolometer.

The case of noise in astronomical interferometric spectroscopy can be very complex, but always it is signal noise which predominates. Fluctuations may be observed in the atmospheric transparency that affect the entire band of detected radiation uniformly (that is, grey or achromatic fluctuations); they may affect different parts of the band differently (chromatic fluctuations) or else they may cause variations in the refractive index of the atmosphere which may disturb the signal wavefront as well as causing disturbances in the total detected flux.

The effects of the two kinds of noise in sequential spectroscopy are easy to visualize. Additive noise is independent of signal strength and one gets the kind of spectrum shown in Figure 9.2(a). Multiplicative noise is most obvious where the signal strength is high, as shown in Figure 9.2(b). In Fourier spectrometry the situation is considerably more complicated since the noise which is manifest in the *spectrum* in the sequential case is manifest in the *interferogram* in the Fourier case. The noise in the transformed interferogram is not simply related to the noise in the original record except in certain simple cases, but, since the cases can be so different, it is best to consider a number of them separately. We shall begin with the analysis of noisy single-sided interferograms and go on to consider the more complicated double-sided case later. In each case just one kind of noise will be assumed to be dominant. The necessary mathematics for dealing with random noise signals was developed in Chapter 2. However, before fulfilling this programme, it is worthwhile to analyse in some detail the nature of the noise reduction operations whose practice was described in the previous chapter.

9.2. THE EFFECTS OF SMOOTHING ON THE RECORDED SIGNAL

As was mentioned in the previous chapter, the detector is followed by an electronic smoothing chain made up of a tuned amplifier, a phase-synchronous rectifier (PSR) and an *RC* filter network. The tuned amplifier is not strictly necessary since, provided the amplitudes of the noise fluctuations are insufficient to exceed the dynamic range of the phase-synchronous

266

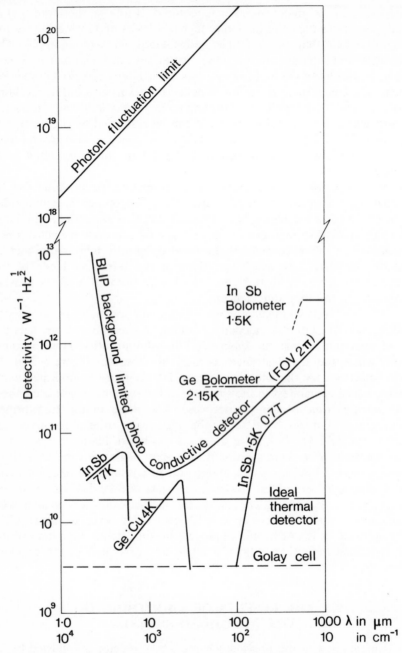

Figure 9.1. Performance of some typical detectors in relation to the photon and background noise limits

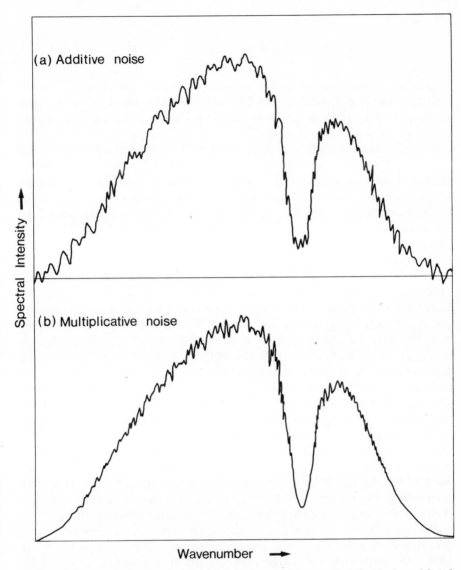

Figure 9.2. Additive and multiplicative noise in the case of a simple broad-band emission spectrum

rectifier/*RC* filter combination, this latter will perform the noise reduction task just as well. In fact, the conclusion reached earlier that the phase-synchronous rectifier will increase the signal-to-noise ratio merely by a factor of two applies only when the signal and noise are both confined to a very narrow pass band and the noise is therefore merely phase noise. In the broad-band case the phase-synchronous rectifier/filter combination will reject noise signals at frequencies remote from the modulation frequency and the signal-to-noise ratio will be given by

$$(S/N) = \frac{\Delta f_{in}}{\Delta f_{out}}, \qquad (9.3)$$

where Δf_{in} is the noise frequency bandwidth and Δf_{out} the response bandwidth of the rectifier/filter. An identical result is obtained if a tuned amplifier is used when Δf_{out} is the overall bandwidth of the amplifier/phase-synchronous rectifier/filter combination. In practice, and especially when non-routine spectroscopy is involved, it is still convenient to have a tuned amplifier. This is because one has to have an amplifier anyway so that the output d.c. voltage span can be adjusted to match the range of the recording system and also because at high gain levels the noise spikes may drive the rectifier into a non-linear region. When this is the case, restriction of the amplifier band pass by a suitable amount will reduce the excursions and produce a resultant signal which the rectifier can handle.

The phase-synchronous rectifier is nearly always followed by an RC smoothing network. This, because of its finite response time, $\tau = (RC)^{-1}$, limits the frequency response of the combination and is usually the principal factor determining Δf_{out}. The reduced response at high frequencies smooths out the modulation frequency and reduces the noise. In sequential spectroscopy the need for and the operation of the filter are intuitively obvious; in multiplex spectroscopy however the matter is rather more subtle. The elimination of the modulation frequency is now even more important—after all, the last thing we want to do is to digitize a cosinusoidal artefact! However, it might be thought at first sight that it is unnecessary to attenuate the residual noise since the process of Fourier transformation will assign each audio frequency component of the analogue record to its corresponding place in the spectral frequency domain. This process will therefore separate the noise remote from the meaningful audio frequency band, and what is left in this band connot be attenuated without, of course, also attenuating the wanted signal. The fallacy here lies in the necessity to digitize the analogue record. Because of the discrete sampling, the spectrum will be aliased and, if noise is assigned by the Fourier transformation to a region occupied by one of the higher aliases, it will promptly be 'folded' back and reappear in the fundamental region. It is therefore necessary for the smoothing chain to attenuate strongly all noise components lying above the folding frequency (i.e. the audio frequency analogue of the folding wavenumber). The folding or replicating wavenumber is set by the path difference interval at which digitization occurs (see Figure 6.8) and is usually a fixed quantity. The folding frequency is therefore set just by the speed at which the mirror moves, and experimentally the maximum permissible time constant of the filter network is determined by the scan speed. If the analogue record is very noisy and heavy damping is required, then a very slow scan speed will be indicated. Of course with very heavy damping the scan speed would have to be exceedingly slow if attenuation of the signal audio frequencies is to be avoided, but in practice this solution is not adopted and some attenuation of the signal frequencies is endured in order that the experiment be completed in a reasonable time. With monochromatic radiation there is not much hardship since, in the limit of large τ, the spectral S/N reaches a maximum value where signal and noise are being attenuated equally.† This happy state of affairs remains until the unavoidable electronic noise in the subsequent equipment used to 'blow-up' the signal back to an acceptable level starts to obtrude and the overall S/N then starts to fall. With broad-band radiation there is a spectrally dependent attenuation of signal, the higher frequencies being attenuated more than the lower. In 'single-beam' absolute spectroradiometry this would be a serious matter, but in 'double-beam' spectroscopy, where one takes a ratio, the various spectral features will be in the correct proportion to one another and the S/N would again be approximately constant across the band. Of course it is essential that the time constant be kept the same for specimen and background run! Therefore the conclusion which emerges is that in Fourier transform spectroscopy much more heavy damping is permissible than could be tolerated in sequential spectroscopy. In the latter, spectral distortion over small frequency ranges would occur if the time constant were too long for the chosen scan speed, and this would not be compensated for by the usual 'double-beam' techniques (e.g. attenuating combs) used in dispersive spectrometers. In FTS the distortion would be spread evenly over a wide band and the effect on a sharp feature would be very small.

The spectral effects of the filter can be analysed in a more quantitative manner by using the concepts developed in section 4.2.3 and defining an impulse response function $R(t)$ which, for the elementary filter networks so far discussed, takes on the simple form

$$R(t) = \tau^{-1} \exp\left[-t/\tau\right]. \tag{9.4}$$

Other more complicated filter networks are often discussed; one sometimes sees, for example, the 'box-car' filter mentioned. This has the characteristic that it eliminates all frequencies up to a lower bound $\nu_0 - \Delta$ and all those above the upper bound $\nu_0 + \Delta$. Between the two bounds the filter has unit response. In these discussions the assumption is generally made (and nearly always implicitly!) that such a filter causes no phase shifts. By the causality arguments sketched out in Chapter 2, this is, of course, impossible. When one bears in mind that to make even a reasonable approximation to a box-car filter is extremely difficult in practice, it will be realized that further discussion of this concept would not be very fruitful. Therefore we shall use only the response function (9.4) for the rest of this development.

Thus, in both sequential and multiplex spectroscopy, we have a time varying signal (the spectrum in the former, the interferogram in the latter)

---

† The interferogram S/N falls as $\tau^{1/2}$ in the limit of large τ, but this is because the signal, at a single frequency, is being compared with noise whose frequency spectrum covers the whole band.

which is the convolution of the true signal with the response function

$$V_s(t) = V_0(t) * R(t). \tag{9.5}$$

In sequential spectroscopy, the only requirement is that the scan speed be slow enough to make the effects of the convolution negligible. In multiplex spectroscopy a deeper analysis is required, and we begin as usual by discussing the monochromatic case and then go on to derive the broad-band results either exactly by integration or else approximately by heuristic arguments. We therefore have

$$V_0(t) = V_0 \cos 2\pi ft \tag{9.6}$$

as the original time-dependent signal, where $f = v\sigma$ and v is the mirror velocity. The output voltage is found from (9.5) and the operation of convolution is illustrated in the middle inset of Figure 9.3, from which it will be seen that

$$V_s(t') = \tau^{-1} V_0 \int_{-\infty}^{t'} \cos 2\pi ft \exp\left[(t - t')/\tau\right] dt; \tag{9.7a}$$

that is,

$$V_s(t') = \frac{1}{1 + 4\pi^2 f^2 \tau^2} \left[\cos 2\pi ft' + 2\pi f\tau \sin 2\pi ft'\right] \tag{9.7b}$$

$$= \frac{1}{\sqrt{1 + (2\pi f\tau)^2}} \cos\left[2\pi ft' - \phi\right] \tag{9.7c}$$

where $\tan \phi = 2\pi f\tau$. The dummy variable t' can then be replaced by t to give the observed time-varying signal.

Two things emerge from studying (9.7b). First, as τ increases the amplitude of the cosine component decreases and, second, since there is now a sine component present, the interferogram is no longer symmetrical about $t = 0$. We therefore lose information (for $2\pi f\tau$ much larger than unity, the amplitude of the wave varies inversely with τ) and produce an interferogram which will need to be transformed by one of the phase correction procedures discussed in Chapter 6. For small values of $2\pi f\tau$ the phase shift increases linearly with the frequency, as emerges clearly from (9.7c). Therefore, if we make it negligible at the maximum frequency present, a fortiori, it will be also negligible at all lower frequencies. The necessary condition is that

$$\tau \ll (2\pi f_{max})^{-1}. \tag{9.8}$$

As an illustration of this inequality, imagine that a far infrared interferometer is irradiated by a band of radiation whose maximum wavenumber is $500\ cm^{-1}$. Suppose further that the mirror scan speed produces a path difference rate of change of $0.002\ cm\ s^{-1}$. The highest audio frequency at the detector will then be 1 Hz and τ must be less than 0.1 s. This result may come as a surprise to many spectroscopists, and it is regrettably clear that

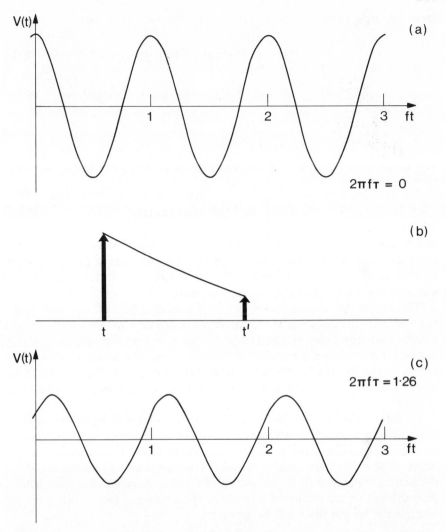

Figure 9.3. The signal at time t' is the sum of the decaying echoes from all previous times t. A pure cosinusoid input (a) therefore becomes distorted and shifted (c) after passing through a filter with $\tau > 0$

many spectra have been recorded in the past with the time constant of the smoothing network set at far too high a value. However, for the reasons given above, if 'double-beam' spectroscopy with adequate phase correction were being carried out, the effects would not have been too serious.

The performance of the filter can also be analysed in the frequency domain—a procedure which throws much light on the process involved. To do this we follow the treatment given in section 2.9.7.1 and calculate the transfer function $\hat{T}(f)$, which is of course the full complex Fourier transform of

$R(t)$. We thus have

$$\hat{T}(f) = \int_0^\infty R(t) \exp(-2\pi i f t)\, dt \tag{9.9a}$$

$$= \tau^{-1} \int_0^\infty \exp(-t/\tau) \exp(-2\pi i f t)\, dt \tag{9.9b}$$

$$= \frac{1}{1 + 2\pi i f \tau}. \tag{9.9c}$$

The effect of the filter on the spectrum in audio frequency terms, $S(f)$, will then be given by

$$\hat{V}_s(t) = \int_0^\infty S(f)\hat{T}(f) \exp(2\pi i f t)\, df, \tag{9.10}$$

which, since it is the Fourier transform of a product (section 2.3), is the equivalent (via the convolution theorem) in complex form of (9.7a). Readers familiar with modern forms of the Debye theory of dielectric loss will find much of the above mathematics very familiar.

(The reason for the similarity lies in the equivalence of the basic problems, namely the question of the forced response of a system which exhibits a finite response time. In the Debye theory it is the response of viscously damped dipoles to a driving electromagnetic wave. The modern forms of the theory, which use the idea of a correlation function, show that *any* response which has the characteristics of (9.4) will have the spectral response of (9.9c).)

We can now go on to analyse practical spectroscopy using this formalism. If we have a monochromatic wave, it will be represented in the frequency domain by a delta spike at f_s (i.e. the audio frequency produced in the interferometer by the scanning mirror). The spectrum will be aliased about the folding frequency f_f. If we have 'white' noise, its power spectrum will be proportional to the square of its voltage spectrum and the power spectrum emerging from the filter will be given by

$$P_N = \mathscr{A}|\hat{T}(f)|^2, \tag{9.11}$$

where $\mathscr{A}$ is a measure of the input noise power. The situation is illustrated in Figure 9.4. When τ is very small, the noise is strongly aliased and the S/N is at its lowest value. As τ is increased, the aliasing is reduced but attenuation of the signal begins. Finally, for large τ, there is little aliasing and the fundamental noise and the signal are being attenuated equally. It will be seen from Figure 9.4 that the bigger the separation of f_s and f_f the more effective is the operation of the filter. Of course if f_f is greater than f_s the interferogram is being over-sampled and one can look at the improvement in S/N in a different light, namely that by information theory the S/N must improve when the number of observations of a steady signal in the presence of truly random noise is increased.

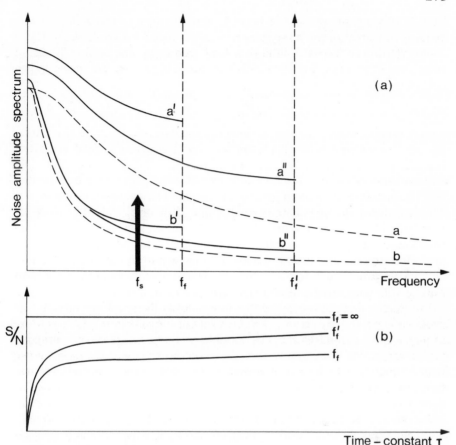

Figure 9.4. The effects of a moderate (a) and a large (b) time constant on the noise amplitude spectrum emerging from an *RC* filter: the input is assumed to be white. When the noise being discussed is on the interferogram signal and this is transformed to give a spectrum, the effects of aliasing must be considered. For infinitely fine (that is, continuous) sampling the signal-to-noise ratio is independent of the time constant since signal and noise are equally attenuated. For finite sampling, the noise lying above the folding frequency (f_f or f_f') is 'folded back' into the fundamental region to give curves (a', b') or (a'', b''). The signal-to-noise ratio therefore deteriorates and is now dependent on τ. When the folding frequency becomes larger, f_f', i.e. when the sampling becomes finer, the signal-to-noise ratio improves and the dependence on τ is reduced

RC smoothing networks reduce the fluctuations due to the noise and hence improve the S/N most effectively when the noise power spectrum is truly 'white'. The amplified voltage fed to the phase-sensitive detector will then be a signal at the modulation frequency accompanied by band-limited white noise—the band limitation occurring either deliberately in the amplifier or else inadvertently, due to component limitations, in other elements of the electronic chain. The operation of rectification transfers the signal plus its accompanying band of noise to zero frequency. The RC filter then reduces the 'area' of the noise spectrum as τ is increased, as shown in Figure 9.4(a). However, an RC filter is a high frequency attenuation device. If reciprocal frequency (or $1/f$) noise is present, where the noise power spectrum has the form

$$P_N \sim 1/f^n \qquad (9.12)$$

with n ranging from $\frac{1}{2}$ to 1, as is generally the case, we find that this part of the noise spectrum falls mainly on the low frequency side of the signal frequency and, whereas for low values of τ the RC filter is effective, for larger values it becomes less so and eventually becomes worse than ineffective (!) since it discriminates against the signal. This is shown in Figure 9.5. The reason for this is that the signal is not, of course, truly d.c., since the interferometer mirror is scanning and the signal in fact occupies an audio-frequency band. When the value of τ is so high that the signal frequencies are being strongly attenuated, the low frequency part of the $1/f$ noise will be being attenuated much less and the interferogram S/N will fall much more rapidly with increasing τ than is the case with white noise.

In sequential spectroscopy, where one records the spectrum directly, the problem is potentially serious. The best way of mitigating its effects is to use as high a modulation frequency as possible, so that one will be operating in a region where the noise is low and where the noise power is varying slowly with frequency. In addition, it is highly desirable to use bandwidth restriction prior to the PSR. With these two provisions, the observed noise will appear to be 'pseudo-white' since it will be a narrow slice of high frequency noise transferred to the low frequency region. Another approach is to use a 'box-car' filter. Now, as was mentioned above, it is not practical to make analogue box-car filters, but one can readily acquire digital devices whose performance when analysed in the frequency domain is essentially similar.

The best of these is based on the multichannel analyser or one of its computer equivalents. One digitizes the noisy signal and, using an internal clock, distributes the digital values to a set of addresses (i.e. memories). The clock continues to go round and round the addresses and, when one of them becomes full, the value of the least is subtracted from each of them. After a period equal to the pre-set integration time, a running average is taken and this may be put through a digital/analogue converter (if desired) to give the smoothed spectral ordinate at that particular (average) value of the scanning variable.

In FTS the problem is not in principle so serious since, as remarked above, the very act of Fourier transformation distributes the noise power components to their proper place along the frequency axis, and it is only aliasing of the noise which need worry us. In the case of $1/f$ noise, aliasing is not a serious problem since the majority of the noise power is concentrated at low frequencies. However it may be necessary to reduce the interferogram noise excursions for the reasons outlined earlier and, in this case, the

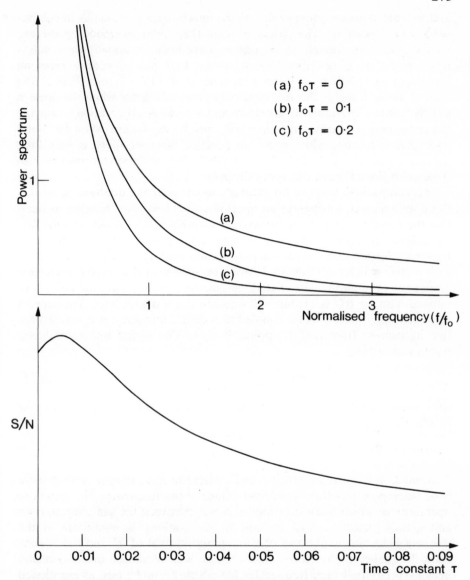

Figure 9.5. The effect (upper inset) of *RC* filtering on the power spectrum of one-over-f' noise. In the lower inset is shown the effect on the signal-to-noise ratio of the *RC* filtering, where f_{min} is 0.001 Hz, $f_s = 10$ Hz, and the values of τ are as marked

use of as high a modulation frequency as possible coupled with some bandwidth restriction is again very desirable. Alternatively, as before, we can use a digital device instead of the *RC* filter to smooth the interferogram record. In this connection the rapid scan instrument has, in principle, an advantage since it is the practice to record a large number of interferograms and to store corresponding values of the interferogram ordinates in computer memory locations. The repetitive scans then build up smoothed average values and, when enough interferograms have been 'co-added', the result is transformed. It is not too difficult to see that this procedure gives an improvement in S/N by a factor proportional to $(T_{obs})^{1/2}$ regardless of the type of noise, whereas the RC filter only gives this factor when the noise is strictly 'white'. Therefore an aperiodic and a periodic observation occupying the same length of observing time will only be equivalent when the noise encountered is pure white noise. In practice however there is very little difference because of the use of reasonably high modulation frequencies plus bandwidth limitation in the aperiodic case.

This comparison leads us on naturally to consider the question of what is the optimum way to operate an aperiodic interferometer bearing in mind that the signal will inevitably be accompanied by noise. Connes was the first to analyse this and, noting that aliasing is the principal noise problem in FTS, she showed that continuous observation through an RC filter coupled with a (theoretical) continuous Fourier transform of the filtered interferogram gave optimum results. The discontinuous, or step-scan, method is not optimal with an RC filter, but is if one can use a digital filter. Therefore a step-scanning interferometer coupled to a digital integrator is a very attractive instrument from a (S/N) point of view. The digital integrator has a response function

$$R(t) = \frac{1}{T'} \sqcap \left(\frac{t}{2T'} \right), \qquad t > 0, \tag{9.13}$$

since its operation is a simple running sum from $t = 0$ to $t = T'$. The transfer function is therefore

$$\hat{T}(f) = \frac{\sin \pi f T'}{\pi f T'}. \tag{9.14}$$

This function has zeros where $f = m/T'$, where m is an integer, and so there is no noise power at these particular values of the frequency. The step-scan operation, in which the mirror moves rapidly between the sampling stations and spends virtually all of its time at the stations, is equivalent in the frequency domain to the use of a sampling interval of T' and the folding audio frequency is therefore $(1/2T')$. Noise will be aliased into the region near $f = 0$ from all high frequencies for which $f \approx m/T'$, but, as mentioned above, there is no noise power at these values and so the aliasing will not be significant. The continuous scan method with continuous transformation is not practical, because the computer can only deal with a finite set of sampled ordinates, so the step-scan plus digital integrator emerges as the best possible way of observing practical interferograms.

One can, however, achieve essentially the same result without using any integrating hardware by having recourse to software modifications in the computer. Suppose that we wanted to multiply our spectrum by a box-car filter which had unit gain between σ_{min} and σ_{max} but had zero gain

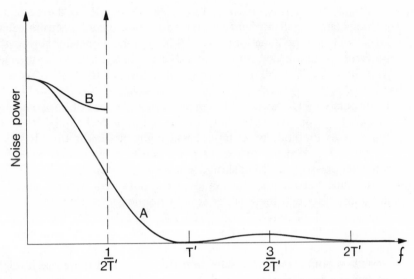

Figure 9.6. The noise power spectrum produced by the digital integrator is of the $(\sin x/x)^2$ form, as shown in curve A. After discrete Fourier transformation, the noise power lying above the folding frequency $f_f = (2T')^{-1}$, will be aliased back, as shown in curve B. However, for low frequencies under economical sampling, or for all frequencies with deliberate oversampling (i.e. $(2T')^{-1}$ large), the additional noise will be very small and a good approximation to ideal behaviour is secured

elsewhere. The Fourier transform of this modified spectrum would, by the convolution theorem, be the convolution of the original interferogram with a sinc function. Therefore if the original unsmoothed interferogram is convolved with a suitable sinc function and then transformed, all noise at frequencies remote from the signal band will be eliminated and therefore cannot be aliased back. This procedure is called *mathematical filtering* and can be very useful. It is an adaptation of a similar technique which was introduced by Connes and Nozal[62] to enable them to reduce the number of samples to be taken and the amount of computing to be done when only a small region of a much larger spectrum was desired to be studied.

To summarize the important points which have emerged from the analysis, we have:

(a) An interferogram observed, with damping introduced to reduce noise, will have either its grand maximum (AM) or its zero crossing (PM) displaced from the true zero path difference position and the interferogram will no longer be perfectly symmetrical (AM) or perfectly antisymmetrical (PM).

(b) Computation of the spectrum by either a two-sided transform or else by the use of a phase-corrected single-sided transform will compensate for the distortion, but the higher frequencies will be attenuated.

(c) In double-beam spectroscopy, the attenuation will be the same for both specimen and background and the *transmission* spectrum calculated will be the true one. The S/N will be constant across the spectrum to first order, but to second order, and due to noise injected post filter, the S/N will tend to deteriorate towards the higher frequencies.

(d) In dispersive FTS there will be an additional amount of 'instrumental' phase shift appearing, but any of the modern techniques (the two thicknesses method for example) will compensate for this. It is not (fortunately!) realistic to expect phase shifts greater than π due to filter action, so branching should not be encountered.

(e) If the time between samples is long, the noise will become uncorrelated and although there is still a high frequency 'roll-off', due to the filter, this will be masked in the transform and the noise will appear to be 'white'. The S/N will then deteriorate towards high frequencies.

(f) In FTS the operation of carrying out the transform will sort out the noise components according to their frequency and therefore it might be thought that smoothing of the interferogram is unnecessary. However, aliasing will always fold noise back into the signal region and some form of interferogram smoothing is required. The best way of observing the interferogram is to use the step-scan method with a digital integrator.

The physical concepts developed above will now be applied to the possible cases which emerge in practice—that is, where we have one-sided or double-sided interferograms recorded via either AM or PM and with either additive or multiplicative noise.

9.3. THE ONE-SIDED AM INTERFEROGRAM WITH ADDITIVE NOISE

9.3.1. The calculated spectrum

It will be assumed that the noise-free interferogram before filtering is perfectly symmetrical and that it has been obtained using amplitude modulation (AM) and aperiodic registration. The actual interferogram may therefore be written

$$I_N(vt) = I_0(vt) + \varepsilon(t), \tag{9.15}$$

where $\varepsilon(t)$ is the additive noise contribution. The spectrum evaluated from this by Fourier transformation,

$$B_e(f) = 2 \int_0^\infty [I_0(vt) + \varepsilon(t)] \cos 2\pi ft \, dt, \tag{9.16}$$

can be written as the sum of two terms

$$B_e^N(f) = B_{e0}(f) T_e(f) + N_e(f) \tag{9.17}$$

where the first represents the ideal (filtered) spectrum and the second the calculated noise, the Fourier transform of the original interferogram noise. We see, therefore, that since Fourier transformation is a linear process, the noise in the calculated spectrum superimposes additively on the required spectral power. Moreover, the noise in the spectrum is given by the cosine Fourier transform of the noise in the interferogram. The noise and the signal power may be treated completely separately not only in the interferogram where, by the definition of additive noise, this is necessarily so, but also in the spectrum.

9.3.2. The interferogram noise

The study of noise is the study of the behaviour of a random variable, of which the mean value is zero but the root mean square value is finite. Full details of the mathematical treatment of noise are given in the standard texts[63] and much of the necessary mathematics has been already developed in Chapter 2. The important point to reiterate is that $\varepsilon(t)$, for example, is not a function of time in the ordinary sense. At a given time t (counting from $t = 0$ at $x = 0$) it will have a given value, but at an equal time lapse later on it will have some quite different and independent value. If a large number of observations of $\varepsilon(t)$ is made, these repeated records constitute an *ensemble*. Similar arguments apply to $N(f)$. If one observes a large number of interferograms and transforms them to give a large number of spectra, the noise values at a fixed value of f will form a random independent *ensemble*. We have every reason to believe that the noise is stationary and ergodic and therefore the mean square value

$$\overline{[\varepsilon(t)]^2} = \lim_{M\to\infty} \frac{1}{M} \sum_{k=1}^{k=M} [^k\varepsilon(t)]^2 \qquad (9.18)$$

across an *ensemble* $k = 1, 2 \ldots M$ of records (that is the *ensemble* variance) will be the same as the mean square value

$$\varepsilon^2 = \langle[\varepsilon(t)]^2\rangle = \lim_{T\to\infty} \frac{1}{2T} \int_{-T}^{+T} [\varepsilon(t)]^2 \, dt \qquad (9.19)$$

along any single record contributing to the *ensemble* (that is, the temporal variance). The variance of the interferogram noise, ε^2, can therefore be evaluated from a single record.

9.3.3. The variance of the interferogram noise

The variance ε^2 is a special case of the autocorrelation

$$\gamma_{11}(\theta) = \langle\varepsilon(t)\varepsilon(t+\theta)\rangle, \qquad t > 0, \qquad (9.20)$$

about the mean $\langle\varepsilon(t)\rangle = 0$. By the use of the Wiener–Khinchin theorem (section 2.5), we can deduce that the spectral density of the autocorrelation

$\gamma_{11}(\theta)$ is the Fourier transform

$$\mathfrak{P}(f) = \int_{-\infty}^{+\infty} \gamma_{11}(\theta) \exp(-2\pi i f \theta) \, d\theta. \tag{9.21}$$

It is more convenient to deal with the even part of $\mathfrak{P}(f)$, which is given by the cosine transform

$$\mathfrak{P}_e(f) = \int_{-\infty}^{+\infty} \gamma_{11}(\theta) \cos 2\pi f \theta \, d\theta. \tag{9.22}$$

Inverting this, we have

$$\gamma_{11}(\theta) = \int_{-\infty}^{+\infty} \mathfrak{P}_e(f) \cos 2\pi f \theta \, d\theta, \tag{9.23}$$

and quoting the special case $\theta = 0$ we have

$$\varepsilon^2 = \gamma_{11}(0) = \int_{-\infty}^{+\infty} \mathfrak{P}_e(f) \, df. \tag{9.24}$$

The variance of the interferogram noise is therefore equal to the integral over all f of the even part of the noise power spectral density. (Strictly $\mathfrak{P}(f)$ does not have the dimensions of a power spectral density, but since it is related to the true noise power spectral density merely by a constant it is simpler to use this term with this meaning.)

If the spectral density of the interferogram noise can be represented by a constant density of white noise modified by the smoothing filter, we can write

$$\varepsilon^2 = \int_{-\infty}^{+\infty} \mathfrak{P}_e \, |\hat{T}(f)|^2 \, df = A \mathfrak{P}_e, \tag{9.25}$$

where $\mathfrak{P}_e$ is the even part of the spectral density $\mathfrak{P}$ of the white noise and

$$A = \int_{-\infty}^{+\infty} |\hat{T}(f)|^2 \, df \tag{9.25a}$$

is the 'area' of the filter function. Clearly, therefore, we wish to make A as small as possible consistent with the restraints imposed by the necessity to minimize the spectral distortion. In the case of the RC filter,

$$|\hat{T}(f)|^2 = [1 + (2\pi f \tau)^2]^{-1} \tag{9.26}$$

and

$$A = (2\tau)^{-1}. \tag{9.27}$$

In this case the variance of the noise is

$$\varepsilon^2 = \mathfrak{P}_e / 2\tau \tag{9.28}$$

and we deduce quantitatively the result we derived heuristically earlier: that the larger the time constant τ, the less noisy is the interferogram. This is because the noise power spectrum has been band limited by the filter. In the same way, if we have strictly $1/f$ noise, then the variance of the noise becomes

$$\varepsilon^2 = 2\mathfrak{P}_e \int_0^\infty \frac{1}{f[1+(2\pi f\tau)^2]} \, df, \qquad (9.29)$$

which, on making the substitution $u = 2\pi f\tau$, becomes

$$\varepsilon^2 = 2\mathfrak{P}_e \int_0^\infty \frac{du}{u(1+u^2)} = 2\mathfrak{P}_e \left[\ln \frac{u}{\sqrt{1+u^2}} \right]_0^\infty. \qquad (9.30)$$

The full interpretation of this equation is complicated by the divergence of the integral (a well-known difficulty), but if one evaluates it from some ultra-low frequency f_{min} instead of from 0, then it will be seen that ε^2 is virtually independent of τ. This underlines what was said earlier (Figure 9.5) about the generally ineffective qualities of RC networks when faced with $1/f$ noise. If the noise is of the more generalized $1/f^n$ type, with n less than one, then some amelioration is possible, but usually not much. A rectangular band pass filter is much more effective since here we evaluate

$$\varepsilon^2 = 2\mathfrak{P}_e \left[\ln \frac{u}{\sqrt{1+u^2}} \right]_{f_{min}}^{f_{max}}, \qquad (9.31)$$

which can become arbitrarily small as $f_{max} - f_{min}$ goes to zero. Analogue filters of this type are very difficult to make, but low-pass filters coupled to digital integrators can give frequency responses of a similar kind and their use is strongly recommended when $1/f$ noise is significant and needs to be reduced. This could arise, for example, when using the fast scanning technique. It would, however, not usually be a problem in aperiodic work, for unless one was restricted to the use of a slow detector, one could use a sufficiently fast chopping speed to be essentially clear of the $1/f$ region. After transformation into the spectral domain, residual $1/f$ noise will not be a problem since it is only aliasing of high-frequency noise that gives difficulties and for $1/f$ noise, of course, most of the power lies at low frequencies.

9.3.4. The autocorrelation function of the interferogram noise

So far we have assumed that the noise samples are strictly independent no matter how closely they are taken. To investigate this point more closely we need to develop some further formalism.

The autocorrelation function of the interferogram noise is, in the case of white noise,

$$\langle \varepsilon(t)\varepsilon(t+\theta) \rangle = \mathfrak{P}_e \int_{-\infty}^{+\infty} |\hat{T}(f)|^2 \exp 2\pi i f\theta \, df. \qquad (9.32)$$

By noting that $\hat{T}(f)$ is Hermitian, i.e. that

$$|\hat{T}(f)|^2 = \hat{T}(f)\hat{T}^*(f) = \hat{T}(f)\hat{T}(-f), \tag{9.33}$$

and by noting that

$$\hat{T}(f) = \int_{-\infty}^{+\infty} R(\theta) \exp{(-2\pi i f\theta)} \, d\theta \tag{9.34a}$$

and

$$\hat{T}(-f) = \int_{-\infty}^{+\infty} R(-\theta) \exp{(-2\pi i f\theta)} \, d\theta, \tag{9.34b}$$

it follows that

$$\langle \varepsilon(t)\varepsilon(t+\theta) \rangle = \mathfrak{P}_e R(\theta) * R(-\theta) \tag{9.35}$$

because of the convolution theorem. Now by virtue of the relation between convolution and autocorrelation (2.60), we can write (9.35) as

$$\langle \varepsilon(t)\varepsilon(t+\theta) \rangle = \mathfrak{P}_e \langle R(t)R(t+\theta) \rangle = \mathfrak{P}_e R(\theta) \star R(\theta). \tag{9.35a}$$

Applying Parseval's (Rayleigh's) theorem to (9.25a), we also have

$$A = \int_{-\infty}^{+\infty} [R(t)]^2 \, dt = \langle R(t)R(t) \rangle. \tag{9.36}$$

We can therefore express the autocorrelation as

$$\langle \varepsilon(t)\varepsilon(t+\theta) \rangle = \mathfrak{P}_e A\gamma_{RR}(\theta) = \varepsilon^2 \gamma_{RR}(\theta), \tag{9.37}$$

where

$$\gamma_{RR}(\theta) = \frac{\langle R(t)R(t+\theta) \rangle}{\langle R(t)R(t) \rangle} \tag{9.38}$$

is the *autocorrelation coefficient* of the *filter impulse response*.

The extent to which the interferogram noise samples are independent depends on the impulse response of the output filter. We can only regard a pair of noise samples as independent if the value of (9.38) is zero or, in practice, very small. It is desirable, therefore, that the autocorrelation coefficient have a very sharp peak at $\theta = 0$. The RC filter which we have discussed so extensively as a standard example has the autocorrelation function

$$\gamma_{RR}(\theta) = \exp{[-|\theta|/\tau]}, \tag{9.39}$$

as will be found by substituting (9.4) in (9.38). This is not particularly sharply peaked unless τ is very small—a situation that accords with intuition, which would suggest that the noise observed at two instants along an interferogram is increasingly independent the smaller the response time. However, A increases as τ is reduced and the variance of the noise increases. In the practical case there may be quite strong correlation of the noise signals and the noise in the smoothed interferogram will not represent a truly independent set of random signals. In other words, it will not be strictly 'white'.

9.3.5. Qualitative treatment of the spectral noise problem

The detailed treatment of the relationship between the noise in the interferogram and the resulting signal-to-noise ratio in the calculated spectrum will take us into some of the more esoteric branches of the subject. The reader may find it useful therefore to have a more qualitative and more pictorial treatment to begin with so that the broader features of the scenario are familiar landmarks during the navigation of the trickier fine points.

The interferogram plus additive noise may be resolved into its two components as shown in Figures 9.7 and 9.8. In the monochromatic case (Figure 9.7), we can consider two alternatives. The first is to let the scanning mirror continue to run (at speed v) until the total observing time T is used up. The second is to choose a resolution R and to set the mirror speed such that one just covers the appropriate amount of path difference. In the first régime, the signal will build up in direct proportion to T, the noise will build up in direct proportion to $\sqrt{T}$, and the signal-to-noise ratio in the calculated spectrum will therefore increase in proportion to $\sqrt{T}$. In the second, the signal in the calculated spectrum will be fixed but the noise will vary inversely as $\sqrt{T}$ since we will, of course, set the time constant τ at a value appropriate to the speed of scan. Again the S/N will go as $\sqrt{T}$.

In the broad-band case, Figure 9.8, the signal will increase up to the point where the oscillations are damped out, but beyond this there will be no further increase. The noise will, however, continue to increase as the mirror scans and the signal-to-noise will *fall* in proportion to $\sqrt{T}$. In the broad-band case, therefore, there is a large advantage in setting the range of the mirror appropriately to the span of the interferogram where there is meaningful oscillation. One can then set the time constant appropriately to give the optimum S/N in the calculated spectrum.

In practical spectroscopy an additional point arises, namely that there is a lower limit to the discrimination of the digital system. Oscillations less than this will not be recorded and each ordinate that is recorded (even if its value is noise free) will be uncertain to within the discrimination limit. This gives rise to an additional type of noise—sometimes called 'digitization noise'. In the past this was a serious matter, but nowadays, with modern digital systems which divide the total range into 10000 parts, it has become negligible. The question of how to do spectroscopy under this limitation used to be answered by the advice to set the time constant such that the amplitude of the noise just equalled the discrimination limit (i.e. one 'bit'), to set the range of scan to just cover the region of the interferogram where there is detectable information, and then to set the scan speed appropriate to the chosen time constant. With modern interferometers and sensitive

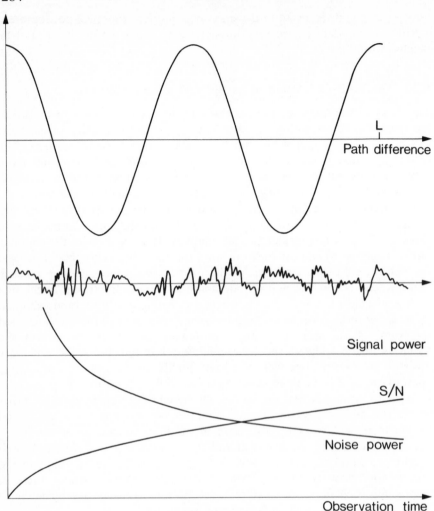

Figure 9.7. The interferometric observation of monochromatic radiation can be done in two ways (see text). The diagrams above illustrate the mode where the total path difference scanned is fixed. The signal power is thus constant but the noise power falls as $\sqrt{T}$, giving a signal-to-noise ratio which increases as $\sqrt{T}$. This is the same result as that for the fixed scan speed mode

cryogenically operated detectors it is not too easy to follow this advice since the noise will not usually be visible on the chart record of the interferogram. However, one can usually judge the magnitude of the noise excursions by observation of the digital voltmeter display and set the time constant appropriately. Fleming, at the National Physical Laboratory, who has carried out most of the high-resolution studies of gases in the far infrared to be reported in recent years,[64] operates in this way. He is therefore working with an interferogram signal-to-noise ratio of $10^4 : 1$. In Figure 9.9 is shown one of his interferograms corresponding to pure rotational absorption by nitric oxide. After the zero path region is past, the gain is increased one hundredfold and the interferometric signatures are shown to be still being enregistered in the digitized version. In fact, signatures with amplitudes of at least thirty times the noise level are still present at the limit of his mirror travel (i.e. total path difference introduced = 98 mm).

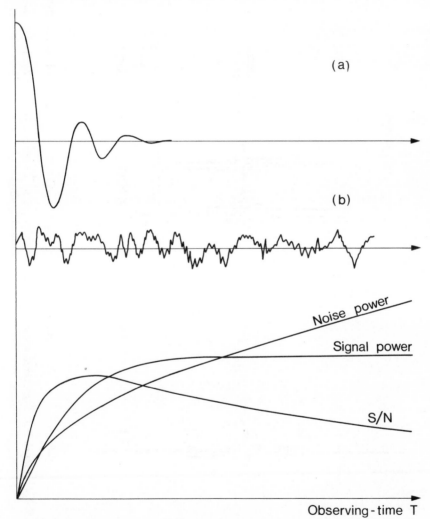

(a)

(b)

Noise power

Signal power

S/N

Observing-time T

Figure 9.8. In broad-band FTS, the noisy interferogram can be resolved into its signal (a) and noise (b) components. The noise power increases monotonically as $\sqrt{T}$, whereas the signal power approaches a limit asymptotically. There is therefore a point where a maximum signal-to-noise ratio is obtained

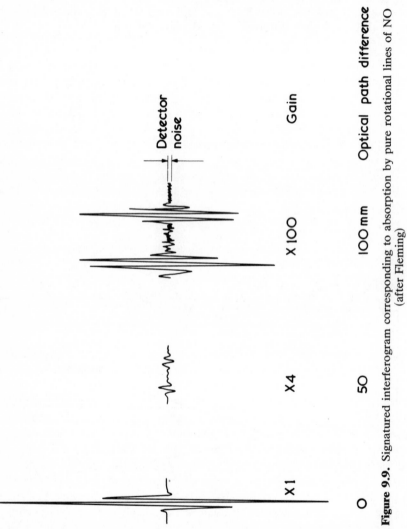

Figure 9.9. Signatured interferogram corresponding to absorption by pure rotational lines of NO (after Fleming)

Discussing the matter of signatures leads us to consider one of the puzzling conundrums of interferometric spectroscopy, i.e. that as one scans a signatured interferogram one goes through a burst of oscillation (at a signature) then a featureless region before one has a burst again at the next signature. The resolution will not apparently increase as one scans over a featureless region, but it will increase very rapidly as one scans over the short region where the signature is located. Thus one arrives at the surprising result that the resolution increases discontinuously, or at least non-linearly with increasing path difference. This conclusion has caused much confusion in the past and has even on occasion been labelled a paradox. Mathematically one can resolve any arbitrary spectral profile into mono-chromatic components, and, since the resolution of each of these should increase linearly with path difference, it is at first rather difficult to see why this is not true of the *ensemble*. The reason, of course, is that the theoretical resolution (i.e. the width of the computed spectral feature corresponding to *hypothetical* strictly monochromatic input) and the actual resolution (i.e. that produced from the *real* input) are very different things. Thus if one has an isolated feature of finite width, it will produce an interferogram which might have the form

$$I(x) = \exp\left(-2\pi k x\right) \cos 2\pi\sigma_0 x. \tag{9.40}$$

This, on transformation over the infinite range of x, will give a Lorentzian line shape whose component at positive wavenumbers will be

$$S(\sigma) = \frac{2\pi k}{4\pi^2 k^2 + 4\pi^2(\sigma_0 - \sigma)^2}. \tag{9.41}$$

The full width at half height of this feature (a measure of the resolution) is $2k$. One sees therefore that if one were to transform the interferogram over a *finite* range the resolution would at first increase linearly but would then begin to tail off and would ultimately approach the limiting value of $2k$ asymptotically. This result suggests a useful approach to the problem, an approach based on the concept of 'visibility of fringes' that was developed in Chapter 5. The modulus envelope of the interferogram will have small values between the signatures and large values at the signatures (see Figure 9.10). The information available from the interferogram will be proportional to the area under the modulus envelope and this will of course not increase linearly. Thus both the resolution and the signal-to-noise ratio will vary non-linearly with path difference whenever there is more than one component present in the spectrum since both quantities are related to the amount of information gleaned and this depends on the area under the modulus envelope. In the damped monochromatic case, for example, the total area under the envelope is $(2\pi k)^{-1}$ and the limiting resolution should be, as it is, proportional to the reciprocal of this limiting area.

With more than one component, one needs a new definition of resolution,

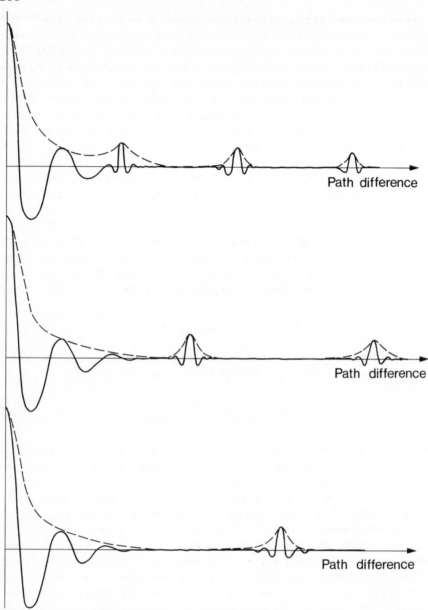

Figure 9.10. As a set of regularly spaced sharp lines in a spectrum move closer and closer the corresponding signatures move out more and more from the ZPD region. In the limit the first signature is at infinity and we have a continuous featureless broad-band spectrum. The area under the modulus envelope (dashed curve) is a measure of the resolution and it clearly does not vary linearly with path difference

but the area under the modulus envelope will always be a good measure. However we need to distinguish between two senses of the word 'resolution', which are mathematically equivalent but spectroscopically very different. Thus spectroscopists use resolution to mean the separation of close sharp components, whereas mathematically one could use the term, without contradiction, to mean the width of a broad feature. Hence, in the limiting case of a very large number of quasimonochromatic components blending into a continuum, there would be nothing to resolve and the concepts developed above would not be useful. If there were *sharp* components subtracted from a continuum—as one would get in absorption spectroscopy—the correct approach would be to split the interferogram into two parts: a zero path difference region, whose envelope area tells something of the width of the overall pass band, and all the rest, whose modulus envelope area tells something of the widths of the sharp components. One can continue this analysis and consider a set of equispaced emission lines: the closer the lines are set the further out does the first signature become. The area under the modulus envelope will reach a steady value before rapidly increasing and will then be steady again until the second signature is reached. In the limit when the signatures have moved out to very large values of the path difference one has essentially broad-band radiation and the envelope soon falls to a steady zero value.

The same approach can be used to discuss the question of the variation of S/N in the computed spectrum as a function of path difference introduced. The signal can again be identified with the amount of information gleaned, that is, the area under the modulus envelope, and provided the noise is uncorrelated we can make the assumption that the noise in the computed spectrum will depend just on the square root of the path difference introduced. The resulting dependence of S/N upon path difference is shown schematically in Figure 9.11. One can see, therefore, that the resulting spectral S/N will deteriorate steadily once observation is continued beyond a certain point in the interferogram. In the practical case, where discrimination is limited, it will decrease even more rapidly. The practical question of deciding where the mirror movement should be stopped depends, therefore, on how much attenuation of S/N can be tolerated. If possible, the smoothing should be set such that the noise excursions are about equal to one unit of the encoding system, but if the interferogram fringes are larger than one unit at the limit of mirror travel this recipe can be relaxed considerably. In general this will not be the case and one can then say that there is no point in continuing observation beyond the point where the interferogram oscillations have dropped below one unit of the encoding system. Having decided this, the experimentalist will choose the best compromise of smoothing and scan speed that his particular requirements dictate.

One can go over the same ground but using a rather different approach and thereby gain still more insight. This is to consider the process in reverse, i.e. to analyse the noise that will be produced in the interferogram by the

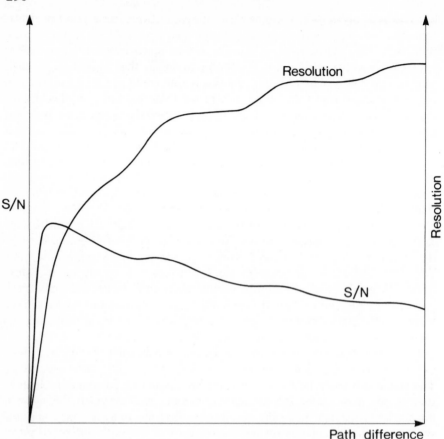

Figure 9.11. Variation of spectral S/N ratio and resolution with path difference for the case of a signatured interferogram such as that shown in Figure 9.10

noise present in the spectrum. If one defines the spectral and interferogram signal-to-noise ratios as $(S/N)_s$ and $(s/n)_I$ in terms of the maximum amplitudes divided by the average noise amplitudes in the vicinity of these maxima, one can see that a monochromatic spectrum in which all the power is piled up in a narrow region will give a much noisier interferogram, in which the signal amplitude is spread out over the whole range of path difference. In other words,

$$(S/N)_s/(s/n)_I > 1. \tag{9.42}$$

On the other hand, a broad-band spectrum which has finite amplitude over the whole spectral range will give an interferogram which is essentially confined to the zero path region. A noisy spectrum will therefore give a *less* noisy inteferogram and

$$(S/N)_s/(s/n)_I < 1. \tag{9.43}$$

Fleming[65] has considered the question of how one might estimate the signal-to-noise ratio to be expected, in broad-band spectrometry, from the transformation of a given interferogram. He shows first, following Connes (see later), that the variance in the transform is related to the variance in the interferogram by the equation

$$\varepsilon_\sigma^2 = 2N\Delta x^2 \varepsilon_x^2 = 2L\Delta x \varepsilon_x^2, \tag{9.44}$$

where N is the total number of points observed, Δx is (as usual) the sampling interval, and L is the total path length observed. The interferogram variance ε_x^2 has dimensions of (watts),^{2} but the spectral variance has dimensions of watts per wavenumber squared. The next point is to estimate the size of the maximum ordinate. One can do this roughly by assuming a triangular form for the spectrum with peak height B_m (watts per wavenumber) and total width σ_{max}, where σ_{max} is the folding wavenumber. One has then that $F_0 = 2(B_m/2)\sigma_{max}$ and therefore that the spectral signal-to-noise ratio will be

$$\frac{B_m}{\varepsilon_\sigma} = (S/N)_\sigma = \frac{F_0}{\varepsilon_x} \cdot \frac{1}{\sigma_{max}} \cdot \left[\frac{1}{2L\Delta x}\right]^{1/2} = (S/N)_x \cdot \left[\frac{1}{L\sigma_{max}}\right]^{1/2}. \tag{9.45}$$

The last step follows, of course, because $2\sigma_{max}\Delta x = 1$. Fleming considers the more general case where the spectrum does not fill the range up to the folding wavenumber, but instead is confined to a narrower region $\Delta\sigma$. He shows that the equivalent of (9.45) is then

$$(S/N)_\sigma = (S/N)_x \left(\frac{\sigma_{max}}{L(\Delta\sigma)^2}\right)^{1/2}. \tag{9.46}$$

These equations merit careful study, but one thing that stands out immediately is that the S/N in the spectrum varies as $L^{-1/2}$ whereas in the monochromatic case it varies as $L^{1/2}$. This argument is a trifle heuristic, but it is basically sound and highlights once again the folly of continuing observation in broad-band spectrophotometry beyond the point where the interferogram ripples have dropped below the noise level. It is well worth pointing out that the simple exposition of the multiplex advantage in FTS, namely that the S/N improves as $\sqrt{K}$, where K is the number of spectral elements, can only be true if these elements can be resolved from one another. If they cannot, as for instance would be the case in continuous broad-band spectrometry, then there is no multiplex advantage. When the elements can be resolved, then observing more interferogram, i.e. taking a longer time over the experiment, will gain you more information because the interferogram will still be oscillating. When they cannot, taking observation out to further and further values of path difference merely adds noise!

9.4. THE NOISE IN THE COMPUTED SPECTRUM

9.4.1. Continuous scanning and transformation

Ignoring for the moment the operations of sampling, the noise to be added to the spectrum is given by

$$\mathcal{N}_e^c(f) = 2 \int_0^\infty \varepsilon(\tau_D) W\left(\frac{\tau_D}{T}\right) \sqcap\left(\frac{\tau_D}{T}\right) \cos 2\pi f \tau_D \, d\tau_D, \qquad (9.47)$$

where τ_D is the delay time in the interferometer, *not* the time constant of the RC filter. The variance of the noise will be found from an expression such as (9.18), but because the noise is stationary and ergodic it may also be found from the integral

$$[\eta^c(f_0)]^2 = 4 \int_0^\infty d\tau_D \int_0^\infty d\tau_D' \varepsilon(\tau_D)\varepsilon(\tau_D') W\left(\frac{\tau_D}{T}\right) W\left(\frac{\tau_D'}{T}\right) \sqcap\left(\frac{\tau_D}{T}\right) \sqcap\left(\frac{\tau_D'}{T}\right)$$
$$\times \cos 2\pi f_0 \tau_D \cos 2\pi f_0 \tau_D', \quad (9.48)$$

where f_0 is the spectral frequency of interest. Much of the mathematics necessary to the elucidation of integrals such as (9.48) was spelled out in detail in Chapter 6. Using this one finds

$$[\eta^c(f_0)]^2 = \int_{-\infty}^{+\infty} df' \mathfrak{P}_e(f')\{[A((f_0+f')T) + A((f_0-f')T)]^2$$
$$+ [U((f_0+f')T) + U((f_0-f')T)]^2\} \quad (9.49)$$

where A and U are the symmetrical and unsymmetrical apparatus functions, respectively. Expanding the brackets, setting the integrals of the cross terms equal to zero, noting that $\mathfrak{P}_e(f')$ is either constant absolutely (white noise) or virtually constant over a resolution width, and finally invoking Parseval's theorem gives

$$[\eta^c(f_0)]^2 = \mathfrak{P}_e(f_0) \int_{-\infty}^\infty \left[W\left(\frac{\tau_D}{T}\right) \sqcap\left(\frac{\tau_D}{T}\right)\right]^2 d\tau_D \qquad (9.50a)$$

$$= \mathfrak{P}_e(f_0) \int_{-T}^{+T} \left[W\left(\frac{\tau_D}{T}\right)\right]^2 d\tau_D = 2TQ\mathfrak{P}_e(f_0), \quad (9.50b)$$

where

$$Q = \frac{1}{2T} \int_{-T}^{+T} \left[W\left(\frac{\tau_D}{T}\right)\right]^2 d\tau_D \qquad (9.50c)$$

is the average value of the square of the weighting function. In the case of RC filtering, making use of (9.28), we may also write (9.50b) as

$$[\eta^c(f_0)]^2 = 4TQ \cdot \tau \cdot \varepsilon^2 \qquad (9.51)$$

where τ is the time constant of the RC filter. From (9.50b) it will be seen that the spectral noise variance does not depend on the filtering. This is of course necessary since there is no aliasing under conditions of continuous observation and continuous transformation. Equation (9.51) expresses the difference in dimensions between η^2 and ε^2. The latter, being an interferogram variance, has the dimensions of $(\text{watts})^2$, whereas η^2 has the dimensions of $W^2 \, Hz^{-2}$. The fluctuations in the spectrum will be given by

$$\eta^c(f_0) = \sqrt{4TQ\tau}\, \varepsilon. \tag{9.52}$$

This appears to vary with τ but of course as τ is increased ε will vary accordingly to make $\tau\varepsilon^2$ constant. The important variation is that the amplitude of fluctuation increases with the square root of the observation time. This result was anticipated in section 9.3.5.

9.4.2. Finite sampling and transformation

Mme Connes, in her thesis,[16] has given the equivalent treatment for the case where the interferogram is sampled and the transform calculated at a finite number of ordinates. The noise will now be aliased and the problem is to sum the infinite number of contributions to the noise, at any particular frequency, which come from the folding back of the infinite number of high frequency contributions. A noise component at a frequency f_n and power b_n will give rise on Fourier transformation to a set of aliases at frequencies $f_n + m2f_f$ and at frequencies $-f_n + m2f_f$, where f_f the folding frequency is given by $f_f = (2h)^{-1}$, h being the sampling interval in time. This is illustrated essentially in Figure 6.8(a). Each alias will have power $\frac{1}{2}b_n$ because it is the even part of the power spectrum which is calculated. It follows that if we are considering a frequency f_0 in the fundamental region (i.e. $0–f_f$), then the total noise will be the sum

$$b(f_0) + b(-f_0 + 2f_f) + b(f_0 + 2f_f) + b(-f_0 + 4f_f) + \text{etc.}$$

Because the even noise power spectrum is symmetrical about $f = 0$, the ordinate at the positive frequency $(-f_0 + 2f_f)$ is identically equal to that at the negative frequency $(f_0 - 2f_f)$. The summation therefore can be considered to involve only the solid line aliases in Figure 6.8(a) and we may write it

$$\sum_{m=-\infty}^{m=+\infty} b(f_0 + m \cdot 2f_f).$$

In the particular case of RC filtering, the spectral noise variance becomes

$$[\eta^c(f_0)]^2 = 2QT\mathfrak{P}_e \sum_{m=-\infty}^{m=+\infty} \frac{1}{1 + [2\pi(f_0 + m/h)\tau]^2}. \tag{9.53}$$

Connes[16] has denoted the sum in (9.53) by the symbol $\phi_{f_0}(h, \tau)$ and she shows that its value is

$$\phi_{f_0}(h, \tau) = \frac{1}{2}\left(\frac{h}{\tau}\right)\frac{\sinh(h/\tau)}{\cosh(h/\tau) - \cos 2\pi f_0 h}. \tag{9.54}$$

It may be shown that for h/τ small the function has the value unity and has also a horizontal tangent: for h/τ large the function rapidly takes on the simple form

$$\phi_{f_0}(h \gg \tau) = \frac{1}{2}\left(\frac{h}{\tau}\right).$$ (9.55)

The situation with $h/\tau \ll 1$ corresponds to continuous scanning and transforming, so in this limit (9.53) reduces to (9.50b) as it should. The situation with $h/\tau \gg 1$ corresponds to very undamped observing (i.e. with heavy aliasing) and we obtain

$$[\eta^c(f_0)]^2 = QT\mathfrak{P}_e \cdot \left(\frac{h}{\tau}\right) = 2QTh\varepsilon^2.$$ (9.56)

Equation (9.55) applies strictly only in the limit where $\cos 2\pi f_0 h = 1$, but this will nearly always be true in practice. Equation (9.56) on the other hand is true for any value of f_0 since it applies in the limit where both $\sinh(h/\tau)$ and $\cosh(h/\tau)$ are very large. Fleming[65] has given an interpretation of equation (9.56). When h is large compared to τ, the noise in the interferogram samples becomes uncorrelated and the noise power spectrum therefore appears to be white when computed digitally. The dependence on h of equation (9.56) is merely a consequence of the finite sampling and hence of the aliasing.

One might now ask the question that, if one is carrying out discontinuous (sampled) interferometry with an RC filter, what is the best value of h/τ? As usual the answer is a compromise. If h/τ is small the S/N is high, but there is a mammoth data recording and handling problem; if h/τ is much greater than one, the S/N is poor and there is little data handling. Connes has shown that if h/τ is set equal to unity the S/N is only 5% worse than the optimum value and therefore one can recommend this setting as the answer to the question. Of course there will be quite severe phase shifts and phase correction techniques will be necessary, but if one has a noise problem these will just have to be employed to deal with it unless one can afford to oversample.

Normally the delay in the interferometer will be measured in terms of optical path difference Δx. Under these circumstances one is interested in the spectral noise power variance in units of watts^2 wavenumber^{-2} rather than in watts^2 Hz^{-2}. The conversion is readily achieved by multiplying throughout by the mirror velocity (v) squared; one then has

$$v^2[\eta^c(f_0)]^2 = [\eta^c(\sigma)]^2 = 2N\Delta x^2\varepsilon^2$$ (9.57)

since $N\Delta x = Tv$ and $hv = \Delta x$.

9.5. THE ONE-SIDED PM INTERFEROGRAM WITH ADDITIVE NOISE

The treatment of the (ideally perfectly antisymmetric) PM case is just the same as that for the AM case except that sine transformations replace the cosine transformations. The noise variance in the interferogram is found to be identically the same as that in AM operation (equations (9.53)) but for the reasons mentioned in section 8.4 the signal, defined as the maximum excursion near the zero path difference position, may well be larger and the PM interferometer have an advantage in that it has a larger signal but identically the same noise. There will also be an enhancement in the spectral domain for those frequencies which lie near the peak of the phase modulation characteristic function. This argument applies of course to a discussion of *interferogram* signal and source independent *random* noise. The PM signal-to-noise ratio will always be better than this in practice because the interferometrically modulated high-frequency radiation is encoded by the chopper in the AM case and fluctuations of this power will add to the noise whilst, of course, contributing nothing to the signal.

9.6. THE TWO-SIDED INTERFEROGRAM WITH ADDITIVE NOISE

In the absence of noise, the power spectrum that is recovered from the full Fourier transformation of an aperiodically recorded two-sided interferogram is virtually identical with that which would have been obtained from a single-sided transform. The small differences which arise in practice and the relative merits of the two approaches were discussed in Chapter 6. When noise is present however, this is not the case, because the power spectrum in two-sided FTS is obtained by taking the modulus of the complex spectrum, and this is a non-linear operation. The spectrum and noise are no longer independent as they are in the single-sided case. The noisy cosine and sine transforms may be written

$$P_N(f) = P(f) + \mathcal{N}(f), \tag{9.58a}$$

$$Q_N(f) = Q(f) + \mathcal{M}(f). \tag{9.58b}$$

The noisy calculated spectrum will therefore be

$$B_N(f) = \{P^2(f) + Q^2(f) + 2P(f)\mathcal{N}(f) + 2Q(f)\mathcal{M}(f) + \mathcal{N}^2(f) + \mathcal{M}^2(f)\}^{1/2}, \tag{9.59}$$

and this depends in a complicated way on the desired spectrum and the noise. In particular it should be noted that there will be a steady contribution to the power at any particular frequency from the 'rectified' noise since, of course, the average value of both $\mathcal{N}^2(f)$ and $\mathcal{M}^2(f)$ is not zero. In the extreme case, where there is no true signal $B_N(f)$ will not be zero but will be a positive quantity of the order of the square root of the variance of the original noise. Detailed calculation shows, in fact, that the calculated signal under these conditions consists of noise made up of a constant component with mean value 1.19η and a fluctuating component of variance $0.59\eta^2$, where η^2 is the variance of the original noise.

Connes[16] has shown that if b_i is the spectral power that one would get in ideal single-sided operation, η^2 the variance of the noise in that spectrum, and m_i the spectral power in the two-sided spectrum, then

$$m_i = [b_i^4 + 2b_i^2\eta^2 + 2\eta^4]^{1/4}. \tag{9.60a}$$

If we write

$$\mu^2 = b_i^2 + 2\eta^2 - (b_i^4 + 2b_i^2\eta^2 + 2\eta^4)^{1/2}, \tag{9.60b}$$

the expected value of m_i is then given by

$$\xi(m_i) = m_i \pm \mu. \tag{9.60c}$$

The signal-to-noise ratio in the modulus spectrum will therefore be m_i/μ. The variation of m_i/b_i with single-sided noise-to-signal ratio η/b_i is shown in Figure 9.12. At values of noise-to-signal ratio of less than 0.1 (that is, signal-to-noise ratios of greater than 10 to 1), the two-sided modulus calculation gives virtually the same answer as the ideal single-sided transform. In Figure 9.13 is plotted the noise-to-signal ratio in the modulus spectrum as a function of that in the equivalent single-sided computation. Again it will be seen that for S/N values greater than 10 the behaviour is acceptable, but in regions where S/N falls to lower values the resulting modulus S/N is distorted by the proportionally larger contribution of the rectified noise to the 'signal' as compared with the noise.

The conclusion one comes to is that provided the equivalent S/N is greater

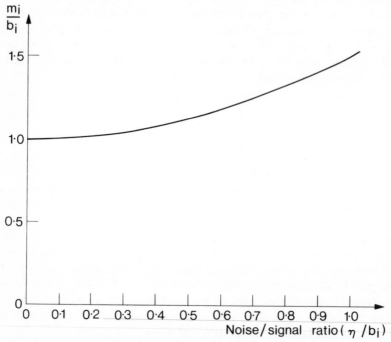

Figure 9.12. Variation of the ratio of double-sided to single-sided power (m_i/b_i) as a function of single-sided noise-to-signal ratio (η/b_i)

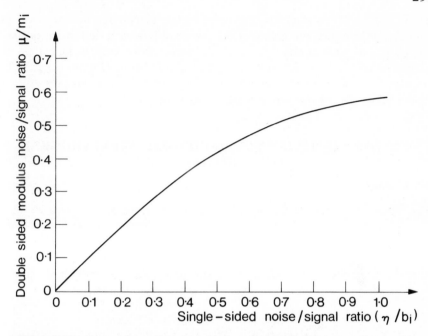

Figure 9.13. Comparison of double-sided to single-sided noise-to-signal ratios

than 10, the two-sided transform gives virtually the same result as the ideal single-sided transform. However, in applying this conclusion one must remember that by far the commonest use of two-sided transforms is the derivation of transmission spectra, where one takes the ratio of two modulus spectra. If the 'sample' interferogram has a region of very low signal (due to the use of a sample which is too thick), the computed ordinates will be in error because of the contribution of the rectified noise and the absorption coefficients will not be trustworthy. When both modulus spectra have low ordinates, as would for example happen at the extremes of the spectral band pass, the ratio can be completely unreliable. Under normal circumstances the signal-to-noise ratio in a transmission spectrum will be related to that in, say, the background spectrum by the relation[65]

$$(S/N)_T = 2^{-1/2}(S/N)_{BS}, \tag{9.61}$$

but in the case of two-sided transforms at the extremities of the range much worse behaviour might be anticipated. The error of course contains a large *systematic* component and this must be borne in mind when comparing spectra derived by different workers under different conditions.

The other principal use of two sided transformations is the deduction of phase spectra. Now when noise is present the phase spectrum which is calculated will become uncertain. In regions where the spectral power is low the phase spectrum will fluctuate wildly since noise dominates. In regions where the power is high, on the other hand, the phase spectrum will be smooth. At intermediate values of the power one can say that the root mean square phase fluctuations will be of the same order (and in fact approximately equal to) the noise-to-signal ratio (η/b_i) in the equivalent single-sided transform. In other words, the root mean square phase error is proportional to the reciprocal of the spectral signal-to-noise ratio.

9.7. THE CONTINUOUSLY RECORDED INTERFEROGRAM WITH PHOTON NOISE

9.7.1. Photon noise

The electromagnetic oscillations found naturally invariably consist of a superposition of waves of many different frequencies ν lying within some range $\Delta\nu$. There is usually no definite phase relationship between these different frequency components and the resultant real wave (see Chapter 3) must be regarded as a *random* time function. This can be expanded as a Fourier series in a long time period T_0 in the form (section 2.11)

$$V^{(Re)}(t) = \sum_{m=0}^{\infty} \left(a_m \cos \frac{2\pi mt}{T_0} + b_m \sin \frac{2\pi mt}{T_0} \right), \qquad (9.62)$$

where a_m and b_m are Gaussian random variables (section 2.15.4). This was first mentioned by Rayleigh. Recent work by Janossy has confirmed earlier demonstrations by Einstein that a_m and b_m are statistically independent Gaussian random variables with the same variance if T_0 is sufficiently long and the random process is stationary (section 2.13).

When the radiation field is weak, as it is for thermal radiation from a black-body-type source, it may alternatively be considered to be composed of particles (i.e. photons), the randomness being associated with the random variations in the time intervals between consecutive arrivals of the particles at the detector. These intervals are too small to be detected; in practice, the randomness can be discerned by recording the variations in the total number of photons collected in a given time interval. This record is not made directly but by way of counting the electrons liberated from a photocathode which is bombarded by the photon beam. The photoelectric current consists of a sequence of discrete pulses which give a measure of the rate at which electrons are emitted. However, the derivation of the properties of the photon beam from those of the electrons is not straightforward. If the radiation falling on the cathode were perfectly steady (a hypothetical situation) the photoelectric current would show fluctuations—the probability of a certain number of electrons being emitted in a certain time being given by a Poisson distribution. These fluctuations of current are called the *shot noise*. The photoelectric current consists therefore of a steady term, the mean rate of photoelectron emission, which is proportional to the intensity of the incident beam (that is, the number of photons incident per unit time) and a random term. The variance of the electron fluctuations is also proportional to the intensity of the incident beam and, therefore, to the mean rate of photoelectron emission. When a real, fluctuating, quasi-monochromatic beam falls on the photocathode, the intensity fluctuations are Gaussian and the probablity of a certain number of electrons being emitted in a certain time is given by a Bose–Einstein distribution. The variance of the electron fluctuations now consists of the shot noise term and an 'excess noise' term. Mandel *et al.*[66] have concluded that this general description of the variance of the number of photoelectrons ejected reflects the fluctuation properties of the radiation itself in so far as they are accessible to measurement and is, in fact, applicable to thermal and non-thermal, stationary and non-stationary sources.

9.7.2. The use of the photomultiplier in Fourier spectrometry

In a photomultiplier, the incident radiation passes through a suitable window and is incident on a photocathode. The resulting primary photoelectrons are accelerated to a positive electrode (called a dynode) where they produce an increased number of secondary electrons by bombardment. These are accelerated to another dynode where the process is repeated. There are commonly at least ten dynodes, so a few incident electrons can be multiplied up to quite a significant final photocurrent. The process of secondary electron production is itself subject to random factors which introduce shot noise, but it is usually the characteristics of the photocathode which predominate in photomultiplier noise.

The photocathode will invariably show a characteristic dependence of responsivity on frequency (see Figure 9.14). This is because at incident photon energies lower than the work function of the cathode material (the minimum energy required to extract an electron from the surface), no electrons will be released and the responsivity will fall abruptly to zero, whereas at higher frequencies, generally in the ultraviolet, photons may be reflected or transmitted by the cathode coating; or interact with the cathode in other non-electron-ejecting ways. Moreover, there is a fundamental reason why the power responsivity must ultimately fall as frequency increases. For a given quantum efficiency, the power detected for each electron released at the cathode increases linearly with $h\nu$, i.e. directly with frequency, and the spectral responsivity expressed in milliamperes per watt must therefore inevitably fall. However, the limiting factor on the high frequency side is generally the window material of the photomultiplier tube. For short-wave band pass, lithium or magnesium fluoride may be used, while silica, sapphire, or various glasses may be used to give a longer-wave cutoff. Figure 9.14(a) illustrates some typical spectral response curves for three combinations of cathode and window, while Figure 9.14(b) shows the power transmission of some window materials.

300

We therefore have to use a series of photomultipliers to cover the available range from 100 to 1100 nm. At the ultraviolet end the noise is predominantly shot noise, but at the infrared end, where low work function materials have to be used, thermal emission of electrons becomes significant and one may again encounter detector noise limitation. Cooling of the cathode to as low as −80 °C may be used to reduce this thermal noise to the shot noise level, and we continue this analysis assuming that thermal noise is negligible.

We are using, by definition, a very sensitive detector and will therefore be experiencing serious limitations by shot noise. If we compare the performance of a Fourier spectrometer with a sequential spectrometer using a photomultiplier, we find a very different situation to that which prevails when thermal detector noise is the limitation. In the case of the Fourier spectrometer, because the whole band of radiation of, say, M elements falls

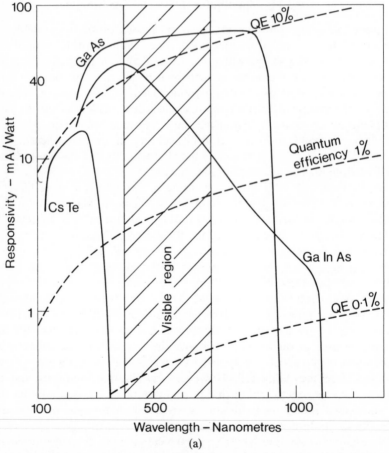

Figure 9.14. (a) Typical photocathode response curves. (b) Transmission of photomultiplier window materials: A, LiF; B, sapphire; C, Pyrex

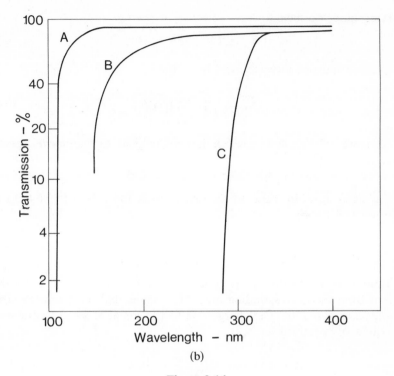

Figure 9.14.

on the detector during the measurement instead of only one element, the noise level, being proportional to the square root of the power, is $\sqrt{M}$ greater than in sequential spectrometry. This increase in noise just offsets the multiplex advantage and indicates that, when photon noise dominates, a Fourier spectrometer need only be used if the luminosity advantage is important.

Kahn[67] has given a simplified classical account of the influence of photon noise in Fourier spectrometry, the basis of which he subsequently elaborated to describe the correlation experiments of Hanbury–Brown and Twiss.[68] If the spectrum being observed is made up of M elements of width δf in a band $\Delta f = f_M - f_m$, then the spectral current flowing at an audio frequency f is given by

$$i(f) = e\alpha N_B(f)\delta f \quad \text{[A]}, \quad (9.63)$$

where α is the quantum efficiency of the photomultiplier, e the electronic charge, and $N_B(f)\delta f$ is the spectral intensity in terms of number of photons per second. The spectral current is usually accumulated in a very high impedance storage system (for example, a capacitor) and the resulting voltage is read by a valve voltmeter. This voltage is the spectral signal which is recorded. We therefore need to know the total *charge* accumulated in the integrating time τ and, since this will usually be set equal to T, the total observing time, we have

$$q(f) = e\alpha T N_B(f)\delta f \qquad [C]. \qquad (9.64)$$

The average spectral intensity may be written

$$\bar{N}_B = \frac{1}{\Delta f} \int_{f_m}^{f_M} N_B(f)\, df. \qquad (9.65)$$

The variance of the noise (which is proportional to the *mean* spectral intensity) then becomes

$$\eta^2 = 2Te^2\alpha\bar{N}_B\Delta f \qquad [C^2] \qquad (9.66)$$

and the signal-to-noise ratio in the spectrum is the ratio of (9.64) to the square root of (9.66):

$$(S/N)_{\text{multiplex}} = \frac{1}{\sqrt{2}} \sqrt{\alpha T} \frac{N_B(f)\delta f}{\sqrt{\bar{N}_B \Delta f}}. \qquad (9.67)$$

This ratio improves as the square root of the observation time T, but is little different from the corresponding ratio for a sequential spectrometer *observing the same spectrum*. The charge due to the signal at some frequency in the sequential spectrum is

$$e\alpha[N_B(f)\delta f]\frac{\delta f}{\Delta f} T \qquad [C], \qquad (9.68)$$

where $\delta f/\Delta f$ is the fraction of the total time spent observing the power $N_B(f)\delta f$. The corresponding noise is

$$e\sqrt{\alpha N_B(f)\frac{(\delta f)^2}{\Delta f} T} \qquad [C] \qquad (9.69)$$

and the signal-to-noise ratio is

$$(S/N)_{\text{sequential}} = \sqrt{\alpha T} \sqrt{N_B(f)\frac{(\delta f)^2}{\Delta f}}. \qquad (9.70)$$

To compare (9.67) and (9.70) we find the ratio

$$\frac{(S/N)_{\text{multiplex}}}{(S/N)_{\text{sequential}}} = \frac{1}{\sqrt{2}} \sqrt{\frac{N_B(f)}{\bar{N}_B}}. \qquad (9.71)$$

This is identical to the expression found by Kahn. The multiplex advantage is virtually cancelled; the value of $(S/N)_{\text{multiplex}}$ is greater than $(S/N)_{\text{sequential}}$ in those regions of the spectrum where $N_B(f) > 2\bar{N}_B$ and less where $N_B(f) < 2\bar{N}_B$. This finding means that strong emission lines are best observed with a Fourier spectrometer, weak lines in between are best seen with a sequential instrument.

9.8. THE CONTINUOUSLY RECORDED INTERFERENCE RECORD WITH MULTIPLICATIVE NOISE

When the total received power fluctuates, the level of the interference signal fluctuates also. From the point of view of the Fourier transform process these variations cannot be distinguished from true interference variations and the resultant spectrum contains spurious features which, in the limit, are random. One gets a similar result if the variations are imposed *after* the detector for example if the amplifier gain drifts. Since the observed interferogram is given by

$$F_{obs}(v\tau) = F_{theor}(v\tau)\varepsilon(\tau), \tag{9.72}$$

where $\varepsilon(\tau)$ is a random factor of mean 1 and variance about the mean of

$$\varepsilon^2 = \overline{[\varepsilon(\tau) - 1]^2} \tag{9.73}$$

(this is also equal to $\overline{[\varepsilon(\tau) - 1]^2}$ since $\varepsilon(\tau)$ is stationary), it follows that the observed spectrum will be given by the convolution

$$B_{obs}(f) = B_{theor}(f) * N(f), \tag{9.74}$$

where $N(f)$ is the Fourier transform of $\varepsilon(\tau)$. The consequences of this convolution with noise take us into some of the more obscure branches of the topic and we will not therefore pursue these but instead will refer the reader to the original literature—for example, Pinard.[69] From the viewpoint of experimental spectroscopy, multiplicative noise is an unmitigated nuisance since it brings a multiplex disadvantage in its trail, so the concern of the experimental Fourier transform spectroscopist is always to reduce the effects of this type of noise as much as he can.

9.8.1. Experimental methods of reducing the effects of multiplicative noise

It is clear that, when multiplicative noise is dominant, the worst way of doing experiments is to use straightforward AM interferometry. Assuming symmetrical interferograms for simplicity, the interferogram is

$$F_{obs}(v\tau) = F_\infty\varepsilon(\tau) + F_e(v\tau)\varepsilon(\tau), \tag{9.75}$$

and we see that we have a large noisy term $F_\infty\varepsilon(\tau)$ which contributes about half the noise (roughly speaking) but which contributes nothing to the information content of the signal! Shifting now to the spectral domain, it may readily be shown that if

$$N(f) = 2\int_0^\infty \varepsilon(\tau) \cos 2\pi f\tau \, d\tau \tag{9.76}$$

and

$$N^c(f) = 2\int_0^\infty \varepsilon(\tau) W\left(\frac{\tau}{T}\right) \sqcap\left(\frac{\tau}{T}\right) \cos 2\pi f\tau \, d\tau, \tag{9.77}$$

then the calculated spectrum is

$$B^c(f) = I_\infty[N^c(f) - \mathcal{A}(fT)] + B_e(f) * N(f). \qquad (9.78)$$

$\mathcal{A}(fT)$ oscillates rapidly to zero as f increases or alternatively we may put in mean values for the ordinates, but in either case, to a high degree of accuracy, (9.78) becomes

$$B^c(f) \simeq I_\infty N^c(f) + B_e(f) * N(f). \qquad (9.79)$$

The distortion due to the convolution of the spectral window with the noise function is unavoidable, but the large additive term $I_\infty N^c(f)$, which is proportional to the background power I_∞ can be reduced or eliminated and it must be if one is to approach the optimum performance.

The first way of reducing the effect of the constant term noise is the use of both interferometer outputs. Usually, of course, no use is made of the complementary interferometrically modulated beam which returns towards the source, but, if one can extract it, then it may be used to reduce the noise effects. The two signals may be written:

source, $\qquad F_S(v\tau) = (R^2 + T^2)I_0^N - 2RTI_0^N F'(v\tau); \qquad (9.80a)$

detector, $\qquad F_D(v\tau) = 2RTI_0^N + 2RTI_0^N F'(v\tau); \qquad (9.80b)$

where I_0^N is the (noisy) incident signal and $F'(v\tau)$ is the interferogram function which one would get for unit incident power. R and T are the moduli of the complex reflection and transmission characteristics of the beam divider. If one could record the *difference* of (9.80a) and (9.80b), then one would bet

$$F_D(v\tau) - F_S(v\tau) = I_0\varepsilon(\tau)(2RT - R^2 - T^2) + 4RTF_{\text{theor}}(v\tau)\varepsilon(\tau), \quad (9.81)$$

where we have written $F_{\text{theor}}(v\tau)$ for $I_0F'(v\tau)$. On comparing (9.81) with 9.75) it will be seen that the constant term has been greatly reduced (eliminated if one can achieve the ideal $R = T = 0.5$ situation) and that the modulation has been increased by a factor of two. Thus one is much better off than if one had used straight AM interferometry. Still more intriguing is to consider the sum of (9.80a) and 9.80b):

$$F_D(v\tau) + F_S(v\tau) = I_0\varepsilon(\tau)(R + T)^2 = I_0\varepsilon(\tau). \qquad (9.82)$$

From this one can, in principle, extract $\varepsilon(\tau)$ and feed it back into (9.80) to produce a *noise-free* interferogram. This is unfortunately very difficult and has not so far been done. Nevertheless the method is, in principle, promising. The obvious drawback is that two channels are required and amplifier drift will not be compensated. There are ways of using just a single channel (by observing alternately), but modern amplifiers are so good that drift is not a serious problem anyway.

The second method of reducing the constant term noise is to use phase modulation. This automatically eliminates the constant term and there is thus not the problem of making near-ideal beam splitters and of having ultrastable amplifiers. One still has the noise, of course, which arises from the convolution of the (odd) spectrum with the (odd) noise function. Thus with only a single detector one can virtually eliminate the constant term noise. PM therefore offers a considerable advantage (in addition to those listed in Chapter 8) when multiplicative noise is dominant. One could if one wished use *two* detectors likewise and one would get twice the signal from the difference, but the sum would be identically zero and there would be no way of making an exact allowance for noise effects as one can (at least in principle) in the AM case.

A third method of reducing the effects of multiplicative noise and, in fact, of eliminating it altogether is to use ratio recording. In this one uses a monitor detector to record the incident power. One then takes the ratio of the interferometer signal to the monitor signal and, since these will have identical noise factors, all the noise should be eliminated in the ratio. This technique has recently become very common in a non-multiplex branch of spectroscopy, namely the determination of absorption spectra by the use of tunable pulsed lasers. These lasers give out a stream of very brief (10 ns–1 µs) pulses with repetition rates of between 0.1 and 10^4 Hz. The peak height usually varies widely from pulse to pulse, so if a straightforward transmission experiment using RC integration and a slow detector is carried out, rather noisy spectra are obtained. If one can, however, use *two* detectors in a double-beam arrangement and if these detectors are fast enough, it becomes possible to ratio each pulse and then one gets essentially noise-free spectra. In the Fourier case we can imagine, for example, the source suddenly dropping to half its power, the ratio will however not be affected since both the interferometer and the monitor signals will equally drop by half. In sequential spectroscopy this would always be true provided both the wanted signal and the monitor signal had passed through the monochromator, but in Fourier spectroscopy it is only true if the multiplicative noise is strictly achromatic, that is, all spectral frequencies behave similarly. Fortunately this is usually the case. When it is not, the situation becomes very complex indeed. Chromatic multiplicative noise is touched on later in the chapter.

9.8.2. General treatment of signal-to-noise ratio and the multiplex advantage

Now that we have considered the three common forms of noise in some depth, we can re-examine the arguments which established the multiplex advantage and extend them to all three cases. It is assumed throughout that we are dealing with either a structured spectrum made up of sharp lines or else, if the lines are not sharp, that we are not extending the computation beyond the point where the interferogram oscillations have fallen below the digital discrimination. In this way we will ensure at least the possibility of a multiplex advantage.

In all cases we will have identical *signal* and we may simplify the treatment, without losing its essential points, by assuming that all the spectral features have the same spectral intensity $\bar{B}(\text{W Hz}^{-1})$ and that they all have the same spectral width $\delta f(\text{Hz})$. If the total observing time is T, then the *energy* accumulated in the spectrum per spectral element, in other words the *signal*, will be given by:

$$\text{multiplex,} \quad S_M = \bar{B}\delta f T; \tag{9.83}$$

$$\text{sequential,} \quad S_S = \bar{B}\delta f(T/M); \tag{9.84}$$

where M is the number of spectral elements. For the calculation of the noise, we need to know the time constant of the detecting system. In the multiplex case we will set τ equal to the reciprocal of the highest audio frequency present, i.e. f_M, and will have also therefore that $f_M = M\delta f$. Now invoking (9.52) and for simplicity setting $Q = 1$ we have for the three cases

$$\text{multiplex detector noise,} \quad N_M^D = 2\varepsilon_D(T/f_M)^{1/2}; \tag{9.85a}$$

$$\text{multiplex photon noise,} \quad N_M^P = 2K(M\bar{B}\delta f)^{1/2}(T/f_M)^{1/2}; \tag{9.85b}$$

$$\text{multiplex source noise,} \quad N_M^S = 2\varepsilon_S M\bar{B}\delta f(T/f_M)^{1/2}; \tag{9.85c}$$

$$\text{sequential detector noise,} \quad N_S^D = 2\varepsilon_D(T/Mf_M)^{1/2}; \tag{9.86a}$$

$$\text{sequential photon noise,} \quad N_S^P = 2K(\bar{B}\delta f)^{1/2}(T/Mf_M)^{1/2}; \tag{9.86b}$$

$$\text{sequential source noise,} \quad N_S^S = 2\varepsilon_S\bar{B}\delta f(T/Mf_M)^{1/2}. \tag{9.86c}$$

In these equations ε_D has the dimensions of watts, K is a constant of dimensions of watts$^{1/2}$, and ε_S is dimensionless. All are root-variance measures of the respective noise. One therefore has that for detector noise,

$$\frac{(S/N)_M^D}{(S/N)_S^D} = M^{1/2}, \tag{9.87}$$

which is the familiar multiplex advantage, but for photon noise,

$$\frac{(S/N)_M^P}{(S/N)_S^P} = \text{constant} \tag{9.88}$$

and there is no advantage, whilst for source noise,

$$\frac{(S/N)_M^S}{(S/N)_S^S} = M^{-1/2} \tag{9.89}$$

and there is a *disadvantage*. High-resolution spectroscopy from a highly structured interferogram will therefore be expected to give disappointing results when source noise is significant. Since a common use of high-resolution spectroscopy is in atmospheric or astronomical applications where source noise (poor 'seeing') is a commonplace, it is essential that use be made of the means detailed earlier for reducing or eliminating the noise.

9.8.3. Chromatic multiplicative noise

So far the multiplicative noise has been assumed to be both 'white' (i.e. the power spectrum of the noise at any observing spectral frequency is constant) and achromatic (i.e. the fluctuations of the amplitude of the power spectrum are also independent of observing spectral frequency). When the perturbations are chromatic (i.e. the power spectrum amplitude fluctuations do depend on spectral frequency) the situation is extremely complicated. Instead of the whole spectrum being perturbed by the same fluctuation, different component frequencies are perturbed differently. One can of course gain an advantage by using PM to eliminate the constant term, but one is then left with an interferogram whose ordinates depend both on σ and x and therefore, strictly speaking, we no longer have a proper Fourier integral. It follows from this that chromatic noise cannot be eliminated by ratio recording or any other simple means.

In practice, one does not know the nature of the limiting noise until investigations have been made. If source noise is found to be the limiting perturbation, an estimate of whether or not it is chromatic can be made by comparing a large number of the ratioed AM records with their average. It has been suggested that the normalized deviation of the records from their average will be greatest at intermediate path differences if the noise is chromatic, whereas it will be similar for all path differences for achromatic noise. The argument is somewhat subjective but is explained on the grounds that the interferogram term gives an integrated effect near zero path difference that fluctuates as the broad-band incoming signal does and at large path differences the interferogram term is zero, leaving a term that is again proportional to the total input.

The spectrum calculated from a record of this kind has superimposed noise that is very frequency dependent and, according to Connes and Connes,[70] increases enormously towards low frequencies. As they point out, when sampled recording is used (as opposed to continuous recording) and when a high frequency spectral band is under study, the size of the sampling interval is of great significance. If the interferogram is not oversampled, there is a grave danger of the low frequency noise being aliased into the spectral region of interest. This oversampling results in the production of more samples for a given interferogram, which can, in turn, limit the available resolution if the limit of the computer capacity is approached.

When the spectral density of the noise increases towards low frequencies, the corresponding fluctuations in the interference record are slow and at normal aperiodic drive speeds become confused with the true interference fringes arising from power in the range $0 \leqslant f \leqslant f_M$. This causes the lower frequencies in the power spectrum to be very noisy. The calculated spectrum can be shifted relative to the noise by increasing the drive speed. Since this also demands reduction of the smoothing filter time constant, the interferogram becomes very noisy and many interferograms need to be superimposed to recover the signal-to-noise ratio. The superimposition is only effective if the interferograms are exactly superimposed or, as Mertz puts it, *co-added*. The combination of rapid scanning and co-adding 'beats' the low frequency noise and represents a technique which is a mixture of periodic generation (rapid scanning) and aperiodic generation (Fourier transformation of the co-added interferograms). If an on-line computer is used the co-adding can proceed until the signal-to-noise ratio in the calculated spectrum is adequate.

This process of co-adding and final transformation of the co-added interferogram when it has reached the desired S/N is a uniform feature of all the sophisticated automated interferometric spectrometers which have been developed commercially for the mid-infrared region. In the simplest theory the process of co-adding interferograms and transforming only one final result is equivalent to transforming each and averaging the resulting spectra. This latter procedure is, however, very expensive of computer time, so the practical alternative to co-adding and transforming is to observe only a *single* interferogram and to use all the available observing time for smoothing it. The co-adding method has the disadvantage that one has the practical problem of ensuring that all the interferograms are observed starting from *exactly* the same place, but it not only has the advantage mentioned above but also others which in some circumstances make its use mandatory. Principal amongst these is that it provides a simple solution to the dynamic range problem. In the far infrared the dynamic range problem only obtrudes for the most careful high resolution work, but in the mid-infrared, where one will be covering thousands of cm^{-1}, the signal range over the whole interferogram will be enormous. Detectors just cannot be expected to be linear over dynamic ranges of $10^5 : 1$ or greater, so the method of observing a single virtually noise-free interferogram and transforming it is going to fail from the beginning. If on the other hand one observes a set of noisy interferograms, the dynamic range in each will be less and the detector will be working always on the linear part of its characteristic. The required dynamic range is then built up in the digital store capacity of the built-in computer. These components can, of course, handle virtually any value of dynamic range. In this method the signal is 'riding on' the noise and is therefore always being observed on a linear part of the detector's curve. One is therefore making optimum use of the detector and the digital circuits by splitting the work between them.

9.8.4. Atmospheric fluctuations

In the laboratory, the most common causes of multiplicative noise are amplifier drift and actual fluctuation of the source output. The former is minimized by use of high quality electronics, the latter by use of stabilized power supplies, preferably current stabilized. It is in the observation of extraterrestrial sources such as the sun, the planets, the stars, or even the sky itself that multiplicative noise presents the greatest problems, and these do not appear to be fully conquerable. Detailed discussions of some of the aspects are given by Connes and Connes[70] and Connes, Connes, and Maillard;[71] Bowers[72] has attempted a mathematical treatment of the effects of atmospheric turbulence by introducing a time-dependent factor into the atmospheric transmission and calculating the consequent effect on a two-line spectrum. At about the same time, James[73] gave a more general account of time-dependent atmospheric transmission by taking a model in which the atmosphere is split into layers of fluctuating thickness. In this account the dependences of $E(\sigma, t)$ on σ and t are separated. This permits a solution to be arrived at, but even for this simple model the solution is extremely complicated. More realistic models are not amenable to closed-form solution.

9.9. DIGITIZING NOISE

Except when analogue computational methods are employed, the continuous signals representing the record of the interference are converted to digital form. The digital scale consists of finite levels. The smaller these levels the better the approximation of the digital record to the analogue. But however small the levels there is always a limiting uncertainty as to which of two adjacent levels a given input may be converted. The output from the analogue-to-digital converter corresponds to any level of the analogue input in the range $V - \frac{1}{2}V'_{A/D}$ to $V + \frac{1}{2}V'_{A/D}$ and the error takes any value between $-\frac{1}{2}V'_{A/D}$ and $+\frac{1}{2}V'_{A/D}$. These errors are a random function of path difference which we call the *digitizing noise*. Let the root mean square value of this be ε_d. This noise represents an effective limit to the resolution of the power (or voltage) scale. If the maximum voltage allowed to enter the digitizer is V_{max}, we follow Mertz in calling

$$DRF = V_{max}/\varepsilon_d \qquad (9.90)$$

the *dynamic range factor*. This factor is extremely important in considerations of digitizing noise, for the dynamic range required to record the interferogram due to a broad-band source is very large, especially when weak or narrow absorption features are superimposed. This is because all frequencies superimpose to give a grand maximum equivalent to the *total* recorded power at zero path difference, while small variations in power dependent on the absorption features occur at large path differences. This digitizing noise may be the limiting factor—a situation that is quite different from sequential recording. Vanasse and Sakai[74] have shown that if the interferogram signal-to-noise ratio is $(S/N)_I$, then there is no digitizing noise if

$$DRF > (S/N)_I. \qquad (9.91)$$

When $(S/N)_I$ is very large and DRF looks like providing the overall limitation, *scale expansion*, as proposed for example, by Connes *et al.*,[71] may be employed. Near zero path difference, a low gain is used and then, after a few fringes, the gain is step-increased by a known amount to make the signal again fill the digital scale. This proceedure may be repeated several times until detector noise provides the limitation. Before Fourier transformation the record of the interferogram is compensated for the gain changes. This technique is, however, less used nowadays since reliable digital voltmeters having a discrimination of better than 10^5 are available.

9.10. SAMPLING NOISE

When the digital version of a detector-noise-limited interferogram record is obtained, samples are taken from that record at finite intervals rather than continuously. In section 9.4.2 we have already seen that the size of the sampling interval in relation to the response time of the filter has an effect on the signal-to-noise ratio in the spectrum unless the noise frequencies are arranged to coincide with the spectral frequencies. We have also seen earlier that a *periodic* error in the sampling interval leads to 'ghost' features accompanying every frequency in the spectrum. It would seem therefore that random sampling errors will lead to random ghosting or *noise* in the spectrum. This type of error in the spectrum is known as *sampling noise*. It has been discussed by Pinard.[69] Small errors in the interferogram sample produce errors in the spectrum which are proportional both to the original error and to the *slope* of the interferogram at the sampling point. Thus the effect of random errors of a certain amplitude gets worse and worse as the frequency rises. For this reason, mid-infrared interferometers seldom feature mechanical sampling. The almost universal practice is to have the sample commands derived from fringes produced using laser radiation, either by the same interferometer or else by a small 'slave' interferometer sharing the same drive. In this way random sampling errors can be eliminated and the sampling noise will vanish.

CHAPTER 10

Computational aspects

10.1. INTRODUCTION

The voltage (i.e. the analogue) output from the RC smoothing network has to be sampled, in the manner indicated in Chapter 6, and the set of samples has then to be Fourier transformed, in a suitable computer, to yield the desired spectrum. The sampling signal can be provided in various ways: thus, if a stepper motor drive is being used, the current pulse to the motor can also be used to provide the sampling command signal. If a continuous drive is being used, then either a Moiré fringe or else a laser interferometer system can be employed to provide the desired sample command pulses. The encoding of the analogue signal can likewise be done in several ways: thus in the past shaft encoders (i.e. discrete rotary potentiometers) were used connected to the main drive of a chart recorder, but in nearly all modern work the analogue-to-digital converter is a digital voltmeter. The digitized output (often in binary coded decimal form, that is, each digit of a decimal number is recorded as a binary number) can be on paper tape, punched cards, or else on magnetic tape. There were some early experiments with the use of analogue computers to do the Fourier transformation, and in many ways this does seem a very natural way of carrying it out, but the power of digital techniques has been increasing by leaps and bounds and all modern work features digital computation exclusively. The analogue computers despite their charm are now only of historical interest.

The digital computer is therefore fed a series of numbers $I_0 \ldots I_{N-1}$ (the interferogram ordinates) and is also given a set of parameters—the value of the stepping interval Δx, the minimum and maximum wavenumbers to be computed, etc. However, the machine has to be programmed, i.e. told what to do, and it is in the writing of the programme (the 'software' in computer jargon to distinguish it from the 'hardware' of the computer and its ancilliaries) that several matters of interest arise. Generally speaking, computer time is either expensive, at a premium, or both, and the programmer will aim to write a program which carries out the desired computation in the minimum time. Additions can be carried out very quickly indeed inside the machine. Multiplications are naturally slower since they involve repeated additions, but the computation of transcendental functions by summation of series is very slow indeed and costly in computer time. The programmer will therefore write his program in such a way as to avoid the repeated computation of transcendental functions and to reduce the number of multiplications and additions to a minimum. The program itself is, as remarked above, a sequence of instructions to the computer. It can be written in a manner directly relevant to the machine actually being used, that is in 'machine or assembler language', but the mastery of the machine language is usually a very lengthy process and very few operators bother to acquire this mastery since it is peculiar to the one machine and the machine language for a different machine would show considerable variations. Instead all modern large computers feature a *compiler*, that is, a software package which translates a program written in a *higher language* into one written in the appropriate machine language. There are numerous higher languages, for example FORTRAN (in its several versions), ALGOL, BASIC, etc., but they all have the feature that program instructions written in them look like conventional mathematical instructions and these languages are therefore easy to learn. A Fourier transform program written in FORTRAN IV for example can be used on any machine with a FORTRAN IV compiler (that is, the vast majority of computers) so it will be of general utility. Such a program can therefore be duplicated and circulated amongst spectroscopists or even made available commercially for sale. Once the program has been written, it can be transcribed via a teletype to any of the input media, paper tape, punched cards, magnetic tape, etc., which the computer can read. The ease of writing and the interchangeability are the great advantages of higher language programs, but the disadvantages are the need to have a compiler, the necessity to use a large number of the machine's memory locations to store the compiler, and the relatively inefficient use of further memory locations to store the transliterated program. If a small computer only is available and a particular form of calculation has to be carried out repeatedly (e.g. Fourier transformation), then it may well be worth the considerable task of writing an efficient machine language program.

10.2. FOURIER TRANSFORMATION IN A HIGH-SPEED DIGITAL COMPUTER

In numerical Fourier transformation one replaces, as mentioned earlier, the integral

$$S(\sigma) = \int_{-x_{max}}^{+x_{max}} F(x) \cos 2\pi\sigma x \, dx \tag{10.1}$$

by the summation

$$S(\sigma) = \Delta x \sum_{m=(-M/2)}^{m=(+M/2-1)} F(m\Delta x) \cos 2\pi\sigma m\Delta x \tag{10.2}$$

over a total number of M interferogram samples. If $F(x)$ is symmetrical about $x = 0$, i.e. $F(+x) = F(-x)$, then this summation can also be written

$$S(\sigma) = 2\Delta x \left[\tfrac{1}{2}F(0) + \sum_{m=1}^{m=(M/2-1)} F(m\Delta x) \cos 2\pi\sigma m \Delta x \right.$$

$$\left. + \tfrac{1}{2}F\left(\frac{M}{2}\Delta x\right) \cos 2\pi\sigma \frac{M}{2} \Delta x \right]. \tag{10.3}$$

At first sight it might be thought that the summation of this series would be a slow process in the machine since it involves the computation of a large number of cosines. The cosine function has to be calculated each time it is required from the basic series

$$\cos \theta = 1 - \frac{1}{2!} \theta^2 + \frac{1}{4!} \theta^4 \ldots + (-1)^k \frac{1}{2k!} \theta^{2k}, \tag{10.4}$$

though of course one has, in principle at least, only to calculate it in the fundamental range 0–2π since the values in all other regions can be found by first subtracting an appropriate number of 2π's from the argument to reduce it into the fundamental region. The series will then converge much more rapidly. Large computers usually have a permanent set of small programs (called subroutines) for the calculation of commonly occurring functions and such subroutines are also frequently a permanent feature of the compiler. The programmer therefore does not have to include the series of instructions for calculating cosines in his program, but of course there is no way of cutting short the actual time that the machine will take in doing the operation of summing the series (10.4) to as many terms as are necessary to ensure the required precision.

Fortunately, the interferogram is sampled at regular intervals Δx and one is faced, for each wavenumber σ, with the problem not of calculating a set of unrelated cosines but of calculating a set, each member of which has an argument which is an integral multiple of that of the first member. Thus one has to calculate $\cos (2\pi\sigma\Delta x)$, $\cos 2(2\pi\sigma\Delta x)$, $\cos 3(2\pi\sigma\Delta x)$, etc. A result from elementary trigonometry, namely

$$\cos A + \cos B = 2 \cos (A+B)/2 \times \cos (A-B)/2 \tag{10.5}$$

can be rephrased for the integral multiple case to read

$$\cos (n+1)\theta + \cos (n-1)\theta = 2 \cos n\theta \cos \theta, \tag{10.6}$$

that is,

$$\cos (n+1)\theta = 2 \cos n\theta \cos \theta - \cos (n-1)\theta, \tag{10.6a}$$

and therefore it is necessary to calculate *only* $\cos \theta$ from the series, all the rest following from simple algebraic manipulations. Equation (10.6) is of such importance that it is usually distinguished by a special name and called the *Chebyshev recurrence relation*. If the output wavenumbers are also to be regularly spaced (as will nearly always be the case), then a still further saving of time occurs because now we have

$$S'(n\Delta\sigma) = \Delta x \sum_{-M/2}^{(M/2-1)} F(m\Delta x) \cos mn(2\pi\Delta\sigma\Delta x). \tag{10.7}$$

In this case *all* the cosines that will be required for the entire computation can be generated from the one† primitive value $\cos 2\pi\Delta\sigma\Delta x$. They can therefore be displayed as a matrix initially symmetrical about its leading diagonal, thus

		m			
	1	2	3	4	5
1	$\cos u$	$\cos 2u$	$\cos 3u$	$\cos 4u$	$\cos 5u$
2	$\cos 2u$	$\cos 4u$	$\cos 6u$	$\cos 8u$	$\cos 10u$
n 3	$\cos 3u$	$\cos 6u$	$\cos 9u$	$\cos 12u$	$\cos 15u$
4	$\cos 4u$	$\cos 8u$	$\cos 12u$	$\cos 16u$	$\cos 20u$
5	$\cos 5u$	$\cos 10u$	$\cos 15u$	$\cos 20u$	$\cos 25u$

etc., (10.8)

where $u = (2\pi\Delta\sigma\Delta x)$ and where we have omitted for the sake of brevity the trivial row and column when m or n is zero. Of course if $M = N$ then the matrix is completely symmetrical about its leading diagonal. One next needs to compute the cosine of every integral multiple of $(2\pi\Delta\sigma\Delta x)$ up to $\cos MN(2\pi\Delta\sigma\Delta x)$, where M and N are the maximum values of m and n and distribute these at the appropriate matrix location in (10.8). If this matrix is designated C_{mn}, then the interferogram ordinates can be assembled into a column vector F_m and the spectral ordinates follow (also as a column vector) from the multiplication

$$S_n^c = C_{nm}F_m. \tag{10.9}$$

Of course all that has been said above applies, *ceteris paribus*, with equal force to the case where one is calculating a sine transform. The sine equivalent of the cosine Chebyshev recurrence relation is

$$\sin(n+1)x = 2 \sin nx \cos x - \sin(n-1)x \tag{10.10}$$

and one assembles a matrix of sines S_{mn} and carries out the matrix multiplication

$$S_n^s = C_{nm}F_m. \tag{10.11}$$

It is important to note that there is no restriction here on the *number* of output wavenumbers. All that is required is that they be regularly spaced.

---

† In real machines there will always be cumulative errors due to 'rounding off' of numbers. It is therefore wise occasionally to use the series (10.4) to generate a few cosines rather than to rely on a very long chain of applications of the recurrence relation. This will make only a very small increase in the total time needed.

Of course if one is calculating spectral ordinates at finer intervals than the sampling theorem indicates, then no additional information is forthcoming. All that one gets from such a procedure is help with drawing smooth profiles to pass through the computed ordinates. Excessively fine computation of ordinates is therefore merely a form of interpolation and one would get the same result by convolving a minimum set of ordinates with the spectral window function.

10.3. THE FAST FOURIER TRANSFORM—BASIC IDEAS

The total number of multiplications involved in Fourier transformation by this method (assuming $M = N$) will be $M(M+1)/2 + M^2$, where the first term represents the multiplications required to generate the cosines and the second the number of multiplications required to carry out the matrix multiplications (10.8). The total number is therefore $\frac{3}{2}M^2$ when M is large. This approach is nowadays called the *slow Fourier transform* (Mme Connes[37] has reported that the transformation of a 12000-point interferogram using the slow Fourier transform in an IBM 7040 took 12 h!) because it was soon realized that by choosing an appropriate order for the processing and by restricting the number of points to be transformed to be a power of two, a routine could be devised for use in a computer, which was very much faster. Bell[58] has given a good account of this and has also commented on the history of this *fast Fourier transform*. The early work of Good[75] became much better known through the advances introduced by Cooley and Tukey,[31] who stressed the practical advantages and showed that the technique was not restricted to base two calculations but could be expressed in any base. Cooley and Tukey described a repetitive mechanical operation (that is, an algorithm) by means of which a full complex Fourier transformation could be rapidly carried out in an electronic computer. Their algorithm has found extensive use in many branches of physics and engineering since Fourier transformation is a very common operation and small hard-wired (i.e. permanently programmed) computers can be bought which do the algorithmic transformation very rapidly indeed. However these do not usually permit the operator to vary the number of points to be transformed. Spectroscopists too have made extensive use of the algorithm, but since they usually wish to vary the number of points (for example to vary the resolution) it is more common for them to use a general purpose digital computer plus conventional software. It will be shown later that the Cooley–Tukey algorithm reduces the number of operations from $\sim M^2$ to approximately $M \log_2 M$, and if M is large this can represent an enormous saving in time. Connes[37] and Bell[58] have referred to this procedure as 'decimation in time', but the exact meaning to be ascribed to the word 'decimation' is not clear. It appears to allude to the ancient Roman punishment of putting to death every tenth man in a line and therefore refers to the procedure of selecting from a group of ordinates alternate members in a repetitive fashion. The word in vernacular usage has come loosely to mean 'drastic reduction' and it is sometimes assumed that this is what is meant when the word is used in connection with the FFT. That this is not the case can be seen from the existence of an alternate way of writing the algorithm which is referred to as 'decimation in frequency'.

The Cooley–Tukey algorithm is nowadays a very powerful technique indeed, but its basic ideas can be developed in quite simple language and the important results derived by relatively unsophisticated arguments. The crucial point is the introduction of the resolution theorem requirements (see Chapter 6) at the start of the procedure. Thus one requires that

$$\sigma_{max} = \frac{M}{2} \Delta\sigma = \frac{1}{2\Delta x}; \tag{10.12}$$

therefore $\Delta\sigma\Delta x = M^{-1}$. If now we make $N = M$ and make both of them of the form 2^p, where p is integral, then some considerable simplifications ensue. Thus, because the cosines are now of the form $\cos mn(2\pi/M)$ they represent a set which is disposed very symmetrically on the cosine wave— see Figure 10.1. Therefore all the cosine factors which occur in a large transform can be reduced to a very much smaller set, namely that which occurs in the region of abscissa values from 0 to $\pi/2$ together with appropriate changes of sign. The whole situation is very symmetrical and the number of multiplications and additions which is needed is greatly reduced.

Before developing the fast Fourier transform algorithm in general terms, it is helpful to consider an illustrative example. We recall that

$$\wedge(\sigma) \supset \left(\frac{\sin \pi x}{\pi x}\right)^2, \tag{10.13}$$

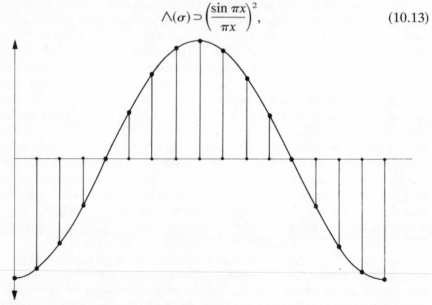

Figure 10.1. Ordinates of the cosine curve sampled at the abscissa values $(2\pi m/16)$

and we will therefore demonstrate the reconstruction of the triangle function $\Lambda(\sigma)$ from a set of sampled ordinates of the $\text{sinc}^2(x)$ function. The maximum value of σ for $\Lambda(\sigma)$ is unity and therefore, by the sampling theorem, the sinc^2 function has to be sampled at intervals $\Delta x = 0.5$. It follows therefore that all the even ordinates (with the exception of I_0, which is unity) will be zero. If we take M $(=2^p)$ ordinates, then the total excursion of x will be $0.5M$ and the resolution interval in the σ domain will be $2M^{-1}$. The total scan of σ for an equal number of output points (i.e. M) will be 2. We expect therefore from the sampling theorem that the transform will show mirror symmetry about $\sigma = 1$ (the folding wavenumber); that is,

$$S(n) = S(M-n), \tag{10.14}$$

and only $M/2$ of the calculated ordinates will be independent. The basic transform equation (10.2) becomes

$$S(n) = \Delta x \sum_{-M/2}^{(M/2-1)} F(m) \cos\left(\frac{2\pi}{M} mn\right). \tag{10.15}$$

The negative values of m are inconvenient. They can be eliminated either by the use of (10.3), which applies in the present symmetric case, or else by the use of a more general stratagem. In this we first note that

$$\cos\left[\frac{2\pi n(-m)}{M}\right] = \cos\left[\frac{2\pi n(M-m)}{M}\right]. \tag{10.16}$$

Now if the interferogram segment running from $-x_{max}$ to $+x_{max}$ is translated along a distance $2x_{max}$ in the positive direction, the summation (10.15) can identically be written

$$S(n) = \Delta x \sum_{0}^{M-1} F(m) \cos\left[\frac{2\pi}{M} mn\right] \tag{10.17}$$

and the negative values of m have been eliminated. We have, in essence, added M to all the negative values of m and arrived at a postive only set. However it must always be borne in mind that the ordinates $F(m)$ only have unique meaning for $0 < m < M/2$. The ordinates for m in the range $M/2 < m < M$ are found by the symmetry relation

$$F(M-m) = F(m), \tag{10.18}$$

which follows from the reflection/translation symmetry of Figure 10.2(a).

It is interesting to pause to consider what has been done here in a little more depth. An interferogram which is only known from $-x_{max}$ to $+x_{max}$ and a calculated spectrum which is only defined from $\sigma = 0$ to $\sigma = \sigma_{max}$ can both be regarded as segments of endlessly replicated functions. This is shown in Figure 10.2 insets (a) and (b), respectively. It will be seen that in this light the Fourier transform operation takes on complete symmetry; the transform of either endlessly replicated function gives the other endlessly replicated

318

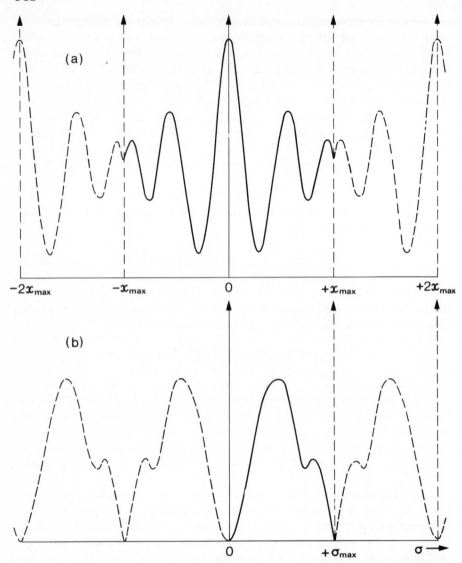

Figure 10.2. A symmetrical interferogram (a) can be endlessly replicated in the x domain and evaluation of the Fourier integral from $-x_{max}$ to $+x_{max}$ replaced by integration from 0 to $+2x_{max}$. The corresponding replicated spectrum in the σ domain is shown below (b) to illustrate the full symmetry of the situation

function. It is therefore most important that there be no discontinuities in the interferogram since these would generate disturbances spread over wide regions of the spectrum. The joining up of the translated segment in Figure 10.2(a) must therefore not result in a discontinuity. This can be best ensured by suitable apodization that reduces the interferogram oscillations smoothly to zero at $x_{max} = \frac{1}{2}M\Delta x$.

Now, with this theory developed, we return to the reconstruction of the sinc^2 function and, taking $M = 8$, we have the eight ordinates

$$I_0 = 1.0$$
$$I_1 (=I_7) = 0.405\,28$$
$$I_2 (=I_6) = I_4 = 0$$
$$I_3 (=I_5) = 0.045\,03.$$

The possibility of discontinuity is neatly avoided since $I_4 = 0$.

For $M = 8$, the transform equation (10.17) becomes

$$S(0) = 0.5[I_0 + I_4 + I_2 + I_6 + I_1 + I_5 + I_3 + I_7],$$
$$S(1) = 0.5[I_0 - I_4 + (I_1 - I_5 - I_3 + I_7) \cos \pi/4],$$
$$S(2) = 0.5[I_0 + I_4 - (I_2 + I_6)],$$
$$S(3) = 0.5[I_0 - I_4 - (I_1 - I_5 - I_3 + I_7) \cos \pi/4],$$
$$S(4) = 0.5[I_0 + I_4 + I_2 + I_6 - I_1 - I_5 - I_3 - I_7],$$

$$(10.19)$$

and $S(5) = S(3)$, $S(6) = S(2)$, $S(7) = S(1)$. (The reason for the particular grouping of ordinates adopted here and in equation (10.20) will become apparent later.) The calculated values are $S(0) = 0.950$, $S(1) = 0.755$, $S(2) = 0.500$, $S(3) = 0.245$, and $S(4) = 0.05$. This is a rather good reconstruction from only four independent 'interferogram' samples.

Simple though the relations expressed in (10.19) are, they reveal properties which are general for all values of M. Thus if $S(0)$ is written as $A + B$, where A is the sum of the even ordinates and B the sum of the odd ordinates, then $S(4)$, i.e. $S(M/2) = A - B$. Similarly, if $S(1) = C + D$, then $S(3) = C - D$. It is clear therefore that if one has calculated $S(0)$, then only one additional operation is required to generate $S(4)$: a similar remark applies to $S(1)$ and $S(3)$. The second point to emerge is that the ordinates occur in the pairs $I_m \pm I_{m+M/2}$, where the plus sign applies when m is even and the minus when m is odd and that therefore the Fourier transform of, say, the even ordinates can be resolved into combinations involving only one set of these pairs and once again there is a saving in the total number of operations which is required. It is these properties which lie at the heart of the saving in machine time provided by the use of the Cooley–Tukey algorithm. As a further illustration, we would have the following relations for $M = 16$:

$$2S(0) = [I_0 + I_8] + [I_4 + I_{12}] + [I_2 + I_{10}] + [I_6 + I_{14}]$$
$$+ [I_1 + I_9] + [I_5 + I_{13}] + [I_3 + I_{11}] + [I_7 + I_{15}]$$
$$2S(1) = [I_0 - I_8] + [I_4 - I_{12}] \cos \pi/2 + [I_2 - I_{10} - I_6 + I_{14}] \cos \pi/4$$
$$+ [I_1 - I_9] \cos \pi/8 - [I_5 + I_{13}] \cos 3\pi/8 + [I_3 - I_{11}] \cos 3\pi/8$$
$$- [I_7 - I_{15}] \cos \pi/8$$

$$2S(2) = [I_0 + I_8] - [I_4 + I_{12}] + [I_2 + I_{10}] \cos \pi/2 - [I_6 + I_{14}] \cos 3\pi/2$$
$$+ [I_1 + I_9] \cos \pi/4 - [I_5 + I_{13}] \cos \pi/4 - [I_3 + I_{11}] \cos \pi/4$$
$$+ [I_7 + I_{15}] \cos \pi/4$$
$$2S(3) = [I_0 - I_8] - [I_4 - I_{12}] \cos \pi/2 - [I_2 - I_{10} - I_6 + I_{14}] \cos \pi/4$$
$$+ [I_1 - I_9] \cos 3\pi/8 + [I_5 - I_{13}] \cos \pi/8 - [I_3 - I_{11}] \cos \pi/8$$
$$- [I_7 - I_{15}] \cos 3\pi/8$$
$$2S(4) = [I_0 + I_8] + [I_4 + I_{12}] + [I_2 + I_{10}] + [I_6 + I_{14}]$$
$$+ [I_1 + I_9] \cos \pi/2 + [I_5 + I_{13}] \cos \pi/2 - [I_3 + I_{11}] \cos \pi/2$$
$$- [I_7 + I_{15}] \cos \pi/2, \tag{10.20}$$

etc. The remaining ordinates can be found by reversing the signs of the odd ordinate parts. Thus from $S(1)$ one would get $S(7)$, from $S(2)$ one would get $S(6)$, and from $S(3)$ one would get $S(5)$. $S(8)$ is obtained similarly from $S(0)$. The use of equations (10.19) and (10.20) to reconstruct the triangle function is shown in Figure 10.3. The reconstruction is even better with sixteen ordinates than with eight, which is as one would expect. The exceptionally good reconstruction of this particular function arises, as will be seen from the Figure, from the very rapid fall-off of the sinc^2 function as its argument increases. The folding symmetry about $\sigma = 1$ and the close relationship between the resolution and the folding frequency when the interferogram is minimally sampled are well brought out in this figure.

The considerable symmetry manifest in equations (10.19) and (10.20) springs basically from the fact, mentioned already and illustrated in Figure 10.1, that the cosine samples are being taken in a highly symmetrical way, but it can be further illustrated by considering the angles involved. This is shown in Figure 10.4. It will be seen that for, say, $M = 16$, the angles break up into the sets for which the running integer m has a common factor 8, 4, 2, and only 1 with M. The angles themselves which generate the sets are then, of course, π, $\pi/2$, $\pi/4$, and $\pi/8$. To illustrate this, since the I_8 ordinate will only involve cosines of $m\pi$, it will appear in the transform equations merely as plus or minus itself. The basic symmetry revealed in this diagram or in equations (10.19) and (10.20) can be immediately programmed to give a very fast Fourier transform indeed. What one has to do is to read the ordinates into an array, not in the order in which they are observed but in the order of decreasing common factor as outlined above. Each group is then multiplied by the appropriate cosine selected from the very small basis set. Unfortunately as this stands one would need a separate program for each value of M. This is rather inconvenient since most users like to vary the number of ordinates they wish to transform and merely wish to instruct the computer, via a parameter card or tape, to do the transformations using one and only one program.

What is needed is a mechanical way of sorting an arbitrary number, $N = 2^p$, of ordinates into the correct order and of then generating rapidly the

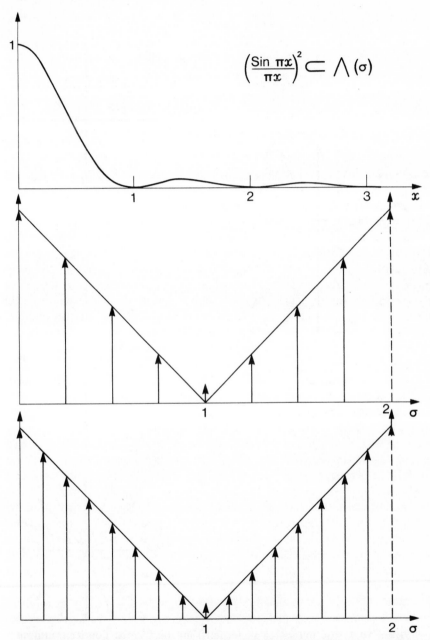

$$\left(\frac{\text{Sin } \pi x}{\pi x}\right)^2 \subset \bigwedge (\sigma)$$

Figure 10.3. Reconstruction of the triangle function from the Fourier transformation of a sampled sinc^2 function

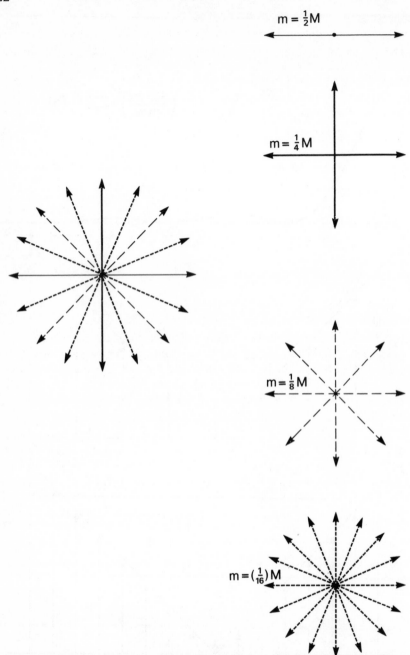

Figure 10.4. The full set of angles (left) involved in the Fourier transform operation can be divided into sets (right) which are characterized by running integers m which have less and less common factors of two with N

required cosines. A mechanical rule or process, either direct or iterative, for performing a calculation is, as mentioned earlier, called an 'algorithm' and the process introduced by Cooley and Tukey for doing fast Fourier transformation is often referred to therefore as the '*Cooley–Tukey*' or FFT *algorithm*. The first step is merely to sort the ordinates into reverse binary order. To see how this works, imagine that we have eight ordinates and that the first eight memory locations of the computer are numbered

$$000, \quad 001, \quad 010, \quad 011, \quad 100, \quad 101, \quad 110, \quad 111$$

as they would be in binary. These binary numbers when reversed will be

$$000, \quad 100, \quad 010, \quad 110, \quad 001, \quad 101, \quad 011, \quad 111$$

and would therefore contain

$$I_0, I_4, I_2, I_6, I_1, I_5, I_3, I_7.$$

If on the other hand we had sixteen ordinates, these would be resorted into

$$I_0, I_8, I_4, I_{12}, I_2, I_{10}, I_6, I_{14}, I_1, I_9, I_5, I_{13}, I_3, I_{11}, I_7, I_{15}.$$

This ordering will be seen to be identical to the occurrence of the ordinates in (10.19) and (10.20). Thus for the $M = 16$ case the pairs $(I_0 \pm I_8)$, $(I_4 \pm I_{12})$, $(I_2 \pm I_{10})$, etc., occur and the $\pm$ signs are chosen in alternating order. The even and odd sets will be seen to be strongly related to one another, thus if one merely adds unity to the running integer of the even set, one arrives at the odd set. At first sight the order of the numbers in reversed binary order looks a little higgledy-piggledy, but it will soon be seen that the even members go

$$0, 1.2^{(p-1)}; \quad 2^{(p-2)}, (1+2)2^{(p-2)}; \quad 2^{p-3}, (1+2^2)2^{p-3}; \quad \text{etc.};$$

and the odd ones are these plus one. The sequence of reversed binary numbers can therefore be generated by a simple algorithm which goes 'multiply the present set by 2 and add on at the end the same set in the same order but with each number augmented by unity'. Thus one has

0

0 1

0 2 1 3

0 4 2 6 1 5 3 7

0 8 4 12 2 10 6 14 1 9 5 13 3 11 7 15

etc. The even numbers always come first in this reverse binary ordering because the least significant binary digit (or 'bit') becomes the most significant on reversal and in the binary representation even numbers have a zero in the units column.

It is always convenient in machine calculations of linear algebra to have the transformation matrices symmetrical. If this is adopted, the spectral ordinates will also appear in reversed binary order and they will have to be finally resorted before the machine outputs the spectrum, but this presents little problem (see later). We will therefore have a symmetric matrix C_{mn} which we can consider divided into four blocks thus:

$$
n \begin{cases}
\quad & \begin{array}{c} \text{Even} \\ \hline \text{Odd} \end{array} & \begin{array}{c|c} ++ & +- \\ \hline -+ & -- \end{array}
\end{cases} \tag{10.21}
$$

Now in the inverted binary order the columns will occur in pairs $m = a$ and $m + 1 = a + M/2$, so for n even the entries in the two columns will be the same in both magnitude and sign since they will be, respectively,

$$\cos na(2\pi/M) \quad \text{and} \quad \cos[na(2\pi/M) + n\pi].$$

Conversely for n odd they will be the same in magnitude but will differ in sign. Thus this inverted binary sorting permits the machine to achieve a further factor of two saving, in the number of multiplications to be carried out. In a similar manner, adjacent elements of pairs within the columns will have the same magnitude and sign for m even and the same magnitude but opposite sign for m odd. Thus within the four blocks of equation (10.21) the elements can be grouped into four blocks of four which have the form

$$
\begin{bmatrix} a & a \\ a & a \end{bmatrix} \begin{bmatrix} a & a \\ -a & -a \end{bmatrix} \begin{bmatrix} a & -a \\ a & -a \end{bmatrix} \begin{bmatrix} a & -a \\ -a & a \end{bmatrix}. \tag{10.22}
$$
$$\;++\qquad\;\;\;+-\qquad\;\;\;-+\qquad\;\;\;--$$

Thus only one element needs to be computed for each block and overall there is a reduction in the amount of computation by a factor of 4.

This being the case, we need only consider the $(1, 1)$ entry in each little 2×2 submatrix and we arrive at an $M/2 \times M/2$ matrix to be calculated. The same process can be applied to this and eventually we shall come down to a primitive 2×2 matrix. By inverting the process one can calculate a matrix of arbitrary size starting with the primitive one. What one does is to create a 2×2 matrix from each element of the preceding matrix by introducing the appropriately changed cosine and then applying the symmetry rules (10.22).

The sorting of the input data into reverse binary order and the resorting of the finally computed spectral ordinates into the conventional sequential form are both readily achieved mechanically in the computer. The data are read into an array and the machine selects alternate elements and repeats this operation within each selected group $(p - 1)$ times. Thus for eight elements we would have

Step 0	(000), (001), (010), (011), (100), (101), (110), (111);
Step 1	(000), (010), (100), (110), (001), (011), (101), (111);
Step 2	(000), (100), (010), (110), (001), (101), (011), (111);

hence achieving the reversed binary order. The resorting of the spectral ordinates is achieved by reversing this process. It should perhaps be mentioned here that it is not usual to compute the output in reversed binary order. This is the most convenient way of presentation to illustrate the basic features of the algorithm and the outline of the FFT procedure given above, in terms of a matrix/vector multiplication, does bring out well the symmetry involved in the operation. However such a procedure would be rather slow in a computer and, worse still, would make poor use of the machine's storage space. Spectroscopists are always striving to get better and better results, either higher resolution, higher radiometric precision or else a wider spectral coverage, and all these things mean more points to be transformed. The principal requirement on the programmer therefore is that his program make optimal use of the available storage capacity. This will be a recurring theme as we now go through an outline of how the Cooley–Tukey procedure is programmed for practical computation.

10.4. SOME PRACTICAL ASPECTS OF FOURIER TRANSFORM PROGRAMMING

When considering the practical aspects of programming a digital computer to do Fourier transformation, we will not dwell on the mechanics of programming and on details of the high level languages which are used to instruct the machine since these are dealt with in exhaustive detail in the computer manuals and since most spectroscopists will either buy a program suitable for their particular computer or else have it included in the standard software package which comes with the machine. FFT routines, for example that of Singleton,[76] are also readily available and the days when a spectroscopist would write his own Fourier transform program are long past. Rather we will try to outline in general terms what is happening inside the machine as the program is executed so that the spectroscopist armed with this knowledge will be in a good position to take appropriate decisions and get the best possible service out of his interferometer computer combination.

10.4.1. The full complex fast Fourier transform

Practical FFT routines are designed to have the widest possible applicability. They are therefore designed to transform one set of complex quantities into another set of complex quantities. Two arrays are used for the data, one to store the real parts and the other to store the imaginary parts. Now, in spectroscopy, the interferogram ordinates are all pure real, so this arrangement is inefficient since one array will have to be filled up with zeros. Likewise the output data will usually be either pure real (with symmetric interferograms) or else pure imaginary (with antisymmetric ones) so the use of store space will again be inefficient. Ways of getting round this and of using the available store to the best advantage have been described by Bell[58] and by Fleming[77] and these will be outlined later in this chapter.

For the moment we will discuss the operation of the algorithm without specifying anything about the ordinates and without dividing store locations into real and imaginary locations. Thus any location can be considered to hold a complex number $A + iB$. The complex equivalent of equation (10.17) is

$$\hat{S}(n) = \Delta x \sum_{0}^{M-1} \hat{F}(m) \exp\left[\frac{2\pi i}{M} mn\right]. \qquad (10.23)$$

The machine will assemble the $F(m)$ into reverse binary bit order as discussed in section 10.3 using the same array that was originally used to store the incoming data. The algorithmic process will then be carried out in a way analogous to that described in section 10.3, except of course that complex arithmetic is implied throughout. One has now, in the general case (i.e. both F_m and S_n complex) rather less obvious symmetry than one had in the previous case where one was transforming purely real data by means of a cosine only transform. In fact if $\hat{S}_n^+$ represents the transform of the even set of F_m and $\hat{S}_n^-$ the transform of the odd set, then one has

$$\hat{S}_n = \hat{S}_n^+ + \exp\left(2\pi in/M\right)\hat{S}_n^-$$

and

$$\hat{S}_{N/2+n} = \hat{S}_n^+ - \exp\left(2\pi in/M\right)\hat{S}_n^-. \qquad (10.24)$$

Thus one has only to do one complex multiplication to get two spectral ordinates, which is of course one of the corner stones of the algorithm, but this is the only simple symmetry available. If however one is transforming purely real data, as one will be in spectroscopy, then further symmetry relations become available because the spectrum must then be Hermitian (see Chapter 2) and its ordinates will be complex symmetric about $N/2$, i.e.

$$\hat{S}_{N/2-n} = \hat{S}_{N/2+n}^*. \qquad (10.25)$$

By taking the real parts of (10.24) and (10.25) all the equalities mentioned in section 10.3 for the real case will be recovered.

10.4.2. The complex FFT algorithm

As remarked earlier, for an automatic calculating machine one needs a method of doing a calculation that is rigorously repetitive, so that the machine will go through a fixed routine a specified number of times then stop and print the answers. In this way the number of cycles can be arbitrary and, therefore, as will transpire below, the number of points which can be transformed can be arbitrarily large. The number of points does have to be of the form 2^p, but if one has in fact observed a number M such that $2^{p-1} < M < 2^p$ the necessary condition can be satisfied by adding $2^p - M$ zero ordinates at the end of the interferogram. The data base so augmented usually needs to be apodized in the manner discussed in Chapter 6, but this is readily achieved by means of a small preliminary program. This is usually built into the main FFT program and some elaborate programs even permit a choice of apodizing function by the choice of an appropriate parameter in the machine instructions.

The data base after apodizing is resorted into the same locations in reverse binary order. After this the algorithm proper begins. It consists, first, of taking the ordinates in pairs, starting at the beginning, and from each pair calculating two other quantities by taking the sum and the difference. The results are then 'overwritten' on to the original array and hence no additional computer storage is required. This first step is illustrated in the second column of Table 10.1. At the fourth step we combine the ordinates I_s and $I_{s+2^{k-1}}$ starting with $s = 0$ and skipping any ordinate already combined. The plus and minus combinations, with the appropriate exponentials included, are

$$I'_s = I_s + I_{s+2^{k-1}} \exp(i\pi s/2^{k-1}),$$
$$I'_{s+2^{k-1}} = I_s - I_{s+2^{k-1}} \exp(i\pi s/2^{k-1}), \tag{10.26}$$

where the prime indicates the contents of the location after the operation. The subsequent steps of the computation (for $M = 8$) are shown in Table 10.1. By taking the real parts of the ordinates, the relations given in equation (10.19) will be recovered.

If we have 2^P original points, then there will clearly be required p cycles to achieve the full calculation. On this basis one can compute roughly the number of multiplications required in the FFT. Each cycle takes $M/2$ multiplications and M additions in addition to the rather faster sorting operations. A fairly good measure therefore, bearing in mind that additions are quicker than multiplications, is to say that each cycle involves $3M/2$ complex operations, that is, $3M$ real operations. The full algorithm takes p cycles, where of course $p = \log_2 M$, and therefore the total number of operations is given by

$$O = 3M \log_2 M. \tag{10.27}$$

This is to be compared with the value

$$O = \tfrac{3}{2}M^2, \tag{10.28}$$

which, as was shown earlier, is the number of operations involved in the slow transform. The FFT is therefore faster than the slow by the factor $(M/2)\log_2 M$. As an example, if M were to equal $1\,048\,576$ (i.e. $p = 20$),

Table 10.1 The computation of the FFT for an eight-point input

Location number s	Initial contents	First step $k=1$	Second step $k=2$	Third step $k=3$	Spectral ordinate
0	I_0	I_0+I_4	$(I_0+I_4)+(I_2+I_6)$	$[(I_0+I_4)+(I_2+I_6)]+[(I_1+I_5)+(I_3+I_7)]$	S_0
1	I_4	I_0-I_4	$(I_0-I_4)+(I_2-I_6)E\left(\dfrac{\pi}{2}\right)$	$\left[(I_0-I_4)+(I_2-I_6)E\left(\dfrac{\pi}{2}\right)\right]+\left[(I_1-I_5)+(I_3-I_7)E\left(\dfrac{\pi}{2}\right)\right]E\left(\dfrac{\pi}{4}\right)$	S_1
2	I_2	I_2+I_6	$(I_0+I_4)-(I_2+I_6)$	$[(I_0+I_4)-(I_2+I_6)]+[(I_1+I_5)-(I_3+I_7)]E\left(\dfrac{\pi}{2}\right)$	S_2
3	I_6	I_2-I_6	$(I_0-I_4)-(I_2-I_6)E\left(\dfrac{\pi}{2}\right)$	$\left[(I_0-I_4)-(I_2-I_6)E\left(\dfrac{\pi}{2}\right)\right]+\left[(I_1-I_5)-(I_3-I_7)E\left(\dfrac{\pi}{2}\right)\right]E\left(\dfrac{3\pi}{4}\right)$	S_3
4	I_1	I_1+I_5	$(I_1+I_5)+(I_3+I_7)$	$[(I_0+I_4)+(I_2+I_6)]-[(I_1+I_5)+(I_3+I_7)]$	S_4
5	I_5	I_1-I_5	$(I_1-I_5)+(I_3-I_7)E\left(\dfrac{\pi}{2}\right)$	$\left[(I_0-I_4)+(I_2-I_6)E\left(\dfrac{\pi}{2}\right)\right]-\left[(I_1-I_5)+(I_3-I_7)E\left(\dfrac{\pi}{2}\right)\right]E\left(\dfrac{\pi}{4}\right)$	S_5
6	I_3	I_3+I_7	$(I_1+I_5)-(I_3+I_7)$	$[(I_0+I_4)-(I_2+I_6)]-[(I_1+I_5)-(I_3+I_7)]E\left(\dfrac{\pi}{2}\right)$	S_6
7	I_7	I_3-I_7	$(I_1-I_5)-(I_3-I_7)E\left(\dfrac{\pi}{2}\right)$	$\left[(I_0-I_4)-(I_2-I_6)E\left(\dfrac{\pi}{2}\right)\right]-\left[(I_1-I_5)-(I_3-I_7)E\left(\dfrac{\pi}{2}\right)\right]E\left(\dfrac{3\pi}{4}\right)$	S_7

The symbol $E(\pi/n)$ is used as a shorthand for $\exp(i\pi/n)$. For convenience the factor Δx has been omitted.

then the improvement would be by about a factor of 25 000!! The routine transformation of million point interferograms does take place in one or two laboratories (for example Aimé Cotton) and the time reduction given by the FFT is absolutely essential to the operation. After all, few people would be prepared to wait several months for a calculation to be completed or be in a position to pay for that much computer time. Even for more humble spectroscopy in which one would be transforming 1000 points, the gain factor would be 50. It is clear that without the FFT much of modern Fourier transform spectroscopy would not be a practical proposition.

10.4.3. Improvements to the FFT algorithm in practical spectroscopy

Dramatic though the improvement brought about by the FFT is, its operation as described above is nevertheless still inefficient since we initially used half the machine's precious storage capacity to store a string of zeros and because we are calculating a spectrum half of whose ordinates are related to the other half by the complex symmetry of equation (10.25). Of course if one is using the two-sided computation one will finish up with the modulus operation (Chapter 6) and one would have $S(N/2+m) \equiv S(m)$, in other words not just complex symmetry but total symmetry.

Connes,[37] Bell,[58] and Fleming[77] have all discussed how this redundancy can be eliminated. The solution is to divide the purely real interferogram into two sets of ordinates, those for which m is even and those for which it is odd. One then has two sets

$$F_e(m'), \qquad m' = 2m,$$

and

$$F_o(m'), \qquad m' = 2m + 1 \tag{10.29}$$

One then sets up the complex interferogram

$$\hat{F}_i(m') = F_e(m') + iF_o(m') \tag{10.30}$$

and carries through the FFT algorithm on this. Of course one merely needs M locations to store $\hat{F}_i(m')$. The algorithm applied to (10.30) does not give the desired spectrum immediately; instead one derives a complex function, called $\hat{E}(n)$, which may be written

$$\hat{E}(n) = \hat{C}(n) + i\hat{D}(n), \tag{10.31}$$

where $C(n)$ and $D(n)$ are the Fourier transforms of $F_e(m')$ and $F_o(m')$, respectively, and are Hermitian since $F_e(m')$ and $F_o(m')$ are pure real. From (10.31) it follows at once that

$$\hat{C}(n) = \tfrac{1}{2}[\hat{E}(n) + \hat{E}^*(-n)] \tag{10.32a}$$

and

$$\hat{D}(n) = \frac{1}{2i}[\hat{E}(n) - \hat{E}^*(-n)]. \tag{10.32b}$$

The desired spectrum may then be calculated from

$$\hat{S}(n) = \hat{C}(n) + \hat{D}(n) \exp(-i\pi n/N). \tag{10.33}$$

The negative values of n may be eliminated from (10.32) readily since $\hat{E}(n)$ is periodic, and hence

$$\hat{E}^*(-n) = \hat{E}^*(N/2 - n). \tag{10.34}$$

This process does take slightly longer than the straightforward transform because of the extra N operations required for (10.33), but the difference becomes rapidly negligible as p increases and the saving of store space when p is large is of great importance.

One would thus arrive at the complex spectrum $\hat{S}(n)$ whose imaginary part would carry all the phase information. From a spectral point of view in which one calculates a modulus spectrum, it really doesn't matter where one starts one's N points: in Figure 10.2(a) any stretch is equivalent to any other stretch of the same length. However, as Fleming[77] has stressed, following Forman,[78] the phase will depend on where one starts, and, since phase is conventionally measured from the point where the two beams are equivalent (i.e. ZPD), it is best to start at $m = 0$. The procedure described above will therefore give the phase correctly according to this convention. Of course one need hardly add that there must not be a discontinuity at the match point $m = N/2$.

10.4.3.1. The special case of the real even interferogram

The ultimate in this process of saving computer store by eliminating redundant information comes when one considers pure real symmetric interferograms. These are, of course, commonly used in non-dispersive work and, even if the actual interferogram observed is not perfectly symmetric, it will be after application of the phase correction convolution described earlier. This kind of interferogram is particularly favoured in high resolution work where, because of the guaranteed symmetry, one can use virtually the whole of the available mirror travel to produce meaningful data. The method stems from the early work of Brenner,[79] but has been described in detail by Connes[37] especially and also by Bell[58] and Fleming[77]. The work of Connes and Bell is unfortunately marred by a series of printing errors so here we will follow, merely in outline, the treatment given by Fleming.[77]

We need consider only perfectly even or perfectly odd interferograms, for which:

$$\text{even functions} \qquad I_m = I_{M-m}; \tag{10.35a}$$

$$\text{odd functions} \qquad I_m = -I_{M-m}. \tag{10.35b}$$

It follows from these equations that $I_{M/2}$ should be zero for an odd function and that there should be no discontinuity for an even function at $M/2$. Also I_0 should be zero for an odd function. This will of course usually have been arranged automatically by the process of subtracting the mean. It is not essential that the mean be subtracted in the even case but this is highly desirable since, as mentioned above, the spectral ordinate S_0 has to be zero for physical reasons.

The interferogram ordinates beyond $I_{M/2}$ are, of course, simply related to those in the range $0-M/2$, so our first step is to define a function which uses only the independent values in the fundamental range $0-M/2$. This is the Hermitian H function, $H(m')$ defined by

even case $\qquad \hat{H}(m=0, 1, \ldots, M/2-1) = I_{2m} + i\frac{1}{2}\{I_{2m+1} - I_{2m-1}\}$, (10.36a)

odd case $\qquad \hat{H}(m=0, 1, \ldots, M/2-1) = \frac{1}{2}\{I_{2m+1} - I_{2m-1}\} + iI_{2m}$. (10.36b)

$H(m)$ is Hermitian (i.e. complex symmetric) about $M = M/4$, so it is only necessary to calculate a total of $(M/4+1)$ samples of H to have it defined over its entire range. The discrete Fourier transform of H, called K, and defined by

$$K_n = \sum_{m=0}^{M/2-1} \hat{H}_m \exp\left(i \cdot \frac{2\pi mn}{N/2}\right) \qquad (10.37)$$

will be real since H is Hermitian. Connes has shown that the required spectral ordinates can readily be extracted from the $K(n)$ by means of the equations

even case $\qquad S_n = \frac{1}{2}\left\{(K_n + K_{N/2-n}) + \frac{(K_n - K_{N/2-n})}{\sin(2\pi n/N)}\right\}$, (10.38a)

odd case $\qquad S_n = \frac{1}{2}\left\{\frac{(K_n + K_{N/2-n})}{\sin(2\pi n/N)} - (K_n - K_{N/2-n})\right\}$, (10.38b)

where n runs from 0 to $N/2-1$. These equations become indeterminate for $n = 0$ and in this case one selects the first term of (10.38a) and writes

$$S_0 = \frac{1}{2}(K_0 + K_{N/2}) \equiv K_0. \qquad (10.39)$$

One likewise selects the second term of (10.38b), but then finds—as expected—that $S_0 = 0$. This result is quite natural since a perfectly odd interferogram must yield a zero spectral ordinate at zero frequency since by definition there is no constant term. Phase modulation therefore eliminates information about the low frequency region—a conclusion reached earlier when the phase modulation characteristic was being discussed.

The method outlined above will work, of course, but really it is no advance on the method discussed earlier for halving the required store. That there is still redundant information present is revealed by the Hermitian character of H_m and, by its consequence, the purely real character of K_n. What we need is to construct another function from H which will have a complex transform and thus eliminate the redundancy, but which can nevertheless be readily used to calculate the original transform of H.

Fleming[77] shows that the required function is

$$F_m = (\hat{H}_m + \hat{H}_{(M/4)-m}) + i \exp{(i4\pi m/M)}(\hat{H}_m - \hat{H}^*_{(M/4)-m}). \quad (10.40)$$

The discrete Fourier transform of F_m (which runs only from $m = 0$ to $M/4$) will give a function E_n whose real part gives the even ordinates of K_n and whose imaginary part will give the odd ordinates. Thus the complete operation may be illustrated schematically.

$$\text{Interferogram} \to I_m \to H_m \to F_m \text{ \{Discrete FFT\} } E_n \to K_n \to S_n. \quad (10.41)$$

The Fourier transform operation is now on a data base only one fourth the size, and there is therefore a large saving in time. This saving is slightly less than would be expected from (10.27) because of the extra calculations involved in the chain (10.41).

An alternative approach to the problem of transforming a perfectly even or odd interferogram has been pointed out by Fleming.[77] He observes that, under normal circumstances, the last ordinate I_M will either be identically zero (because of the application of apodization) or else will be very small. One can therefore ignore the last term of (10.3) and use this equation to perform the transform. The *observed* ordinates are read into the computer's memories except that I_0 is first divided by two. A pure cosine (or sine, as appropriate) transformation is then carried out. This procedure is quick but it suffers from the drawback that the calculated spectral ordinates are spaced twice as coarsely as they are in the orthodox method; in fact the calculated ordinates are the even subset of the original. There is clearly little point in remedying this defect by adding $M/2$ zeros to the data base since one would be back with an M-point interferogram and one might just as well use the orthodox approach.

10.4.4. Spectral interpolation and mathematical filtering

From the theory developed in Chapter 6 it follows that to obtain the desired perfectly symmetric or else perfectly antisymmetric interferogram, which can be transformed by a single-sided transform, one needs to convolve the observed and generally asymmetric sampled interferogram with the phase correction function $\phi(\sigma)$. This latter is derived from a two-sided Fourier transformation carried out for a short stretch on each side of ZPD. One therefore computes (section 6.11.2)

$$F_e(x) = \int_{-\infty}^{+\infty} F_n(x')\chi(x - x')\,dx', \quad (10.42)$$

where

$$\chi(x) = \int_{-\infty}^{+\infty} \exp{(+i\phi(\sigma))} \exp 2\pi i\sigma x\,d\sigma. \quad (10.43)$$

Of course one cannot actually evaluate the continuous integral (10.42) and instead one must approximate it by numerical methods, in which one uses the analogue of the Fourier equations but with the cosine terms set to unity. This procedure nevertheless gives very good results. The resulting symmetrized interferogram can be read out, if desired, on to paper or magnetic tape for off-computer storage. The actual operation is relatively slow and it occupies a sensible fraction of the total time involved.

It was mentioned earlier, in connection with Figure 10.2, that the endlessly replicated spectrum and interferogram functions formed perfect Fourier mates and therefore one could always replace multiplication of one by any chosen function by convolution of the other with the Fourier transform of that function. So the operation of convolution just mentioned for symmetrizing observed interferograms might be expected to have other applications. The first of these is the calculation of spectral ordinates at finer intervals than is allowed by the sampling theorem when one is using the FFT. The point of difficulty here is that, although one can in principle draw a smooth spectral profile through the primary spectral ordinates, one cannot do this using the readily available linear interpolation subroutines (graph plotting). Mme Connes[37] states that at least four secondary (or interpolated) points are necessary before the graph plotting programme may be safely used. Samples at arbitrary values of wavenumber would be available if the spectrum were a continuous function. This can only be so and yet the two functions remain as Fourier mates if the interferogram function is truncated. If one were to apply the rectangle function to the endlessly replicated interferogram shown in Figure 10.2, so as to isolate the solid line period, then the Fourier transform would become continuous. Essentially this is because we have gone from the computation of Fourier series, which are discrete, to the computation of Fourier transforms, which are continuous. Of course we cannot do this in practice, at least we cannot do it this way, since we are constrained to use the discrete FFT to get from the interferogram domain to the spectral domain. However, since the Fourier transform of a product is equal to the convolution of the separate Fourier transforms (see section 2.3), it follows that we can get the same effect, i.e. a continuous transform, by applying a suitable convolution to the discrete function produced by the FFT. The Fourier transform of the rectangle function is the sinc function (section 2.9.2) and so the required convolution is

$$B'(\sigma_0) = \int_{-\infty}^{+\infty} B(\sigma) \frac{\sin 2\pi(\sigma_0-\sigma)D}{2\pi(\sigma_0-\sigma)D} \, d\sigma. \qquad (10.44)$$

This integral cannot be evaluated as it stands since its infinite range implies an infinite time for its computation! We must settle for an approximation and apply a truncation or an apodization to the sinc function so that the integrand is finite only over a restricted range—and preferably we would like this range to be as short as possible. The truncated or apodized sinc function will have a transform in which the sharp discontinuities at $\pm\sigma_{max}$ are replaced by rapidly damped oscillations. (This is essentially the Gibbs'

334

phenomenon discussed in Chapter 2.) The filter function, which is being effectively applied (albeit in the spectral domain via convolution) to the original interferogram, will therefore not isolate a single period since some of the adjoining alias will be involved and this function will also give a distorted account of the interferogram inside the fundamental period close to $\pm\sigma_{max}$. This is illustrated in Table 10.1. If apodization is being applied to the original interferogram, these undesirable effects will be considerably mitigated, but since post-apodization (see below) has much to commend it, we seek a more general stratagem to by-pass the difficulty. Mme Connes suggests setting a small number (~2%) of the final ordinates (i.e. those immediately before σ_{max}) equal to zero. This involves a loss of true resolution, but this is unavoidable when using the FFT because of the constraint that the number of points to be transformed equal a power of two.[†] Bearing this in mind, the true resolution interval in the spectral domain will be larger than the intervals at which $B(\sigma)$ is computed or conversely one can say that $B(\sigma)$ is being sampled more finely than would be indicated by the total path difference over which there is meaningful information. If we were to return to the interferogram domain by means of a Fourier transformation of this oversampled spectrum, then necessarily the aliases in the endlessly replicated $F(x)$ would be slightly spaced from one another with small bands of zeros between meaningful oscillations. Of course this is precisely what one has arranged by the initial setting to zero of the last 2% of the ordinates.

The Gibbs' oscillations of the modified rectangle function are very inconvenient and it is much better practice to apply apodization—for example that represented by the function of equation (6.38)—to the sinc function rather than to chop if off abruptly. This gives a filter function (Table 10.1) which drops smoothly from its unity value down to zero over a small range. The use of the word 'filter' here comes from the analogous, but technically far more important, field of digital data processing. Observed data as functions of time often need to be 'cleaned up' by restricting their frequency spectrum. One could do this by means of the electronic filter networks discussed in the previous chapter, but one can do it equally well by applying suitable convolutions to the actual observed unfiltered data. In the case of FT spectrometry one is applying filtering in the time or space domain, but mathematically the two situations are completely equivalent. The apodized sinc function is a form of apparatus function, so the operation represented by (10.44) can be generalized to give

$$B''(\sigma_0) = \int_{-\sigma_L}^{+\sigma_L} B(\sigma)A[(\sigma_0 - \sigma)D]\,d\sigma, \tag{10.45}$$

[†] If one were using the slow FT one could observe to the limits of one's available mirror travel and add on any arbitrary number of zeros after this, but then of course one would not imagine that one would have an interpolation problem with the slow FT. Oddly enough, in practice, one may well have such a problem because the FT may be much slower than a subsequent convolution of a minimally sampled set of $B(\sigma)$.

where $\pm\sigma_L$ is the range over which the integral is to be evaluated. Here $A(\sigma D)$ is *any* of the possible apparatus or spectral window functions. Thus there is a big advantage to computing *unapodized* spectra because the data so produced can be fitted to any chosen spectral window by the use of (10.45), whereas, if the original interferogram had been apodized, the spectra would have a preordained spectral window after transformation; equation (10.45), although it gives a continuous function $B''(\sigma_0)$, will, of course, in practice only be evaluated for a finite number of samples. The number chosen is determined by the acceptable error involved in fitting the points by means of the standard graph plotter program. Mme Connes suggests that four secondary points be interpolated between each pair of primary points. This would correspond in the interferogram domain to a set of aliases with a periodicity of $10D$. The question of how wide a range the integral (10.45) should be computed over is again determined by the level of acceptable error. Mme Connes states that using ten cycles of the sinc function gives secondary samples with less than 10^{-5} intensity errors. The matter of which procedure is the better, apodization of the interferogram or convolution of the sampled spectrum, when one has decided on the apparatus function, depends to a certain extent on the computer to be used and what subroutines are available. For high resolution spectroscopy with a modern high-speed machine the convolution approach is far preferable.

All that has been said above about filtering the interferogram can be applied just as well in reverse. One could take an apodized interferogram and convolve it with a suitable sinc function and thus arrive at a band-limited spectrum. This would be a desirable approach if one were only interested in a narrow spectral band. The approach is usually called mathematical filtering since the effect is just the same as if the incident radiation had been passed through an equivalent optical filter. For it to be effective, it is essential that the discrimination of the recording digital system be very high. If this cannot be ensured, it is better to rely on conventional optical filtering to reduce the dynamic range of the interferogram.

10.5. RELATIVE MERITS OF THE SLOW AND THE FAST FOURIER TRANSFORMS

At various times in the past, the slow Fourier transform has been considered to have some advantages over the fast and in fact it was often thought that the only advantage of the fast transform was its speed. Most of these alleged advantages have been shown to be illusory and those which remain have been shown to be only of marginal benefit.

The first advantage ascribed to the slow transform was that it permits one to process the data points as they are observed and in the order that they are observed. Thus one can use almost all of the available store for locating spectral ordinates. In essence one runs down the columns of the matrix

(10.8) with each element multiplied by the appropriate (and current) inter-ferogram ordinate and each product is summed into the store location appropriate to the spectral ordinate. Thus, if the mth interferogram ordinate has just been read in and one is calculating $F_m \cos 2\pi nm \Delta\sigma \Delta x$, this quantity will be stored in location n, where S_n is being assembled. Clearly when computer store is limited this feature is very attractive, but since means to increase the store capacity of a computer (floppy discs, magnetic data cartridges, etc.) are readily available and are becoming cheaper and cheaper, this argument is not so compelling as it once was. In addition, the FFT algorithm can be so programmed that each step is 'overwritten' on the preceding step—in other words the store locations used for the incoming data are also the locations where the final spectral ordinates are formed. There is thus no gain in the slow method since all the store not used by the FT program is available for input/output data in either case. Of course in both cases, if one is calculating a two-sided or full complex transformation, one will need $2N$ locations of store, N for the cosine transform and N for the sine transform. One should also point out that one cannot simply transform each ordinate as it is observed since what is required is the ordinate minus the average of all the ordinates. This is because $S(0)$ must, of course, be zero for a physically meaningful spectrum; therefore the sum of all the ordinates which are to be transformed must be zero. Phase modula-tion is very attractive here since phase modulated interferograms oscillate about a true zero and one could consider real time operation of the interferometer/computer combination. With amplitude modulation, the average value needs to be known before the transformation takes place. The best way of doing this is to record all the interferogram ordinates on punched tape, magnetic tape, or cards and then to process these quickly in the computer using a small preliminary program that calculates the average. The tape or cards can then be put into the reader again and slowly processed using the Fourier transform program—the machine automatically subtract-ing the average before doing the transformation.

The second putative advantage of the slow transform, related to the first, is its ability to operate in real time. The FFT has to wait until the last point has been recorded before it can get on with the actual transformation. In real time operation with the slow transform one would observe a short range of interferogram prior to the starting point and use this stretch to calculate the average. This advantage is however of dubious value to the spectroscop-ist because there are so many drawbacks to real time operation that it is seldom used in practice. Thus, generally speaking, the computer is faster than the interferometer and so there is a waste of computer time unless elaborate time sharing arrangements are available. Also, if one is observing single-sided interferograms, there is no obvious way of doing the phase correction, whereas with all the interferogram ordinates recorded on a tape one can do the short two-sided transform about ZPD to establish the phase and hence the function to be convolved with the observed interferogram to give a phase-corrected interferogram. A further disadvantage of real time operation is that if an error occurs, for example a point is omitted, is incorrectly inserted, or else is of the wrong amplitude (for example a 'pick-up' spike), then there is little one can do about it, whereas with a tape the error can be eliminated.

The third advantage claimed for the slow Fourier transform is that one can calculate spectral ordinates at arbitrary values of the wavenumber, whereas in the FFT, of course, the output wavenumbers are prescribed. This is, however, something of an illusory advantage because since the resolution theorem is built into the FFT one cannot get any more real information by calculating the spectrum at finer intervals. In fact if, as was discussed earlier, one were to convolve the discrete spectrum produced by the FFT with the window function, then the ordinate of the continuous function which would thus be produced at any chosen wavenumber would be identical to that calculated at the same wavenumber using the slow Fourier transform. Thus the operation of convolution permits us to interpolate secondary points between the primary points produced by the FFT and we can therefore recover all the flexibility of the slow transform whilst retaining the speed of the fast.

Thus it will be seen from the above discussion that when one is calculating the spectrum over the entire available range (i.e. $0 < \sigma < \sigma_{max}$) all the advantages lie with the FFT. However the slow transform may come into its own on those rare occasions when one is calculating only over a very limited range. This is because the FFT perforce must provide *all* the spectral ordinates. Nevertheless, the saving of time with the FFT is so great that the slow transform would only show a real gain if the number of points to be transformed were rather small and the range to be covered rather narrow.

References

1. FELLGETT, P. B. (1968) *Contemp. Phys.* **9**, 603.
2. CHAMBERLAIN, J. and CHANTRY, G. W. (1973) *High Frequency Dielectric Measurements*, IPC Science and Technology Press, Guildford, England.
3. E.g. HERZBERG, G. (1945) *Infrared and Raman Spectra*, Van Nostrand, New York; TOWNES, C. H. and SCHAWLOW, A. L. (1955) *Microwave Spectroscopy*, McGraw-Hill, New York.
4. CHAMBERLAIN, J. (1972) *Infrared Physics* **12**, 145.
5. MICHELSON, A. A. (1892) *Phil. Mag.* **34**, 280.
6. MICHELSON, A. A. (1891) *Phil. Mag.* **31**, 338.
7. MICHELSON, A. A. (1902) *Light Waves and their Uses*, University of Chicago Press, Chicago.
8. MICHELSON, A. A. (1927) *Studies in Optics*, Phoenix Edition, University of Chicago Press, Chicago, 1962.
9. RAYLEIGH, LORD (1892) *Phil. Mag.* **34**, 407.
10. TERRIEN, J. (1959) in *Symposium on Interferometry*, National Physical Laboratory, HMSO, London.
11. TERRIEN, J. (1967) *J. de Physique et le Radium* **28**, supplement to No. 3–4, C2–3.
12. RUBENS, H, and WOOD, R. W. (1911) *Phil. Mag.* **21**, 249.
13. RUBENS, H. and BAEYER, O. (1911) *Phil. Mag.* **21**, 689.
14. JACQUINOT, P. (1954) *J. Opt. Soc. Amer.* **44**, 761.
15. FELLGETT, P. B. (1951) Thesis, University of Cambridge.
16. CONNES, J. Thesis, University of Paris, 7 October 1960, subsequently published as a special issue of *Revue d'Optique theoretique et instrumentale* Serie A No. 3579 No. d'Ordre 4451, and in *Rev. Opt. Theor. Instrum.*, 1961, **40**, 45, 116; 171; 231.
17. FELLGETT, P. (1958) *J. de Physique et le Radium* **19**, 187.
18. FELLGETT, P. (1958) *J. de Physique et le Radium* **19**, 236.
19. BEEVERS, C. A. and LIPSON, H. (1936) *Nature*, **137**, 825; LIPSON, H. and BEEVERS, C. A. (1936) *Proc. Phys. Soc.* **48**, 477; FELLGETT, P. (1958) *J. Scient. Instrum.* **35**, 257.
20. MERTZ, L. (1958) *J. de Physique et le Radium* **19**, 233.
21. MERTZ, L. (1973) *Opt. Commun.* **6**, 354.
22. STRONG, J. and VANASSE, G. A. (1960) *J. Opt. Soc. Amer.* **50**, 1960.
23. GEBBIE, H. A. and VANASSE, G. A. (1956) *Nature* **178**, 432.
24. STRONG, J. and VANASSE, G. A. (1959) *J. Opt. Soc. Amer.* **49**, 844.
25. GEBBIE, H. A. (1959) in *Symposium on Interferometry*, National Physical Laboratory, HMSO, London.
26. GEBBIE, H. A. (1961) in *Advances in Quantum Electronics* (Ed. J. R. Singer), Columbia University Press, New York.
27. CONNES, J. (1958) quoted in reference 16; see also *J. de Physique et le Radium* (1958) **19**, 197.
28. CONNES, J. and GUSH, H. P. (1959) *J. de Physique et le Radium* **20**, 915.
29. CONNES, J, and GUSH, H. P. (1960) *J. de Physique et le Radium* **21**, 645.
30. FELLGETT, P., (1967) *J. de Physique et le Radium* **28**, supplement to No. 3–4, C2–165.

339

31. COOLEY, J. W. and TUKEY, J. W. (1965) *Math. Comput.* **19**, 297.
32. BRACEWELL, R. (1965) *The Fourier Transform and its Applications* McGraw-Hill, New York.
33. BORN, M. and WOLF, E. (1959) *Principles of Optics*, Pergamon Press, Oxford.
34. HOPKINS, H. H. (1951) *Proc. Roy. Soc. A* **208**, 263.
35. See for example SHURCLIFF, W. A. (1962) *Polarized Light*, Harvard University Press, Cambridge, Mass.
36. See CHANTRY, G. W. and FLEMING, J. W. (1976) *Infrared Physics* **16**, 655, for a discussion.
37. CONNES, J. (1971) in *Aspen International Conference on Fourier Spectroscopy 1970* (Ed. G. A. Vanasse, A. T. Stair, Jr., and D. J. Baker), AFCRL-71-0019, Special Report No. 114, p. 83.
38. LOEWENSTEIN, E. (1963) *Appl. Opt.* **2**, 491.
39. MERTZ, L. (1963) *Appl. Opt.* **2**, 1331; MERTZ, L. (1967) *Infrared Physics* **7**, 17.
40. FORMAN, M. L., STEEL, W. H., and VANASSE, G. A. (1966) *J. Opt. Soc. Amer.* **56**, 59.
41. GIBBS, J. E. and GEBBIE, H. A. (1965) *Infrared Physics* **5**, 187.
41a. WALMSLEY, D. A., CLARK, T. A., and JENNINGS, R. E. (1972) *Appl. Opt.* **11**, 1148.
42. HASWELL, R., MARTIN, A. E., and SHARP, G. (1964) *Appl. Opt.* **3**, 1195; RICHARDS, P. L. (1967) *IEEE Spectrum* **4**, 83.
43. FILLER, A. (1964) *J. Opt. Soc. Amer.* **54**, 762.
44. SAKAI, H. and VANASSE, G. A. (1966) *J. Opt. Soc. Amer.* **56**, 131; see also STEEL, W. H. and FORMAN, M. L. (1966) *J. Opt. Soc. Amer.* **56**, 982.
45. BELL, E. E. (1966) *Infrared Physics* **6**, 57; BELL, E. E. (1967) *J. de Physique et le Radium* **28**, supplement to No. 3–4, C2–18; BELL, E. E. (1971) in *Aspen International Conference on Fourier Spectroscopy 1970* (Ed. G. A. Vanasse, A. T. Stair, Jr., and D. J. Baker), AFCRL-71-0019, Special Report No. 114, p. 71.
46. CHAMBERLAIN, J., GIBBS, J. E., and GEBBIE, H. A. (1969) *Infrared Physics* **9**, 185; THOMAS, T. E., ORVILLE-THOMAS, W. J., CHAMBERLAIN, J., and GEBBIE, H. A. (1970) *Trans. Faraday Soc.* **66**, 2710; CHAMBERLAIN, J., COSTLEY, A. E., and GEBBIE, H. A. (1967) *Spectrochim. Acta A* **23**, 2255.
47. CHAMBERLAIN, J. (1972) *Infrared Physics* **12**, 145.
48. BIRCH, J. R. and BULLEID, C. E. (1977) *Infrared Physics* **17**, 279.
48a. BIRCH, J. R. and PARKER, T. J. (1979) *Infrared Physics* **19**, 103.
49. KISELEV, B. A. and PARSHIN, P. F. (1962) *Opt. i. Spektr.* **12**, 311 (*Optics and Spectroscopy* **12**, 169 (1962)).
50. PARSHIN, P. F. (1962) *Opt. i. Spektr.* **13**, 740 (*Optics and Spectroscopy* **13**, 418 (1962)).
51. GENZEL, L. (1960) *J. Mol. Spectroscopy* **4**, 241.
52. CHAMBERLAIN, J. (1971) *Infrared Physics* **11**, 25; CHAMBERLAIN, J. and GEBBIE, H. A. (1971) *Infrared Physics* **11**, 56.
53. MOSS, D. G. Private Communication.
54. GEBBIE, H. A., HABELL, K. J., and MIDDLETON, S. P. (1962) in *Proceedings of the Conference on Optical Instruments 1961*, Chapman and Hall, London, p. 43.
55. CONNES, J. and CONNES, P. (1966) *J. Opt. Soc. Amer.* **56**, 896.
56. HAPP, H. and GENZEL, L. (1961) *Infrared Physics* **1**, 39.
57. VOGEL, P. and GENZEL, L. (1964) *Appl. Opt.* **3**, 367.
58. BELL, R. J. (1972) *Introductory Fourier Transform Spectroscopy*, Academic Press, London and New York.
59. HANEL, R., CONRATH, B. J., HOVIS, W. A., KUNDE, V. G., LOWMAN, P. D., PEARL, J. C., PRABHAKARA, C., SCHLACHMAN, B., and LEVIN, G. V. (1972) *Science* **175**, 305.
60. GRIFFITHS, P. R., FOSKETT, C. T., and CURBELO, R. (1972) *Appl. Spectroscopy Ref.* **6**, 31.

61. FLEMING, J. W. (1974) *IEEE Trans. Microwave Theory Tech.* **MTT-22,** 1023.
62. CONNES, J. and NOZAL, V. (1961) *J. de Physique et le Radium* **22,** 359.
63. See for example RICE, S. O. (1944) *Bell Syst. Tech. J.* **23,** 282; (1945) **24,** 46; also BENDAT, J. S. (1958) *Principles and Applications of Random Noise Theory,* John Wiley, New York.
64. See for example FLEMING, J. W. (1974) *IEEE Trans. Microwave Theory Tech.* **MTT-22,** 1023; FLEMING, J. W. and GIBSON, M. J. (1976) *J. Mol. Spectroscopy* **62,** 326.
65. FLEMING, J. W. (1977) *Infrared Physics* **17,** 263.
66. MANDEL, L., SUDARSHAN, E. C. G., and WOLF, E. (1964) *Proc. Phys. Soc.* **84,** 435.
67. KAHN, F. (1959) *Astrophys. J.* **129,** 518.
68. HANBURY-BROWN, R. and TWISS, R. Q. (1958) *Proc. Roy. Soc. A* **243,** 291.
69. PINARD, J. (1969) *Ann. Phys.* **4,** 147.
70. CONNES, J. and CONNES, P. (1966) *J. Opt. Soc. Amer.* **56,** 896.
71. CONNES, J., CONNES, P., and MAILLARD, J.-P. (1967) *J. de Physique et le Radium* **28,** supplement to No. 3–4, C2–120.
72. BOWERS, H. C. (1964) *Appl. Opt.* **3,** 627.
73. JAMES, J. F. (1964) *J. Quant. Spectroscopy* **4,** 793.
74. VANASSE, G. A. and SAKAI, H. (1967) *Progress in Optics,* Vol. VI, North Holland, Amsterdam.
75. GOOD, I. J. (1958) *J. Roy Statist. Soc. B* **20,** 361.
76. SINGLETON, R. C., Collected algorithms from CACM: Algorithms 338 and 339.
77. FLEMING, J. W. National Physical Laboratory Report No. Mat. App. **27,** February 1973.
78. FORMAN, M. L. (1966) *J. Opt. Soc. Amer.* **56,** 978.
79. BRENNER, N. W. Quoted by Mme Connes in Reference 37.

Author index

Subject index